Lecture Notes in Mathematics

Edited by A. Dold and B. Eckmann

799

Functional Differential Equations and Bifurcation

Proceedings of a Conference Held at
São Carlos, Brazil, July 2–7, 1979

Edited by A. F

Springer-Verlag
Berlin Heidelberg New York 1980

Editor

Antonio Fernandes Izé
Universidade de São Paulo
Instituto de Ciências Matemáticas de São Carlos
Departamento de Matemática
Av. Dr. Carlos Botelho, 1465
C.P. 668
13560 São Carlos
Brazil

AMS Subject Classifications (1980): 34C05, 34C30, 34C40, 34D20, 34G05, 34J10, 34K15, 35B10, 45K05, 45N05, 47A50

ISBN 3-540-09986-7 Springer-Verlag Berlin Heidelberg New York
ISBN 0-387-09986-7 Springer-Verlag New York Heidelberg Berlin

Library of Congress Cataloging in Publication Data. Conference on Functional Differential Equations and Bifurcation, São Carlos, Brazil, 1979. Functional differential equations and bifurcation. (Lecture notes in mathematics; 799) Bibliography: p. Includes index.
1. Functional differential equations--Congresses. 2. Bifurcation theory--Congresses.
I. Izé, A. F., 1933- II. Title. III. Series: Lecture notes in mathematics (Berlin); 799.
QA3.L28. no. 799. [QA372]. 510s. [515.3'5] 80-14367

Printed in Germany

Printing and binding: Beltz Offsetdruck, Hemsbach/Bergstr.
2141/3140-543210

PREFACE

This volume consists of papers that were presented at the Conference on Functional Differential Equations and Bifurcation, held at the *Instituto de Ciências Matemáticas de São Carlos, Universidade de São Paulo, São Carlos-Brasil*, during July 2-7, 1979. The members of the organizing committee were: A.F. Izé (Coordinator), O.F. Lopes, H.M. Rodrigues and P.Z. Táboas.

The organizers and the editor would like to express their gratitude to the participants for their contributions and for their cooperation for making the conference a sucess and to the Springer-Verlag for its readiness to publish this proceedings.

To many of our colleagues, to several universities and research institute of Brazil and to the Rector of *Universidade de São Paulo*, Waldyr M. Oliva, we express our gratitude for their most valuable help.

We acknowledge the financial support of the Brazilian Agencies CNPq, FAPESP, FINEP and Sociedade Brasileira de Matemática.

A. F. Izé

São Carlos, December 1979.

TABLE OF CONTENTS

PARTICIPANTS

- AKASHI, Wilson Yoshihiro
  Instituto Tecnológico de Aeronáutica
  Departamento de Matemática
  12200 - São José dos Campos - SP - Brasil

- ALONSO, Luiz Antonio Ponce
  Universidade de São Paulo
  Instituto de Ciências Matemáticas de São Carlos
  Departamento de Matemática
  Caixa Postal 668
  13560 - São Carlos - SP - Brasil

- ALTATIM, Ruy Alberto Correa
  Universidade de São Paulo
  Instituto de Ciências Matemáticas de São Carlos
  Departamento de Matemática
  Caixa Postal 668
  13560 - São Carlos - SP - Brasil

- AVELLAR, Cerino Ewerton de
  Universidade Federal de São Carlos
  Departamento de Matemática
  Rdv. W. Luiz, Km 235
  13560 - São Carlos - SP - Brasil

- BALTHAZAR, José Manoel
  Universidade Estadual Paulista "Júlio de Mesquita Filho"
  Instituto de Geociências e Ciências Exatas
  Departamento de Matemática
  Campus de Rio Claro
  13500 - Rio Claro - SP - Brasil

- BAPTISTINI, Margarete T. Zanon
  Universidade Federal de São Carlos
  Departamento de Matemática
  Rdv. W. Luiz, Km 235
  13560 - São Carlos - SP - Brasil

- BARBANTI, Luciano
Universidade Estadual Paulista "Júlio de Mesquita Filho"
Instituto de Geociências e Ciências Exatas
Departamento de Matemática
Campus de Rio Claro
13500 - Rio Claro - SP - Brasil

- BECKER, Ronald I.
University of Capetown
Private Bag
Rondebosch 7700
South Africa

- BOTURA FILHO, Décio
Universidade Federal de São Carlos
Departamento de Matemática
Rdv. W. Luiz, Km 235
13560 - São Carlos - SP - Brasil

- BOULOS, Paulo
Universidade de São Paulo
Instituto de Matemática e Estatística
Departamento de Matemática
Caixa Postal 20570
02019 - São Paulo - SP - Brasil

- CAMARGO, José Luiz Correa
Instituto Tecnológico de Aeronáutica
Departamento de Matemática
12200 - São José dos Campos - SP - Brasil

- CARR, Jack
Brown University
Department of Mathematics
Division of Applied Mathematics
Providence, RI 02912 - USA

- CARVALHO, Luiz Antonio Vieira de
Universidade Federal de Goiás
Departamento de Matemática
74000 - Goiânia - GO - Brasil

- CASSAGO JR., Hermínio
Universidade de São Paulo
Instituto de Ciências Matemáticas de São Carlos
Departamento de Matemática
Caixa Postal 668
13560 - São Carlos - SP - Brasil

- CASTELAN, Walter de Bona
Universidade Federal de Santa Catarina
Departamento de Matemática
88000 - Florianópolis - SC - Brasil

- CERON, Suely do Carmo Siqueira
Universidade Estadual Paulista "Júlio de Mesquita Filho"
Instituto de Biociências, Letras e Ciências Exatas
Departamento de Matemática
Campus de São José do Rio Preto
Rua Cristovão Colombo, 2265
15100 - São José do Rio Preto - SP - Brasil

- CERQUEIRA, Maria Helena L.P. de
Universidade Federal da Bahia
Departamento de Matemática
40000 - Salvador - BA - Brasil

- CHOW, Shui-Nee
Michigan State University
Wells Hall
East Lansing, MI 48824 - USA

- COSTA, Ivo Machado da
Universidade Federal de São Carlos
Departamento de Matemática
Rdv. W. Luiz, Km 235
13560 - São Carlos - SP - Brasil

- CUNHA, Gregorio Maranguape da
Universidade Federal do Ceará
Departamento de Matemática
60000 - Fortaleza - CE - Brasil

- CURSI, José Eduardo Souza de
Universidade de São Paulo
Instituto de Matemática e Estatística
Departamento de Matemática
Caixa Postal 20570
02019 - São Paulo - SP - Brasil

- DARÉ, Carmen Diana Rodrigues
Universidade Estadual Paulista "Júlio de Mesquita Filho"
Instituto de Planejamento e Estudos Ambientais
Departamento de Matemática
Campus de Presidente Prudente
Rua Roberto Simonsen, 305
19100 - Presidente Prudente - SP- Brasil

- DUARTE, João Carlos Sell
Universidade Federal de Santa Catarina
Departamento de Matemática
88000 - Florianópolis - SC - Brasil

- DUTRA, Maurici José
Universidade Federal de Santa Catarina
Departamento de Matemática
88000 - Florianópolis - SC - Brasil

- EGUSQUIZA, Eduardo Alfonso Chincaro
Universidade Federal de Minas Gerais
Departamento de Matemática
30000 - Belo Horizonte - MG - Brasil

- FALEIROS, Antonio Cândido
Instituto Tecnológico de Aeronáutica
Departamento de Matemática
12200 - São José dos Campos - SP - Brasil

- FAVARO, Luiz Antonio
Universidade de São Paulo
Instituto de Ciências Matemáticas de São Carlos
Departamento de Matemática
Caixa Postal 668
13560 - São Carlos - SP - Brasil

- FERNANDES, Davi Teodoro
Universidade de São Paulo
Instituto de Matemática e Estatística
Departamento de Matemática
Caixa Postal 20570
02019 - São Paulo - SP - Brasil

- FERREIRA, José Arminio
Universidade Federal do Espirito Santo
Departamento de Matemática
29000 - Vitória - ES - Brasil

- FIGUEIREDO, Djairo Guedes de
Universidade de Brasília
Departamento de Matemática
70000 - Brasília - DF - Brasil

- FÜRKOTTER, Monica
Universidade Estadual Paulista "Júlio de Mesquita Filho"
Instituto de Planejamento e Estudos Ambientais
Departamento de Matemática
Campus de Presidente Prudente
Rua Roberto Simonsen, 305
19100 - Presidente Prudente - SP - Brasil

- GALANTE, Luiz Fernandes
Universidade Estadual Paulista "Júlio de Mesquita Filho"
Instituto de Planejamento e Estudos Ambientais
Departamento de Matemática
Campus de Presidente Prudente
Rua Roberto Simonsen, 305
19100 - Presidente Prudente - SP - Brasil

- GIONGO, Maria Ângela de Pace Almeida Prado
Universidade Federal de São Carlos
Departamento de Matemática
Rdv. W. Luiz, Km 235
13560 - São Carlos - SP - Brasil

- GUERRA, Fernando
Universidade Federal de Santa Catarina
Departamento de Matemática
88000 - Florianópolis - SC - Brasil

- GUIMARÃES, Luiz Carlos
Universidade Federal de Minas Gerais
Departamento de Matemática
30000 - Belo Horizonte - MG - Brasil

- HALE, Jack K.
Brown University
Department of Mathematics
Division of Applied Mathematics
Providence - RI 02912 - USA

- HATORI, Toshio
Instituto Tencológico de Aeronáutica
Departamento de Matemática
12200 - São José dos Campos - SP - Brasil

- HÖNIG, Chaim Samuel
Universidade de São Paulo
Instituto de Matemática e Estatística
Departamento de Matemática
Caixa Postal 20570
02019 - São Paulo - SP - Brasil

- IZÉ, Antonio Fernandes
Universidade de São Paulo
Instituto de Ciências Matemáticas de São Carlos
Departamento de Matemática
Caixa Postal 668
13560 - São Carlos - SP - Brasil

- KATO, Junji
Tohoku University
Matematical Institute
Sendai 980 - Japan

- LADEIRA, Luiz Augusto da Costa
Universidade de São Paulo
Instituto de Ciências Matemáticas de São Carlos
Departamento de Matemática
Caixa Postal 668
13560 - São Carlos - SP - Brasil

- LANDER, Leslie Charles
Universidade Federal de Minas Gerais
Departamento de Matemática
30000 - Belo Horizonte - MG - Brasil

- LANDER, Maria Livia Mirna H. de
Universidade Federal de Minas Gerais
Departamento de Matemática
30000 - Belo Horizonte - MG - Brasil

- LIMA, Paulo F.
Universidade Federal de Pernambuco
Departamento de Matemática
50000 - Recife - PE - Brasil

- LOPES, Orlando Francisco
Universidade Federal de São Carlos
Departamento de Matemática
Rdv. W. Luiz, Km 235
13560 - São Carlos - SP - Brasil

- LOURENÇO, Mary Lilian
Universidade Estadual de Campinas
Instituto de Matemática, Estatística e Ciência da Computação
Departamento de Matemática
Caixa Postal 1170
13100 - Campinas - SP - Brasil

- MAGNUS, Robert
University of Iceland
Science Institute
Dunhaga 3
107 - Reykjavik - Iceland

- MALLET-PARET, John
Brown University
Departmente of Mathematics
Division of Applied Mathematics
Providence - RI 02912 - USA

- MEDEIROS, Luiz Adauto de Justa
Universidade Federal do Rio de Janeiro
Departamento de Matemática
Caixa Postal 1835
20000 - Rio de Janeiro - RJ - Brasil

- MENDES, Claudio Martins
Universidade de São Paulo
Instituto de Ciências Matemáticas de São Carlos
Departamento de Matemática
Caixa Postal 668
13560 - São Carlos - SP - Brasil

- MENZALA, Gustavo Perla
Universidade Federal do Rio de Janeiro
Instituto de Matemática
Departamento de Matemática
20000 - Rio de Janeiro - RJ - Brasil

- MILOJEVIĆ, Petronije S.
Universidade Federal de Minas Gerais
Departamento de Matemática
30000 - Belo Horizonte - MG - Brasil

- MOCHIDA, Dirce K. Hayashida
Universidade Federal de São Carlos
Departamento de Matemática
Rdv. W. Luiz, Km 235
13560 - São Carlos - SP - Brasil

- MOLFETTA, Natalino Adelmo de
Universidade Federal de São Carlos
Departamento de Matemática
Rdv. W. Luiz, Km 235
13560 - São Carlos - SP - Brasil

- MONTEIRO, Paulo Adão
Universidade Federal de São Carlos
Departamento de Matemática
Rdv. W. Luiz, Km 235
13560 - São Carlos - SP - Brasil

- MOREIRA, Robert Ozório
Universidade Federal de Santa Catarina
Departamento de Matemática
88000 - Florianópolis - SC - Brasil

- NASCIMENTO, Arnaldo Simal do
Universidade Federal de São Carlos
Departamento de Matemática
Rdv. W. Luiz, Km 235
13560 - São Carlos - SP - Brasil

- NOWOSAD, Pedro
Instituto de Matemática Pura e Aplicada
Rua Luiz de Camões, nº 68
20000 - Rio de Janeiro - RJ - Brasil

- OLIVA, Waldyr Muniz
Universidade de São Paulo
Instituto de Matemática e Estatística
Departamento de Matemática
Caixa Postal 20570
02019 - São Paulo - SP - Brasil

- OLIVEIRA, Ivan de Camargo e
Universidade de São Paulo
Instituto de Matemática e Estatística
Departamento de Matemática
Caixa Postal 20570
02019 - São Paulo - SP - Brasil

- OLIVEIRA, Mario Moreira Carvalho de
Universidade Federal do Rio de Janeiro
Instituto de Matemática
Centro de Tecnologia
20000 - Rio de Janeiro - RJ - Brasil

- ONUCHIC, Lourdes de la Rosa
Universidade de São Paulo
Instituto de Ciências Matemáticas de São Carlos
Departamento de Matemática
Caixa Postal 668
13560 - São Carlos - SP - Brasil

- ONUCHIC, Nelson
Universidade de São Paulo
Instituto de Ciências Matemáticas de São Carlos
Departamento de Matemática
Caixa Postal 668
13560 - São Carlos - SP - Brasil

- PALIS JR., Jacob
Instituto de Matemática Pura e Aplicada
Rua Luiz de Camões, nº 68
20000 - Rio de Janeiro - RJ

- PATERLINI, Roberto Ribeiro
Instituto de Matemática Pura e Aplicada
Rua Luiz de Camões, nº 68
20000 - Rio de Janeiro - RJ - Brasil

- PAVLU, Luiz Carlos
Universidade Federal de São Carlos
Departamento de Matemática
Rdv. W. Luiz, Km 235
13560 - São Carlos - SP - Brasil

- PENEIREIRO, João Batista
Universidade Federal de São Carlos
Departamento de Matemática
Rdv. W. Luiz, Km 235
13560 - São Carlos - SP - Brasil

- PEREZ, Geraldo
Universidade Estadual Paulista "Júlio de Mesquita Filho"
Instituto de Geociências e Ciências Exatas
Departamento de Matemática
Campus de Rio Claro
13500 - Rio Claro - SP - Brasil

- PERISSONOTO JR., Anizio
Universidade Estadual Paulista "Júlio de Mesquita Filho"
Instituto de Geociências e Ciências Exatas
Departamento de Matemática
Campus de Rio Claro
13500 - Rio Claro - SP - Brasil

- PETRONILHO, Gerson
Universidade Federal de São Carlos
Departamento de Matemática
Rdv. W. Luiz, Km 235
13560 - São Carlos - SP - Brasil

- PINTO, Roberto Carvalho Engler
Universidade de São Paulo
Instituto de Ciências Matemáticas de São Carlos
Departamento de Matemática
Caixa Postal 668
13560 - São Carlos - SP - Brasil

- PISANELLI, Domingos
Universidade de São Paulo
Instituto de Matemática e Estatística
Departamento de Matemática
Caixa Postal 20570
02019 - São Paulo - SP - Brasil

- RACHID, Munir
Universidade de São Paulo
Escola de Engenharia de São Carlos
Departamento de Estruturas
Caixa Postal 359
13560 - São Carlos - SP - Brasil

- RAUPP, Marco Antonio
Centro Brasileiro de Pesquisas Físicas
Ilha do Fundão
20000 - Rio de Janeiro - RJ - Brasil

- RODRIGUES, Hildebrando Munhoz
Universidade de São Paulo
Instituto de Ciências Matemáticas de São Carlos
Departamento de Matemática
Caixa Postal 668
13560 - São Carlos - SP - Brasil

- RODRIGUES, Paulo R.
Universidade Federal Fluminense
Instituto de Matemática
Departamento de Matemática
24000 - Niterói - RJ - Brasil

- RIBEIRO, Hermano de Souza Ribeiro
Universidade de São Paulo
Instituto de Ciências Matemáticas de São Carlos
Departamento de Matemática
Caixa Postal 668
13560 - São Carlos - SP - Brasil

- SAAB, Miriam
Universidade Federal de São Carlos
Departamento de Matemática
Rdv. W. Luiz, Km 235
13560 - São Carlos - SP - Brasil

- SANTOS NETO, Cristiano dos
Universidade de São Paulo
Instituto de Ciências Matemáticas de São Carlos
Departamento de Matemática
Caixa Postal 668
13560 - São Carlos - SP - Brasil

- SELL, George R.
University of Minnesota
Department of Mathematics
Minneapolis - Minn 55455 - USA

- SILVA, Euripides Alves da
Universidade Estadual Paulista "Júlio de Mesquita Filho"
Instituto de Biociências, Letras e Ciências Exatas
Departamento de Matemática
Campus de São José do Rio Preto
Rua Cristovão Colombo, 2265
15100 - São José do Rio Preto - SP - Brasil

- SILVA, Jairo José da
Universidade Estadual Paulista "Júlio de Mesquita Filho"
Instituto de Geociências e Ciências Exatas
Departamento de Matemática
Campus de Rio Claro
13500 - Rio Claro - SP - Brasil

- SILVEIRA, Mauricio
Universidade Federal de São Carlos
Departamento de Matemática
Rdv. W. Luiz, Km 235
13560 - São Carlos - SP - Brasil

- SINAY, Léon Roque
Universidade Federal do Ceará
Departamento de Matemática
60000 - Fortaleza - CE - Brasil

- SPEZAMIGLIO, Adalberto
Universidade de São Paulo
Instituto de Ciências Matemáticas de São Carlos
Departamento de Matemática
Caixa Postal 668
13560 - São Carlos - SP - Brasil

- TÁBOAS, Carmen Maria Guacelli
Universidade Federal de São Carlos
Departamento de Matemática
Rdv. W. Luiz, Km 235
13560 - São Carlos - SP - Brasil

- TÁBOAS, Plácido Zoéga
Universidade de São Paulo
Instituto de Ciências Matemáticas de São Carlos
Departamento de Matemática
Caixa Postal 668
13560 - São Carlos - SP - Brasil

- TADINI, Wilson Mauricio
Universidade de São Paulo
Instituto de Ciências Matemáticas de São Carlos
Departamento de Matemática
Caixa Postal 668
13560 - São Carlos - SP - Brasil

- TAKENS, Floris
Instituto de Matemática Pura e Aplicada
Rua Luiz de Camões, nº 68
20000 - Rio de Janeiro - RJ - Brasil

- TELLO, Jorge Sotomayor
Instituto de Matemática Pura e Aplicada
Rua Luiz de Camões, nº 68
20000 - Rio de Janeiro - RJ - Brasil

- VENTURA, Aldo
Universidade de São Paulo
Instituto de Ciências Matemáticas de São Carlos
Departamento de Matemática
Caixa Postal 668
13560 - São Carlos - SP - Brasil

- VILA, Antonio Marcos
  Universidade de São Paulo
  Instituto de Ciências Matemáticas de São Carlos
  Departamento de Matemática
  Caixa Postal 668
  13560 - São Carlos - SP - Brasil

- YOSHIZAWA, Taro
  Tohoku University
  Matematical Institute
  Sendai 980 - Japan

- ZEZZA, Pierluigi
  Universita Degli Studi di Firenze
  Istituto di Matematica Applicata
  Via Di S. Marta, 3
  50139 - Firenze - Italy

LIÉNARD EQUATIONS AND CONTROL (*)

by Luciano Barbanti

A. GENERAL THEORY

I. Introduction.

I.1. It is well known ([1], [2]), that Liénard equation

$$\ddot{x} + f(x)\dot{x} + x = 0 \qquad f \in C(R)$$

describes the simple, but fundamental, electric circuit consisting of a capacitor, a inductor and a resistor wired in series.

The Van der Pol equation, a Liénard equation itself,

$$\ddot{x} + \mu(x^2-1)\dot{x} + x = 0$$

was founded in the study about the triod oscillator. For great values of μ, Van der Pol suggests that the equation models certain irregularities in heart pulsations. The same equation, in a different form was studied by Lord Rayleigh, in his investigations of sound theory.

In the last years, it was discovered many applications of the Van der Pol equation to biological systems.

So, it is interesting to study the control of the Liénard equation, with the purpose of controlling the phenomena that these equations models.

E. James [3], using electronic calculators got some results, about control of the VDP equation. In her work, many results have been only stated. Many of them were proved by Gabriele Villari [4].

R. Conti in a fundamental work gave a precise formulation for the theory of the VDP equation with control [5].

N. Alekseev [6], investigating the discontinuities of the function that associates to values of the control, the respective set of null

---

(*) This work was developed in the "Istituto Matematico 'U. Dini'" (Florence) and was partially supported by FAPESP (Fundação de Amparo à Pesquisa do Estado de São Paulo) and "Ministero degli Affari Esteri d'Italia".

controllability, also contributed to the VDP control theory.

It is in the frame of these main lines and closely following Conti's work, that we are going to present this work.

I.2. The Liénard equations, as originally placed by A. Liénard, in "Études des oscillations entreteneus" (Revue Générale de l'Électricité, 23, pp. 901-912 & 946-954, 1928) are of the type:

$$(LI)_o \qquad \ddot{x} + f(x)\dot{x} + x = 0 \qquad\qquad f \in C(R)$$

Later, Levinson-Smith introduced the so called generalized Liénard equations:

$$\ddot{x} + f(x)\dot{x} + f(x) = 0 \qquad\qquad [g \in Lip(R) \quad \text{and} \quad xg(x) > 0]$$

This equation, apparently more general in form, actually can be transformed in the $(LI)_o$ one, by the known transformation of variables due to R. Conti [7]:

$$Z = (2G(x))^{1/2}.\text{sgn } x \quad \text{where} \quad G(x) = \int_o^x g(s)ds.$$

If $x(z)$ is the inverse function and $H(z) = F(x(z))$, where $F' = f$, we have:

$$\ddot{z} + H'(z)\dot{z} + z = 0.$$

The study of control theory to Liénard equations, in a first glance is the study of the processes

$$\ddot{x} + f(x)\dot{x} + x = u(t)$$

where $u \in L^\infty_{loc}(R)$. But the study of those processes is to general for our purposes. Our attention will be directed to the question of transfering in a minimal time, in the phase plane, points of the cycle of $(LI)_o$ (if it is unique) to the origin.

At this point, we are facing our first problem: what conditions would be given on f, for the existence and uniqueness of a non trivial cycle of $(LI)_o$? This question itself proposes a vast field of investigation. Many papers have been written on this subject, (see e.g. U. Staude [8]). It was picked out conditions that derive from Liénard, and are in S. Lefschetz [9].

(A) $\begin{cases} \text{A1) } f \text{ is even and } f(0) < 0 \\ \text{A2) } f \text{ is continuous on } R \\ \text{A3) Let } F(x) = \int_0^x f(s)ds, \text{ then } F(x) \to \pm\infty \text{ with } x \\ \text{A4) } F \text{ has a single positive zero } x = 0_F \text{ and is monotone} \\ \quad \text{increasing for } x > 0_F \end{cases}$

These conditions imply that there exists a lower positive real number 0_f which is a zero of f and such that f is nonnegative in the right hand side of 0_f.

Under the conditions (A), there exists a unique cycle Γ_o which comes from a periodic solution of $(LI)_o$. In addition Γ_o is orbitally stable.

Since Γ_o is stable transfering points of Γ_o to zero, in the phase plane, R^2, is only possible by means of a forcing term $u(t)$, according to the control process

$$\ddot{x} + f(x)\dot{x} + x = u(t)$$

where f satisfies (A) and $u \in L^\infty_{loc}(R)$.

If $f \in C^1$, by the Pontryagin's Maximal Principle the control that transfeers a point of Γ_o to the origin in a minimum time, is a realy control (Lee & Markus [10], p. 427). So, we will restrict ourselves to the study of systems of the type stated above, with the set of admissible controls being now, $u \in L^\infty_{loc}(R)$, such that $|u(t)| = k$ a.e. $t \in R$, $(k \in R)$. (see also Theorem - Lee & Markus [10]).

The orbits of the system

$$\ddot{x} + f(x)\dot{x} + x = k \qquad k > 0$$

are the symmetrical with respect to zero, of the orbits associated to the system

$$\ddot{x} + f(x)\dot{x} + x = -k.$$

So, it is sufficient to consider those processes with nonnegative k.

We fix, then, the processes for which our attention will be directed:

$(LI)_k \qquad \ddot{x} + f(x)\dot{x} + x = k \qquad\qquad k \geq 0$

with $f \in C^1$, satisfying (A) and the hypothesis,

(B) $\quad \exists\ N, M \in R^+ \quad s.t.\{y = \frac{0_f - x}{f(x)}; \quad x \in N\} \subseteq [-M,0].$

I.3. We denote for a fixed strictly positive t, $V(t,k)$ as the set of the points (x_o,y_o) of R^2 for which there exists a solution of $(LI)_k$ $x(t,x_o,y_o,k)$ satisfying:

$$x(0,x_o,y_o,k) = x_o, \quad \dot{x}(0,x_o,y_o,k) = y_o, \quad x(\bar{t},x_o,y_o,k) = 0$$
$$\text{and} \quad \dot{x}(\bar{t},x_o,y_o,k) = 0$$

for some $\bar{t} \in [0,t]$.

The domain of null controllability of $(LI)_k$, is the set

$$V_k = \bigcup_{t>0} V(t,k)$$

For all $k > 0$, V_k is an open connected subset of R^2 ([10],Th. 2, p. 429).

II. <u>The structure of V_k.</u>

If $P \in R^2$, $\gamma_k(P)$ denotes the single orbit of $(LI)_k$ crossing P, $\gamma_k^+(P)$ the segment of $\gamma_k(P)$ that begins at P and advances for increasing t, and $\gamma_k^-(P)$ the segment of $\gamma_k(P)$, that begins at P an advances for decreasing t. The set of points Λ^+-limit (resp. Λ^--limit) of $\gamma_k(P)$ is denoted by $\Lambda^+(P)$ (resp. $\Lambda^-(P)$).

<u>Theorem</u> 1. Consider the process $(LI)_k$. Then $V_k = R^2$, or V_k is bounded. If $V_k \neq R^2$, then V_k is symmetrical with respect to zero, and is bounded by two segments of orbit, one of them coming from the system $(LI)_k$.

<u>Proof</u>. Observing the vector field schema of $(LI)_k$ (Fig. I and II) if D is a positive point of the x-axis, and M, N are the constants that appear in condition (B), then the orbit $\gamma_k^+(D)$ intercepts the line $x = N$ in the lower semiplane $y < 0$, in a point $P(D)$. Since $P(D) \leq -M$, there is a point $P(\infty) = \lim_{D\to\infty} P(D)$.

Let $\gamma_k(\infty)$ be the orbit passing for $P(\infty)$.

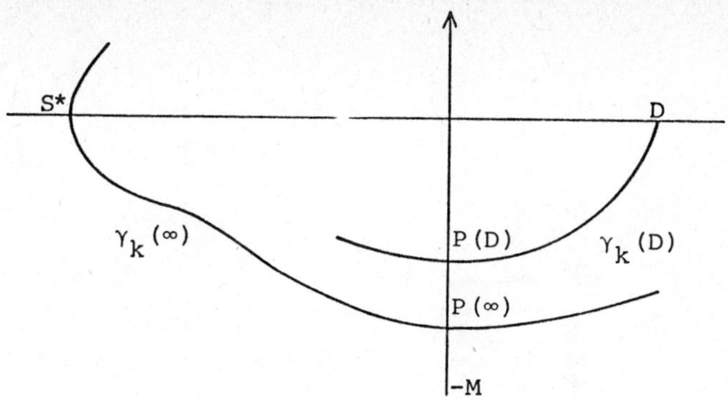

Figure 1

If $\gamma_k(\infty)$ cross the x-axis in a point S^*, whith $S^* < (k,0)$, then S^* has the following property: if Z is a point in the x-axis, then

$Z < S^*$ implies that $\gamma_k^-(Z)$ does not intercept the x-axis after Z

$Z > S$ implies that $\gamma_k^-(Z)$ must intercept the x-axis after Z

If, on the other hand, $\gamma_k(\infty) \xrightarrow[t\to\infty]{} (k,0)$, then all the orbits crossing the x-axis at the right of $(k,0)$ must tend to $(k,0)$ when $t \to \infty$.

Denoting the point $(0,0)$ by θ, let's consider the following sequence of points in the x-axis:

$\theta = 0_o$, and

for a 0_{n-1} $(n \geq 1)$, with $S^* < 0_{n-1}$, let 0_n be the point where $\gamma_k^-(-0_{n-1})$ intercepts the x-axis.

The following holds:

1º) there exists a number n_o for which $-0_{n_o} < S^* < 0_{n_o-1}$.

2º) $-0_n > S^*$ for all n.

In the first case, $V_k = R^2$. In fact: the arc $\gamma_{-k}^-(0_{n_o}) \vee [-0_{n_o}, 0_n] \vee \vee\, \gamma_k^-(-0_{n_o})$ splits the plane into two domains. If P is a point at the right of this arc, we apply the control $-k$ until $\gamma_k^+(P)$ intercepts $\gamma_k^-(-0_{n_o})$ and then we reach θ.

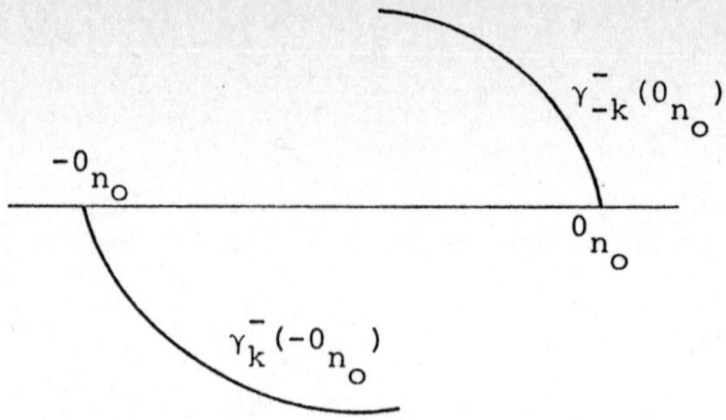

Figure 2

Conversely, if P is at the left of this arc, we begin applying the control k.

In the second case, calling 0_∞ the point $\lim_{n\to\infty} 0_n$ we have $-0_\infty > S^*$, by the continuous dependence theorem. In addition $0_\infty \geq (k,0)$ and $\gamma_{-k}^{-}(0_\infty)$ pass at -0_∞.

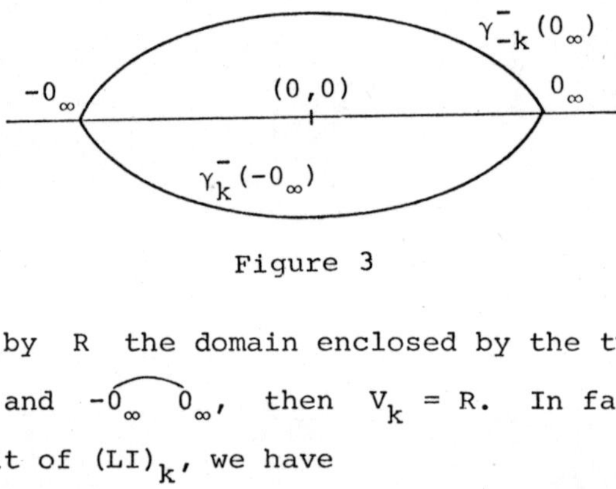

Figure 3

If we denote by R the domain enclosed by the two segments of orbit $\overset{\frown}{0_\infty\ -0_\infty}$ and $\overset{\frown}{-0_\infty\ 0_\infty}$, then $V_k = R$. In fact: if y'_k is the slope of an orbit of $(LI)_k$, we have

$$\text{(II.1)}\qquad y'_{k_1} - y'_{k_2} = -f(x) + \frac{k_1-x}{y} - \left(-f(x) + \frac{k_2-x}{y}\right) = \frac{k_1-k_2}{y}$$

for all $k_1, k_2 \in R^+$.

Then, $y > 0 \rightarrow y'_{-k} < y'_k$

and $y < 0 \rightarrow y'_{-k} > y'_k$

So, at the points (x,y), with $y > 0$, the orbits of $(LI)_k$ always cross out the arc $\overset{\frown}{-0_\infty\ 0_\infty}$ of $\gamma_{-k}^{-}(0_\infty)$. Symmetricaly, we find the same situation in the $y < 0$-semiplane: the orbits of $(LI)_{-k}$ always would cross out the arc $\widehat{-0_\infty 0_\infty}$ of $\gamma_k^-(-0_\infty)$.

At the points -0_∞ and 0_∞, there are no other orbits of $(LI)_k$ or $(LI)_{-k}$, than the $\gamma_k^-(-0_\infty)$ or $\gamma_k^-(0_\infty)$. In conclusion $V_k \subseteq R$.

Let P be a point in R. Then it's possible following $\gamma_{-k}(P)$ or $\gamma_k(P)$ without coming out R, to reach a point in $\gamma_k^-(0_{n_o})$ or $\gamma_k^-(0_n)$ for some n_o. Thus, with more n_o applications of control k or -k alternately we reach the point (0,0).

Concluding, $R \subseteq V_k$.

Moreover, it is possible to establish that R belongs to the interior of Γ_o. But, for this, a preliminary proposition is needed.

<u>Proposition</u> 1. If V_k^u denotes the domain of null controllability of the process

$$(LI)_u \qquad \ddot{x} + f(x)\dot{x} + x = u(t) \qquad |u(t)| \le k \quad \text{a.e.} \quad t \in R.$$

and $f \in C^1$ satisfies (A) and (B), then $V_k^u = V_k$.

<u>Proof</u>. $V_k \subseteq V_k^u$. Suppose $V_k \neq R^2$. The angular coeficient of the $(LI)_u$ orbits, at the point (x,y) is

$$y'_u = -f(x) + \frac{u(t)-x}{y}$$

Then, we have

$$y > 0 \rightarrow y'_{-k} < y'_u$$
$$y < 0 \rightarrow y'_k < y'_u$$

Since $\dot{x} = y$, if $y \neq 0$ the $(LI)_u$ orbits, always get out the domain V_k.

The same happens at -0_∞ and 0_∞, because

$$x''_k(0) < x''_u(0) < x''_{-k}(0),$$

where x(0) is the first coordinate of 0_∞, $x'' = \dfrac{dx^2}{d^2y}$ and the subscripts v are used to indicate that x" is calculate in $(LI)_v$.

<u>Theorem</u> 2. Consider the system $(LI)_k$. If $V_k \neq R^2$, then V_k belongs to the Γ_o interior.

Proof. Denote the domain enclosed by Γ_o as R_{Γ_o} and suppose $V_k \neq R^2$ with $V_k \not\subseteq R_{\Gamma_o}$.

As $V_k^u = V_k$, and the origin for $u(t) = 0$ is repulsive, each point of R_{Γ_o}, can be transfered in a finite time into V_k by the zero control. So $R_{\Gamma_o} \subseteq V_k$.

If $P \in R^2 - R_{\Gamma_o}$, P can be transfered in a finite time, at a point Q, so close of Γ_o, as one wishes. But in a convenient neigborhood of Γ_o, one of the vectors $(y,-f(x)y + k - x)$ or $(y,-f(x)y - k - x)$ is directed to R_{Γ_o}. As $R_{\Gamma_o} \subseteq V_k$, we have then, $V_k = R^2$. This is an absurd. So, if $V_k \neq R^2$, then $V_k \subseteq R_{\Gamma_o}$.

One of the questions that can be posed now is the following: do there exist some value k, for which V_k is R^2 ? The next theorem, gives us an affirmative answer. Despite of this, a slight elaboration on the fact that f satisfies condition (B), allows us to give the same answer (Cor. 1 of Th. 6).

Theorem 3. Consider the system $(LI)_k$ with $k \geq x_{\Gamma_o}$, where $x_{\Gamma_o} > 0$ is the point at which Γ_o cross the x-axis. Then $V_k = R^2$.

Proof. It is sufficient to observe that $0_{\infty_1} \geq k \geq x_{\Gamma_o}$, where $0_\infty = (0_{\infty_1},0)$ is that one obtained in Theorem 1 above.

Fixing f in $(LI)_k$, the mapping which associates V_k to each k, has only one point of discontinuity. This fact is shown by the two next theorems.

Theorem 4. Fix f in $(LI)_k$ and let k be variable. If $k_1 > k_2$, then $V_{k_1} \supseteq V_{k_2}$.

Proof. By (II.1) if $k_1 > k_2$, then $y'_{k_1} > y'_{k_2}$. So, the $(LI)_{k_1}$ orbits always get out the domain V_{k_2}. In conclusion, $V_{k_1} \supseteq V_{k_2}$.

Theorem 5. Fix f in $(LI)_k$ and let k be variable. It exists a number $k_f > 0$ such that

$$k > k_f \text{ implies } V_k = R^2; \quad k \leq k_f \text{ implies } V_k \neq R^2.$$

Proof. The fact that Γ_o is stable implies $V_o \neq R^2$. By Theorem 3 there exists k_o such that $V_{k_o} = R^2$. Then, by Theorem 4, it is possible to conclude the existence of the sup of the numbers k, for which $V_k \neq R^2$. That this sup is the max of such numbers is shown equally as in Conti ([5], n. 9).

By the precedent theorems we conclude that an effective control of the cycle Γ_o, to the origin, by means of relay controls it is only possible when we use $u(t) = \{^{k}_{-k}$ with $k > k_f$. But the problem of finding such a k_f is not yet solved. The next theorems, as in Theorem 3, will be an attempt to fix k_f into certain intervals of R.

Theorem 6. Consider the system $(LI)_k$, when k satisfies

$$f(k+r) + f(k-r) > 0$$

for all $r \geq 0$. Then $V_k = R^2$.

Suppose that for some k, and some $r \geq 0$,

$$f(k+r) + f(k-r) = 0$$

and in addition there exists a semi-open interval $(a_r, k-r]$ if $y < 0$ or $[k-r, b_r)$ if $y > 0$ such that for all $k-s$ $(s \geq 0)$ in it,

$$f(k+s) + f(k-s) > 0,$$

holds. Then $V_k = R^2$.

Proof. Consider the point $P = (0,y)$ with $y > 0$. Verifying the vector field schema for $(LI)_k$, we see that $\gamma_k^+(P)$ intercepts the positive x-axis in a point Q, at the right of $(k,0)$, and so, in the fourth quadrant either tends to $(k,0)$ or cross again the line $x = k$ in a point R.

Let us denote by $\gamma^s(P)$, the symmetrical arc with respect to $x = k$ of $\widehat{PQR}$ of $\gamma_k^+(P)$. Since at the point $x = k$ we have, $y_k' = -f(k) < 0$, where y_k' is the slope of $\gamma_k^+(P)$, then $\gamma_k^+(P)$ after passing at R, get into the domain enclosed by the line $x = k$ and $\gamma^s(P)$.

The slope of $\gamma^s(P)$ at a point $(k-x,y)$, denoted here by $\gamma^{s'}$, and the slope of $\widehat{PQR}$ at $(k+s,y)$ have opposite values. Consequently, at $(k-x,y)$ we have:

$$y^{s'} = f(x+k) + \frac{x}{y} \quad \text{and} \quad y_k' = -f(k-x) + \frac{x}{y}.$$

So, $y^{s'} - y_k' = f(x+k) + f(k-x)$, and then

$$y^{s'} > y_k'.$$

This means that $\gamma_k^+(P)$ always get into the domain bounded by $x = k$, and $\gamma^s(P)$ if eventually touch $\gamma^s(P)$. Therefore, after passing by R, and after crossing the x-axis in a point at the left of $(k,0)$, $\gamma_k^+(P)$ either tends to $(k,0)$ in the second quadrant or intercept again, the line $x = k$ and $y > 0$, at a point P', with $P' < P$.

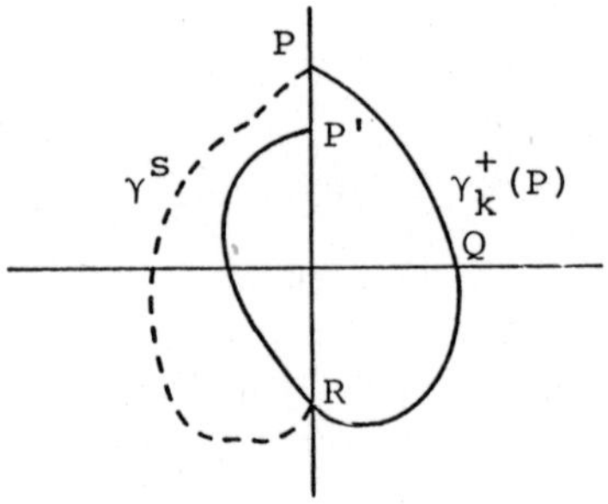

Figure 4

So, it can not exist a cycle for $(LI)_k$, and $\Lambda_k^+(P) = (k,0)$. Then $\Lambda^+(S) = (k,0)$ for all $S \in R^2$.

Suppose that for some $P \neq (k,0)$, $\Lambda^-(P) \neq \emptyset$. $\Lambda^-(P)$ is not a cycle and this implies that $(k,0) \in \Lambda^-(P)$. In conclusion: $\Lambda^-(P) = \emptyset$ for all $P \neq (k,0)$.

The conclusion of this part of the theorem follows as in Theorem 3.1, by Conti [5].

Suppose now that k is such that for some $r \geq 0$,

$$f(k+r) + f(k-r) = 0$$

holds, and for all $s \geq 0$, satisfying $k-s \in (a_r, k-r]$, $f(k+s) + f(k-s)$ is nonnegative. Then, by the Continuous Dependence Theorem, if $\gamma_k^+(P)$

cross away $\gamma^s(P)$ at the point $k-r$, then it is possible to find a curve Δ, so close $\gamma^s(P)$ as one wishes, in such a way that Δ "capture" $\gamma_k^+(P)$, and drives it, into the domain enclosed by $\gamma^s(P)$ and $x = k$, again.

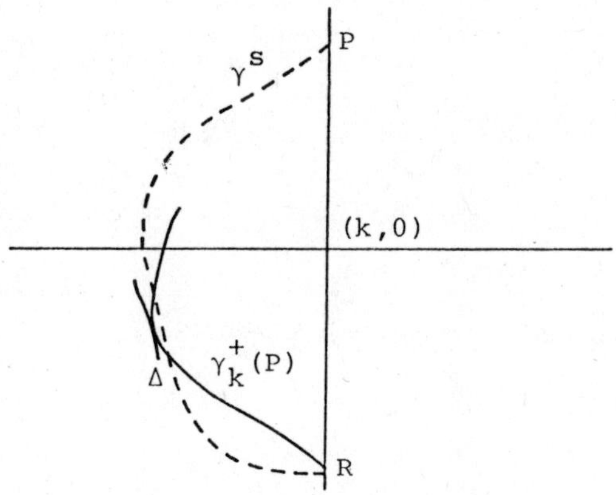

Figure 5

In another context, Gabriele Villari, independently, proved the same theorem [11].

Corollary 1. Condition (B) on f implies the existence of a number k_o, such that for all $k \geq k_o$ and all $r \geq 0$ we have, $f(k+r)+f(k-r)>0$. As a consequence, $V_k = R^2$.

Proof. Condition (B) implies the existence of a number k_o such that for all $k \geq k_o$, $f(k) \geq -\inf f(x)$ holds.

Before proving the next theorems, it is necessary to classify the point $(k,0)$ while a singular point of $(LI)_k$. The criterion (Sansone--Conti [1], p. 48) is the following:

values of $f(k)$	types of singularities
$f(k) < -2$	unstable 2 tg. node
$f(k) = -2$	unstable 1 tg. node
$-2 < f(k) < 0$	unstable focus
$f(k) = 0$	center/focus
$0 < f(k) < 2$	stable focus
$f(k) = 2$	stable 1 tg. node
$f(k) > 2$	stable 2 tg. node

Theorem 7. Consider the system $(LI)_k$ and suppose that exists a function $h : [-k,k] \to R$, satisfying:

(i) $h(k) = 0$ and $h(x) < 0$ if $x \in [-k,k]$,

(ii) h is differentiable,

(iii) $f(x) + \frac{x-k}{h(x)} + h'(x) < 0$ for all $x \in [-k,k]$.

Then we have, $V_k \neq R^2$, and in particular $0_\infty = (k,0)$.

Proof. Denoting $(k,0)$ by K, let us suppose that $\gamma_k^-(-K)$ cross the graph of h, in a point $(x,h(x))$ if $x \in (-k,k)$.

As the slope of $\gamma_k^-(-K)$ in such a point is

$$y_k'(x) = -f(x) + \frac{k-x}{y},$$

then

$$y_k'(x) > h'(x),$$

according (iii).

This implies that $\gamma_k^-(-K)$ can not get out the domain enclosed by $\overline{(-k,0)(k,0)}$, $\overline{(-k,0)(-k,h(-k))}$ and the graph of h. So, $\gamma_k^-(-K) \to K$. Furthermore $\gamma_k^-((0,0)) \to K$.

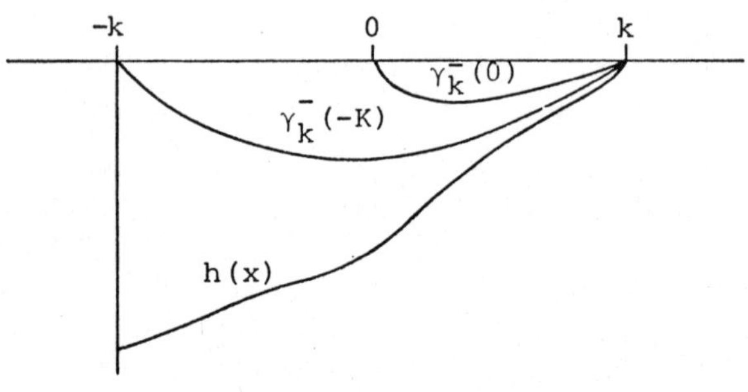

Figure 6

Then, $V_k \neq R^2$, being its boundary $\gamma_k^-(-K) \vee \gamma_{-k}^-(K)$. In particular $0_\infty = K$.

Corollary 1. If $f(k) \geq -2$ (i.e. $(k,0)$ is not an unstable 2 tg. node) then does not exist a function h fulfilling the three conditions asked in the theorem.

Proof. $f(k) < -(h'(k) + \frac{1}{h'(k)}) \leq -2$.

Corollary 2. If $f(x) < -2$, for all $x \in [0,k)$, then there exists a mapping h satisfying the hypothesis of the theorem. Such h, has the form

$$h(x) = \lambda(x-k),$$

for some real positive number λ.

Proof. Since $f(x) < -2$ for all $x \in [0,k)$, then there exists a $\lambda \in R$, such that

$$f(x) + (\lambda + \frac{1}{\lambda}) < 0 \quad \text{for all} \quad x \in [0,k).$$

The eveness of f implies

$$f(x) + \frac{x-k}{\lambda(x-k)} + \lambda < 0 \quad \text{for all} \quad x \in (-k,k).$$

Observation. When in $(LI)_o$, $f(x) = u(1-x^2)$ (VDP equation) the node $(k,0)$ always "capture" the trajectory $\gamma_k^-(-K)$. In general this is not true. For instance, set of the even function.

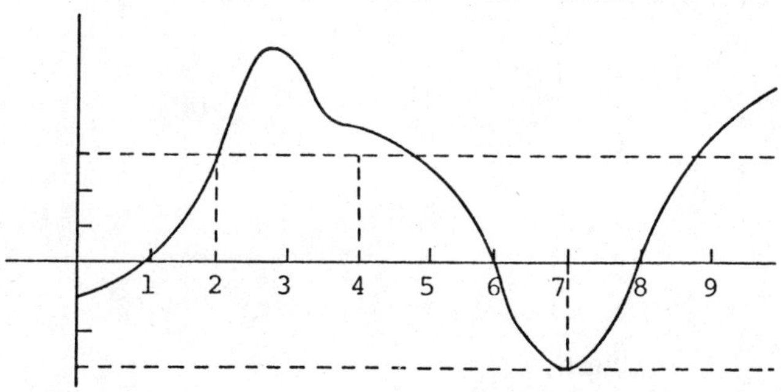

Figure 7

Considering $K = (5,0)$, we have for $k > 5$, that $V_k = R^2$ for Theorem 6 and 5. In particular $(7,0)$ is a node of $(LI)_7$, and $V_7 = R^2$.

Theorem 8. Consider the system $(LI)_k$. The circle with center $(0,0)$ and radius

$$\text{(II.2)} \qquad r_k = \min_{x:f(x)<0} \sqrt{x^2 + \frac{k^2}{(f(x))^2}}$$

lies inside V_k, for all k.

<u>Proof</u>. If (x,y) is a point in a orbit of $(LI)_k$, taking ρ as

$$\rho^2 = x^2 + y^2$$

then

$$\rho\dot{\rho} = x\dot{x} + y\dot{y} = (-f(x)y + k)y.$$

So, $y < 0$ implies that $\dot{\rho} < 0$, when

$$-f(x)y + k > 0.$$

That is, when

$$y > \frac{k}{f(x)}.$$

Then, denoting by r_k, the number

$$r_k = \min_{x:f(x)<0} \sqrt{x^2 + \frac{k^2}{(f(x))^2}}$$

if $V_k \neq R^2$, its boundary does not get into the circle with center $(0,0)$ and radius r_k, otherwise its intersection points with the x-axis could not be symmetrical with respect to zero.

<u>Corollary</u> 1. Suppose that

$$k \geq -f(x_o) \sqrt{r_o^2 - x_o^2} \; ,$$

where x_o is a point at which r_k is obtained, in (II.2), and r_o is the radius of the largest circle with center $(0,0)$ that can be inscribed into Γ_o. Then, $V_k = R^2$.

<u>Proof</u>. If

$$k \geq -f(x_o) \sqrt{r_o^2 - x_o^2} \; ,$$

then

$$\frac{k^2}{(f(x_o))^2} + x_o^2 \geq r_o^2$$

and consequently

$$r_k \geq r_o.$$

Corollary 2. Suppose $V_k \neq R^2$. Denoting by $0_{\infty 1}$ the first coordinate of 0_∞, then:

(i) if x_o is a point for which r_k is obtained as in (II.2), then $|x_o| \leq 0_{\infty 1}$.

(ii) $f(x_o) < -\dfrac{k}{\sqrt{0_{\infty 1}^2 - x_o^2}}$

(iii) if $x_o = 0$, then $f(x_o) < -\dfrac{k}{0_{\infty 1}}$.

Suppose that $0_{\infty 1} = k$, then

(i) $|x_o| < k$

(ii) $f(x_o) < -\sqrt{\dfrac{1}{1 - (\frac{x_o}{k})^2}}$

(iii) if $x_o = 0$ then $f(0) < -1$.

Proof. These results follows from the fact that $r_k < 0_{\infty 1}$.

Consider the system $(LI)_k$ and denote by R_k the number

$$\text{(II.3)} \qquad R_k = \max_{x \in [-0_F, 0_F]} \sqrt{x^2 + \frac{1}{k^2} - (xF(x))^2}$$

where $F(x) = \int_0^x f(s)ds$, and 0_F is the only positive zero point of F. Then, its possible to state:

Proposition 2. Consider the system $(LI)_k$, r_k and R_k as in (II.2) and (II.3). Then:

$$\text{(i)} \quad k_1 > k_2 \implies \begin{cases} r_{k_1} > r_{k_2} \\ R_{k_1} < R_{k_2} \end{cases}$$

$$\text{(ii)} \quad k \to 0 \implies \begin{cases} r_k \to 0 \\ R_k \to \infty \end{cases} \quad \text{and} \quad k \to \infty \implies \begin{cases} r_k \to \infty \\ R_k \to 0 \end{cases}$$

(iii) there is a number k_o, such that $r_{k_o} = R_{k_o}$.

Proof. For each line $x = a$ with $f(a) < 0$,

$$k_1 > k_2 \Longrightarrow \begin{cases} \dfrac{k_1}{f(a)} > \dfrac{k_2}{f(a)} \\ \\ \dfrac{aF(a)}{k_1} < \dfrac{aF(a)}{k_2} \end{cases} \quad \text{holds.}$$

So,

$$k_1 > k_2 \Longrightarrow \begin{cases} \left|\dfrac{k_1}{f(a)}\right| > \left|\dfrac{k_2}{f(a)}\right| \Longrightarrow \left|(a, \dfrac{k_1}{f(a)})\right| > \left|(a, \dfrac{k_2}{f(a)})\right| \\ \\ \left|\dfrac{aF(a)}{k_1}\right| < \left|\dfrac{aF(a)}{k_2}\right| \Longrightarrow \left|(a, \dfrac{aF(a)}{k_1})\right| < \left|(a, \dfrac{aF(a)}{k_2})\right| \end{cases}$$

and this implies (i).

The parts (ii) and (iii) follows from (i).

The Liénard transformation

(II.4) $\quad (x,y) \to (x,y + F(x))$

maps the solutions of $(LI)_u$ (see Proposition 1) into solutions of the process

$$(\mathcal{LI})_u \quad \begin{cases} \dot{x} = \eta - F(x) \\ \dot{\eta} = u(t) - x \end{cases}$$

Since the x coordinate remains invariant by the Liénard transformation, we have V_k and its transformed by means of (II.4) (denoted here by $\mathcal{V}_k$) identical together to R^2 or not. Moreover, the set $\mathcal{V}_k$ will be bounded if and only if there exists and orbit arc of

$$(\mathcal{LI})_k \quad \begin{cases} \dot{x} = \eta - F(x) \\ \dot{\eta} = k - x \end{cases}$$

which remains below the graph of $F(x)$ and intercepts the graph itself in two symmetrical with respect to zero, points.

After this considerations, we can prove the following theorem:

Theorem 9. Set k_o the number for which $R_{k_o} = r_{k_o}$. Then for every

k, such that $k \geq k_o$ we have $V_k = R^2$.

Proof. Let (x,η) be a point in a orbit of $(LI)_k$.

If

$$\sigma^2 = x^2 + \eta^2,$$

then

$$\sigma\dot{\sigma} = -xF(x) + k\eta.$$

Therefore

$$\text{(II.5)} \qquad \eta < \frac{xF(x)}{k} \longleftrightarrow \dot{\sigma} < 0.$$

Suppose now that V_k is different from R^2. We denote by ξ the orbit segment of $(LI)_k$ which is part of the boundary of V_k and pass by the semiplane $\eta < 0$. The arc ξ, must intercept the graph of $F(x)$ at two symmetrical with respect to zero, points. By (II.5), ξ must lies inside the circle with center $(0,0)$ and radius R_k. Thus, either $V_k = R^2$ or it belongs to this circle. Therefore, supposing $V_k \neq R^2$, since the circle with center $(0,0)$ and radius r_k lies inside V_k (Theorem 8), then we must have $R_k > r_k$.

If $k \geq k_o$ then we have an absurd, according Proposition 2, (i). This concludes the theorem.

Despite what have been done untill now, several problems in Liénard equations with control are still unsolved.

This fact can be foreseen by reading Conti's work [5]. At this work, after proposing several open questions, Conti says that those questions are more numerous than the solved ones.

Most of these questions forms themselves great "veins" that are worthy of being developed. That is the case, for example, of the questions about existence and localization of the cycles of $(LI)_k$, when $k < k_f$. Many partial results about the matter have been given by R. Conti and Gabriele Willari, and will be indicated afterwards.

B. THE SYSTEMS $(LI)_k$, WITH f BEING A STRICTLY CONVEX FUNCTION

When f at $(LI)_k$ is strictly convex, the theory presents a great cohesion.

If $k \geq 0_f$, then for every $r \geq 0$ we have, $f(k+r) + f(k-r) > 0$. Even when $k = 0_f$ and $r = 0$, the conditions of Theorem 6, part A, are fulfilled, and so it is possible to conclude that for all $k \geq 0_f$ we have $V_k = R^2$.

On the other hand, if $f(k) \leq -2$, it follows that $V_k \neq R^2$, by Corollary 2, Theorem 7, part A. So, k_f belongs to the interval $f^{-1}((-2,0))$.

The constant r_k in (II.2) will be equal to $-\frac{k}{f(0)}$.

Actually, many results of part A, can be transposed:

1. V_k contains the circle with center $(0,0)$ and radius $-\frac{k}{f(0)}$ (Theorem 8).

2. Denoting by r_o the radius of the largest circle with center $(0,0)$ lying inside Γ_o, then $k_f < -f(0).r_o$. (Corollary 1, Theorem 8).

3. If $V_k \neq R^2$, then $f(0) < -\frac{k}{0_{\infty 1}}$.

For Van der Pol equations we have the well-known result by La Salle, that Γ_o lies outside the circle with center $(0,0)$ and radius $\sqrt{3}$. So, 2. above do not improve Conti's result (in [5]), which states that $k \geq \sqrt{3}\,\mu$ implies $V_k = R^2$.

C. SYSTEMS $(LI)_{k,\mu}$, WITH f STRICTLY CONVEX, DEPENDING ON A PARAMETER $(f_\mu(x) = \mu f(x))$

Here, we slightly change the notations, to get them more suggestive. V_k will be changed, for every k and μ, by $V(k,\mu)$. For every μ, k_f will be changed by k_μ, and r_o will be denoted by r_o^μ.

If $k \geq 0_f$, by Theorem 6, part A, then $V_k = R^2$. If $k < 0_f$, k and μ can be related as:

1º) $0 < \mu < -\frac{2}{f(k)}$ ((k,0) is an unstable focus)

2º) $0 < -\frac{2}{f(k)} \le \mu$ (unstable node)

If the 2nd inequality holds, then $V(k,\mu) \neq R^2$ and $0_\infty = (k,0)$. By other hand, if the 1st holds, then $V(k,\mu)$ may be the whole R^2 or not. But, for every μ, $0 < k_\mu < 0_f$, and consequently the existence of pairs (k,μ) for which $V(k,\mu) = R^2$ or $V(k,\mu) \neq R^2$ are guaranteed.

Since $F_\mu(x) = \int_0^x f_\mu(s)ds = \mu\int_0^x f(s)ds$, then the constant k_o that came from Proposition 2, (iii), part A, have the form $-\mu f(0)\alpha$, for some α in R.

Using Theorem 3 of part A, for each pair (k,μ), such that

$$-\frac{k}{\mu f(0)} \ge \max\{r_o,\alpha\} = \beta \quad (\text{with } r_o = \max_\mu r_o^\mu, \text{ if finite})$$

we have then

$$V_k = R^2.$$

Taking the precedent results all gathered, it is possible to establish a mapping $k^*(\mu)$, such that the following configuration holds:

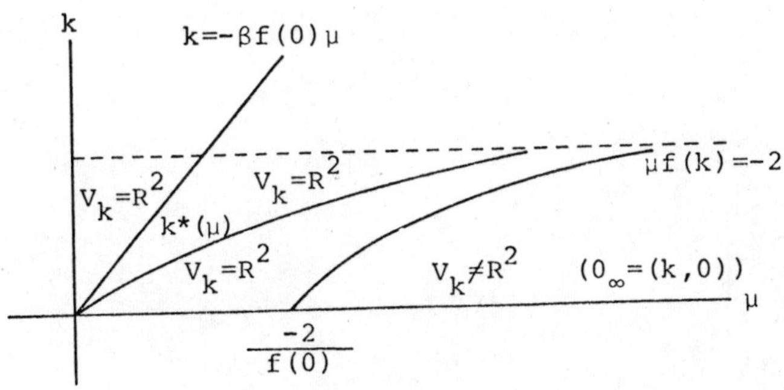

Figure 8

The mapping $k^*(\mu)$ is not well known, but using results by N. K. Alekseev ([6], Theorem 1) it is possible to set up:

<u>Theorem</u> 1. $k^*(\mu)$ is a Lipschitzian function, in any interval $[0,\mu^*]$.

One question related to the matter is the following: is $k^*(\mu)$ an increasing function ? We have not yet an answer to this question, but it is possible to demonstrate a result showing a certain regularity in this sense.

<u>Theorem</u> 2. The pairs for which the $V(k,\mu)$ boundary intercepts the x-axis at $x^* \le 0_f$, form a increasing curve $\overline{K}$ in the plane (μ,k). Moreover, this curve approaches assymptotically the line $k=0_f$, when μ increases.

<u>Proof</u>. Let us consider $\Delta_\mu > 0$, the two points (k_o,μ) and $(k_o,\mu - \Delta_\mu)$ in the (μ,k)-plane.

The slope of a point belonging to a orbit of $(LI)_{k_o,\mu}$ or $(LI)_{k_o,\mu-\Delta_\mu}$ systems, are respectively

$$y' = -f(x) + \frac{k_o - x}{y}$$

$$y'_\Delta = -\mu f(x) + \Delta_\mu f(x) + \frac{k_o - x}{y}$$

Therefore, $y' - y'_\Delta = -\Delta_\mu f(x)$, and this difference is possitive when $x \in (-0_f, 0_f)$. This difference being positive implies that every orbit of $(LI)_{k_o,\mu}$ cross the $(LI)_{k_o,\mu-\Delta_\mu}$ orbits. Consequently,

$$V(k_o,\mu-\Delta_\mu) \supseteq V(k_o,\mu).$$

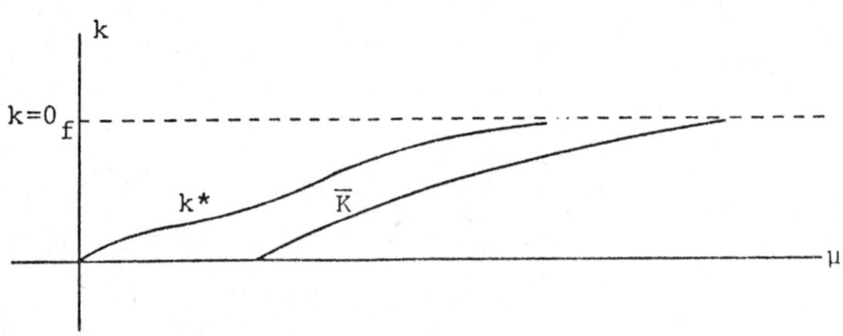

Figure 9

As said above, there are many partial results about existence and localization of the $(LI)_k$ cycles, when f in $(LI)_k$ is $\mu(x^2-1)$ (VDP equations).

When $k < 1 = 0_f$, it was proved the existence of at least one cycle associated to $(LI)_k$. E. James [3] proposes without demonstration that when μ is fixed we have uniqueness too.

R. Conti in [5], n. 10, proposes interesting results, about uniqueness of cycles.

Gabriele Villari [12], stated the following:

"Fixing μ, let x_o be the positive point when Γ_o intercepts the x-axis. Then for all $x \in (1,x_o)$, there is a k, with $0 < k < 1$, for which a $(LI)_k$ cycle pass at x. If in addition it holds uniqueness for $(LI)_k$ cycles, then there is the following 1-1 correspondence between $[1,x_o]$ and $[0,1]$: for every x in $(1,x_o)$ there is a k in $[0,1]$ such that the $(LI)_k$ cycle cross the x-axis at x. Moreover this correspondence is decreasing ($k = 0$ for x_o and $k = 1$ for 1)."

ACKNOWLODGEMENTS: My deep thanks to Prof. R. Conti and Prof. A. F. Izé by their suggestions about this matter.

REFERENCES

[1] - SANSONE, G. and CONTI, R., Equazioni differenziali non lineari, Ed. Cremonesi, Roma (1956).

[2] - NIEMYTSKII, V.V. and STEPANOV, V.V., Qualitative Theory of Differential Equations, Princenton U. Press, (1960).

[3] - JAMES, E., Time optimal control and the VDP oscillator, J. Inst. Maths. Applics, 13(1974).

[4] - VILLARI, G., Controllo del ciclo di Van der Pol, Tesi di in Matematica, Ist. "U. Dini", Univ. Firenze, 76(1975).

[5] - CONTI, R., Equazione di VDP e controllo in tempo minimo, Rapporti del Ist. Mat. "U. Dini", 13(1976/1977).

[6] - ALEKSEEV, N.K., Some questions about controllability for two-dimensional systems, (Russo) Diff. Uravnenya, 13(1970).

[7] - STAUDE, U., Uniqueness of periodic solutions of the Liénard equations, Proc. of "Recent Advances in Differential Equations", ICPT, Trieste (1978) (to appear).

[8] - LEFSCHETZ, S., Differential Equations: Geometric Theory, Interscience Publishers, New York (1957).

[9] - LEE and MARKUS, Foundations of Optimal Control Theory, J. Wiley, New York (1967).

[10] - VILLARI, G., Non esistenza di cicli per equazioni di Liénard, BUMI (1979) (to appear).

[11] - VILLARI, G., Personal Communication, 1978.

PERIODIC SOLUTIONS OF SEMILINEAR FUNCTIONAL DIFFERENTIAL EQUATIONS IN A HILBERT SPACE

by Ronald I. Becker

Abstract.

The existence of solutions in a weak sense of the following boundary value problem is proved:

$$\frac{dx}{dt} = Ax + \sum_{i=1}^{k} B_i(t,\overline{x}(t))x(t+\omega_i) + \int_{-r}^{0} B_o(t,\overline{x}(t),\theta)x(t+\theta)d\theta + f(t,x_t)$$

$$x_o = x_p$$

where $x(t)$ takes values in a Hilbert space H;

$$\overline{x}(t) = (x(t+\nu_1),\ldots, x(t+\nu_m), \int_{-r}^{0} C(t,\theta)x(t+\theta)d\theta) \quad \text{and}$$

$C(t,\theta)$ is a bounded linear operator on H, uniformly bounded in (t,θ); x_t is the function on $[-r,0]$ defined by $x_t(\theta) = x(t+\theta)$; the ω_i and ν_i lie in $[-r,0]$; $p > r > 0$; B_i $(i=0,\ldots,k)$ are bounded linear operators on H, uniformly bounded in their arguments, continuous in the second argument in H^{m+1} and measurable in t,θ; f is a function with values in H, continuous on $[0,p] \times \{$continuous functions on $[-r,0]\}$ and uniformly bounded.

The further hypotheses needed are that for each $z(t)$ continuous on $[-r,p]$, the above equation with $B_i(t,\overline{z}(t))$, $B_o(r,\overline{z}(t),\theta)$ and 0 in place of $B_i(t,\overline{x}(t))$, $B_o(t,\overline{x}(t),\theta)$ and $f(t,x_t)$ has unique solution $x \equiv 0$, and that A is an operator generating a strongly continuous semigroup which is compact for $t > 0$.

0. Introduction.

This paper applies methods of the type introduced by Opial [7] for ordinary differential equations in R^n, and extended in Becker [2] to the case of semilinear equations of evolution of compact type. The results obtained here are for semilinear functional differential equations of retarded type.

In their paper [9], Waltham and Wong consider the boundary value problem

$$(0.1) \qquad x'(t) = L(t,x_t) + f(t,x_t) \qquad (t \in [a,b])$$

$$(0.2) \qquad Mx_a + Nx_b = \psi$$

where L is continuous linear in the second argument, f has slower than linear growth in the second argument and M and N are bounded linear operators on $C([-r,0],R^n)$ (the continuous functions with domain $[-r,0]$ and range R^n). The notation x_t is standard (see section 2). It is assumed that the Cauchy problem for (0.1) has unique solution. Conditions are given that the boundary value problem (0.1) and (0.2) has a solution.

By analogy with the results of Opial [7], it could be expected that conditions of the form:

$$x' = L(t,\phi(t),x_t) + f(t,\phi_t)$$

has a unique solution satisfying the boundary conditions (0.2) for each ϕ (L linear in the third variable) rely that the nonlinear problem

$$x' = L(t,x(t),x_t) + f(t,x_t)$$

has a solution satisfying (0.2). In this paper we prove that this is indeed the case for periodic boundary conditions and somewhat less general quasi-linear portion L, but for equations taking values in a Hilbert space. The proof could be specialized in R^n to include quasi-linear portions which are more general. See section 4 for further comments.

We use nonlinear perturbations of $x' = Ax$ where A generates a semigroup $T(t)$ which is compact for $t > 0$. This type of equation seems to behave more like the R^n case in many ways than do general differential equations in Banach spaces (see Pazy [8] and Fitzgibbon [4]). It should be remarked that even if we consider equations in R^n with $A \equiv 0$, the main results below seem to be new.

Section 1 contains most of the definitions and preliminary results needed later. In particular, there are results stated on the compactness of certain integral operators in the spaces of continuous functions and square integrable functions. Section 2 deals with the existence of solutions of the Cauchy problem for functional differential equations and their continuity with respect to weakly converging perturbations. See Becker [2] for analogous results for equations of evolution. Section 3 has the proof of the existence of periodic solutions using Schauder's theorem and section 4 has concluding remarks outlining some possible directions of extension.

1. Notation and preliminary results.

We denote by H a separable Hilbert space with norm $|\cdot|$ and inner product $(\cdot,\cdot)$, and by $L(H)$ the bounded linear operators on H with norm $|\cdot|$.

For $J \subseteq R$ (the reals) a bounded interval, we denote by $C(J,H)$ the space of continuous H-valued functions on J with norm $||\phi|| = \sup_{t\in J} |\phi(t)|$; we denote by $M(J,H)$ the space of strongly measurable H-valued functions on J and by $L^p(J,H)$ the subset of $M(J,H)$ consisting of those x for which $\int_J |x(t)|^p dt < \infty$ $(1 \le p < \infty)$. The inner product in $L^2(J,H)$ will be denoted by $[\cdot,\cdot]$, and with this inner product, $L^2(J,H)$ becomes a Hilbert space if we take the usual equivalence classes. We write $[x(t),x(t)]^{1/2} = |x|_{L^2}$. Extending the above notation, we denote by $M(J,L(H))$ the functions $B(t): J \to L(H)$ such that $B(t)x$ is strongly measurable for each $x \in H$. For measure-theoretic details see [3].

Note that if $B(t) \in M(J,L(H))$ and $|B(t)| \le M$ a.e. for $t \in J$, then $B(t)$ defines a bounded linear operator $\tilde{B}$ from $L^2(J,H)$ to itself by

$$(\tilde{B}x)(t) = B(t)x(t).$$

Also $B(t,\theta)$ for $(t,\theta) \in J_1 \times J_2$, J_1 and J_2 bounded intervals, defines a bounded operator $\tilde{B}$ from $L^2(J_1,H)$ to itself such that $\tilde{B}$ is bounded and

$$(\tilde{B}x)(t) = \int_{J_2} B(t,\theta)x(\theta)d\theta$$

if $|B(t,\theta)| \le M$ a.e. on $J_1 \times J_2$, and $B(t,\theta) \in M(J_1 \times J_2, L(H))$. In what follows, $\tilde{B}$ denotes this transformation induced by $B(t)$ or $B(t,\theta)$. We denote by $L(L^2(J,H))$ the bounded linear operators on $L^2(J,H)$.

Proposition 1.1.

(a) For $M > 0$, the set

$$S_M = \{\tilde{B} \mid B(t) \in M(J,L(H)), \ |B(t)| \le M \text{ a.e. for } t \in J\}$$

is compact in the weak operator topology in $L(L^2(J,H))$.

(b) For $M > 0$, and J_1, J_2 bounded intervals, the set

$$S'_M = \{\tilde{B} \mid B(t,\theta) \in M(J_1 \times J_2, L(H)), \ |B(t,\theta)| \le M \text{ a.e. for } (t,\theta) \in J_1 \times J_2\}$$

is compact in the weak operator topology in $L(L^2(J_1 \times J_2, H))$.

Proof. (a) is proved in Becker [2] and (b) is proved in the same way with trivial modifications to take into account the two-variable case.

We remark that S_M and S'_M will be used in what follows in the sense of Proposition 1.1, after J and J_1, J_2 have been specified. The following will be useful in the sequel:

Proposition 1.2.

Let $I_r^p = [-r,p]$, $I_r = [-r,0]$, $I = [0,p]$ for some $r \ge 0$, $p \in R$. Let (x^n) be a sequence in $C(I_r^p,H)$ converging to x.

(a) Let $(B^n(t))$ be a sequence in $M(J,L(H))$ for which $(\tilde{B}^n)$ lies in S_M and for which there exists $B(t)$ such that $\tilde{B}^n \to \tilde{B}$ weakly. Then for $\theta \in I_r$, $B^n(t)x^n(t+\theta)$ is bounded a.e.

for $t \in I$, and $\tilde{B}^n x^n(t+\theta) \to \tilde{B}x(t+\theta)$ weakly in $L^2(I,H)$.

(b) Let $(B^n(t,\theta))$ be a sequence in $M(I \times I_r, L(H))$ for which $(\tilde{B}^n)$ lies in S_M' and for which there exists $B(t,\theta)$ such that $\tilde{B}^n \to \tilde{B}$ weakly. Then

$w^n(t) = \int_{I_r} B^n(t,\theta)x^n(t+\theta)d\theta$ is bounded a.e. for $t \in I$ and $(w^n(t))$ converges weakly in $L^2(I,H)$ to $w(t) = \int_{I_r} B(t,\theta)x(t+\theta)d\theta$.

Proof. (a) $x^n(t+\theta)$ converges strongly in $L^2(I,H)$ and $\tilde{B}^n$ converges in the weak operator topology. Hence $\tilde{B}^n x^n$ converges weakly. Boundedness is clear.

(b) Let $y(t) \in L^2(I,H)$. Then

$$[w^n(t),y(t)] = \int_I \left(\int_{I_r} B^n(t,\theta)x^n(t+\theta)d\theta, y(t)\right)dt$$

$$= \int_I \int_{I_r} (B^n(t,\theta)x^n(t+\theta),y(t))d\theta dt$$

$$= \int_{I \times I_r} (B^n(t,\theta)x^n(t+\theta),y(t))d\theta \times dt$$

by Fubini's theorem.

Then

$$[w^n(t)-w(t),y(t)] = \int_{I \times I_r} (B^n(t,\theta)(x^n(t+\theta)-x(t+\theta)),y(t))d\theta \times dt$$

$$+ \int_{I \times I_r} ((B^n(t,\theta)-B(t,\theta))x(t+\theta),y(t))d\theta \times dt$$

The uniform convergence of (x^n), the boundedness of $(|B^n(t,\theta)|)$ a.e. and the weak convergence of $(\tilde{B}^n)$ imply that $[w^n-w,y] \to 0$. The boundedness of $(|w_n(t)|)$ is clear.

In what follows, we will use the following notation:

I_r, I and I_r^p will be as in Proposition 1.2.

$\mathcal{B}$ will denote the set of $(k+1)$ tuples

$(B_o(t,\theta), B_1(t), \ldots, B_k(t))$

where $B_o \in M(I \times I_r, L(H))$, $B_i \in M(I, L(H))$ $i = 1, \ldots, k)$.
We will write members of $\mathcal{B}$ as $B(t,\theta)$.
We will suppose that $-r < \omega_1 < \ldots < \omega_k \le 0$ are given and we write as an abbreviation

$$\int_{I_r} d_\theta B(t,\theta)x(t+\theta) = \sum_{i=1}^{k} B_i(t)x(t+\omega_i) + \int_{I_r} B_o(t,\theta)x(t+\theta)d\theta \qquad (t \in I)$$

for $B \in \mathcal{B}$ and $x \in C(I_r^p, H)$.

While this is a formal abbreviation, the integral $\int_{I_r} d_\theta B(t,\theta)$ could be given a meaning in a different way as a vector-valued Stieltjies integral. A sequence (B^n) in $\mathcal{B}$ is said to converge weakly to B in $\mathcal{B}$ if for some $n > 0$, $(B_o^n) \subseteq S_M'$ converges in the weak operator topology on $L^2(I \times I_r, H)$ to B_o and if $(B_i^n) \subseteq S_M$ converges as $n \to \infty$ in the weak operator topology on $L^2(I,H)$ to B_i $(i = 1, \ldots, k)$. (This is formal convergence - we have introduced no topology). With this notation we may deduce from Proposition 1.2.

Corollary 1.3.

Let (x^n) be a sequence in $C(I_r^p, H)$ converging to x.

(a) Let (B^n) be a sequence converging weakly to B in $\mathcal{B}$. Then $(\int_{I_r} d_\theta B^n(t,\theta)x^n(t+\theta))$ is bounded a.e. for $t \in I$ and converges weakly in $L^2(I,H)$ to

$$\int_{I_r} d_\theta B(t,\theta)x(t+\theta).$$

(b) Let $B \in \mathcal{B}$, $|B_i(t)| \le k(t)$ a.e. $(i = 1, \ldots, k)$ and $|B_o(t,\theta)| \le k'(\theta)$ a.e. for $(t,\theta) \in I \times I_r$, where $k \in L^1(I,R)$ and $k' \in L^1(I_r, R)$. Then $w^n(t) = \int_{I_r} d_\theta B(t,\theta)x^n(t+\theta)$ converges boundedly on I to $\int_{I_r} d_\theta B(t,\theta)x(t+\theta)$.

Proof. (a) follows from Proposition 1.2 and (b) is easily proved.

We will have to do in the sequel with integral operators whose kernels are compact operators. We give two results which we will need on the compactness of these operators in L and C.

Proposition 1.4.

Let $R(t,s) : I \times I \to L(H)$ be compact for almost all $(t,s) \in I \times I$, and let $\int_{I\times I} |R(t,\tau)|^2 dt d\tau < \infty$. Then the map

$$R : L^2(I,H) \to L^2(I,H) \quad \text{defined by}$$

$$(Rx)(t) = \int_I R(t,\tau)x(\tau)d\tau$$

is compact.

Proof. See Laptev [6], Lemma 2.

We will call a map $\Phi : S \to L(H)$, where $S = \{(t,s) \mid 0 \le s \le t \le p\}$, an *evolution operator of compact type* if it satisfies

(A_1) $\Phi(t,s)$ is strongly continuous on S,

$|\Phi(t,s)| \le M$ on S, $\Phi(t,t) = \text{identity}$ and

$\Phi(t,\sigma)\Phi(\sigma,s) = \Phi(t,s) \qquad (0 \le s \le \sigma \le t)$

(A_2) $\Phi(t,s)$ is compact for $t > s$, and is continuous on $S' = \{(t,s) \mid 0 \le s < t \le p\}$ in the uniform operator topology.

Note.

1) If we merely know that $\Phi(t,s)$ is compact $(t > s)$ and that (A_1) holds, then the continuity in the uniform norm as in (A_2) follows. This is essentially due to Lax, for which see Balakrishnan [1], Theorem 4.4.1. (The proof there is for $T(t)$ a semigroup but the proof extends easily to the present case). See also Ward [10], Lemma 1.1.

2) For $\varepsilon > 0$, and s fixed, $s+\varepsilon < p$, (A_2) implies that $\Phi(t,s)$ is uniformly continuous in the uniform operator topology for t in $[s+\varepsilon,p]$.

<u>Proposition</u> 1.5.

Let $k(t) \in L^1(I,R)$ and let P_k be the set

$$P_k = \{x \in L^1(I,H) \mid |x(t)| \leq k(t) \text{ a.e. for } t \in I\}.$$

Let $R : L^1([s,p],H) \to C([s,p],H)$ be defined by

$$(Rx)(t) = \int_s^t \Phi(t,\tau)x(\tau)d\tau.$$

Then RP_k is relatively compact in $C([s,p],H)$. Further the image of a convergent sequence from P_k converges uniformly with respect to s.

<u>Proof</u>. Similar to the proofs in Pazy [8], Theorem 2.1 and Ward [10], Theorem 2.1. The uniformity in s follows from perusal of these proofs.

In what follows, we will assume that A is an (in general unbounded) operator on H which is closed and has dense domain, $\mathcal{D}(A)$. We will write the semigroup generated by A as $\Phi(t,s)$ (it could also be written $T(t)$ where $T(t) = \Phi(t''-t',0)$, $t''-t' = t$).

We will assume that $\Phi(t,s)$ corresponding to A is an evolution operator of compact type so that is satisfies (A_1) and (A_2). Such an A will be said to be of *compact type*.

<u>Proposition</u> 1.6.

Under the above conditions on A, the following holds:

(A_3) For any $f \in L^2(I,H)$ and $x_o \in H$ and $0 < s < p$ there is a unique $x(t) \in C([s,p],H)$ such that for all $y \in \mathcal{D}(A^*)$, $(x(t),y)$ is absolutely continuous on I and

$$\frac{d}{dt}(x(t),y) = (x(t),A^*y) + (f(t),y) \quad \text{a.e. on } I$$

$$x(s) = x_o.$$

The solution is given by

$$x(t) = \Phi(t,s)x_o + \int_s^t \Phi(t,\tau)f(\tau)d\tau.$$

Proof. See Balakrishnan [1], Theorem 4.8.3. Compactness of $\Phi(t,s)$, $(t > s)$ is not necessary for (A_3) to hold.

2. Existence and continuity for functional equations.

We consider equations of the form

$$\frac{dx}{dt} = Ax + \sum_{i=1}^{k} B_i(t)x(t+\omega_i) + \int_{I_r} B_o(t,\theta)x(t+\theta)d\theta + f(t)$$

$(-r < \omega_1 < \ldots < \omega_k \leq 0)$ which we abbreviate as in section 1 to

$$\frac{dx}{dt} = Ax + \int_{I_r} d_\theta B(t,\theta)x(t+\theta) + f(t) \tag{2.1}$$

We use the notation x_t for the function defined on I_r for $t \in I$ and $x(t)$ defined on I_r^p by

$$x_t(\theta) = x(t+\theta) \qquad (\theta \in I_r).$$

We also denote by $x_{\bar{t}}$ the restriction of this function to $I_{\bar{r}} = \{\theta \mid -r \leq \theta < 0\}$. We consider initial conditions

$$\begin{aligned} x_{\bar{o}} &= \phi \\ x(0) &= \xi \end{aligned} \tag{2.2}$$

where $\phi \in C(I_r,H)$ and $\xi \in H$ and we will understand by the first equation that $x_{\bar{o}}$ is equal to the restriction of ϕ to the interval $I_{\bar{r}}$. Throughout this section A will be assumed to have properties as outlined in section 1, and in particular (A_1), (A_2) and (A_3).

We establish the basic existence theorem using the method of Caratheodory.

Theorem 2.1.

Let $B \in \mathcal{B}$ and let $|B_i(t)| \leq K'$ $(i = 1, \ldots, k)$ a.e. for $t \in I$ and $|B_o(t,\theta)| \leq K'$ a.e. for $(t,\theta) \in I \times I_r$. Let $f(t) \in L^2(I,H)$. Then for each $\phi \in C(I_r,H)$ and $\xi \in H$ there exists a unique function continuous on $[s,p]$ and satisfying:

$$(2.3)\qquad \begin{aligned} x(t) &= \Phi(t,s)\xi + \int_s^t \Phi(t,\tau)\left(\int_{I_r} d_\theta B(\tau,\theta)x(\tau+\theta) + f(\tau)\right)d\tau \\ x_{\bar{s}} &= \phi \end{aligned}$$

Further, $|x(t)| \leq K$ where K depends only on K', M of (A_1) and r and p.

Proof. Define the sequence $(x^n(t))$ of functions on $[s-r,p]$ by

$$(2.4)\qquad \begin{aligned} x^n_{\bar{s}}(\theta) &= \phi(\theta) && (\theta \in I_r) \\ x^n(t) &= \xi && (t \in [s,s+\tfrac{p}{n}]) \\ &= \Phi(t,s)\xi + \int_s^{t-p/n} \Phi(t,\tau)\left(\int_{I_r} d_\theta B(\tau,\theta)x^n(\tau+\theta) + f(\tau)\right)d\tau && (t \in [s+\tfrac{p}{n},p]). \end{aligned}$$

It is easily seen, using the method of steps that $x^n(t)$ is well-defined on $[s-r,p]$ for sufficiently large n, and $\in C([s-r,p],H)$. Further, since $x^n_\tau(\theta) = x^n(\tau+\theta)$, we have

$$|x^n(t)| \leq |\Phi(t,s)\xi| + K'M\int_s^{t-p/n} ||x^n_\tau||d\tau + K''||f||_{L^2}.$$

Hence

$$||x^n_t|| \leq (M|\xi| + ||\phi||) + K'M\int_s^t ||x^n_\tau||d\tau + K''||f||_{L^2}$$

and by Gronwall's lemma,

$$(2.5)\qquad ||x^n_t|| \leq (M|\xi| + ||\phi|| + K''||f||_{L^2})e^{K'Mp}$$

Using the analogue of Proposition 1.5 for $[s,p]$, and writing the last equation in (2.4) as

$$x^n(t) = \Phi(t,s)\xi + \int_s^t \Phi(t,\tau)\chi_{[s,t-\frac{p}{n}]}\left(\int_{I_r} d_\theta Bx^n_\tau + f(\tau)\right)d\tau$$

where $\chi_{[a,b]}$ is the characteristic function of $[a,b]$, we see that the boundedness of x^n_τ implies the existence of a uniformly convergent subsequence of $(x^n(t))$ on $[s-r,p]$. Taking limits in

(2.4) and using Corollary 1.3, (b), we obtain the existence of a solution. Uniqueness and the nature of the bound on the solution follows from (2.5). Q.E.D.

We now define evolution-type linear operators associated with the equation (2.1). We call these the *fundamental operators* of (2.1).

Let $T(t,s) : C(I_r,H) \to H$ be the linear operator defined by the requirement that $T(t,s)\phi$ is the value at t of the solution of (2.3) with $f \equiv 0$ and with $\xi = \phi(s)$. We will also use the map $T_t(\cdot,s) : C(I_r,H) \to C(I_r,H)$ defined by $T_t(\cdot,s)\phi = (T(t,s)\phi)_t$ $(0 \le s \le t \le p)$.

Let $U(t,s) : H \to H$ be defined by the requirement that $U(t,s)\xi$ is the value at t of the solution of (2.3) with $f \equiv 0$ and with $\phi \equiv 0$. Note that $U(t,s)\xi$ is continuous on $[s,p]$ and that it is zero on $[s-r,s)$.

Theorem 2.2.

Let the conditions of Theorem 2.1 hold. Then $T(t,s)$ and $U(t,s)$ are strongly continuous in $0 \le s \le t \le p$. Further, $T(t,s)$ is compact for each $t > s$, and $T_t(\cdot,s)$ is compact for each $t > s+r$. T and U are bounded on $0 \le s \le t \le p$ by constants depending only on K, M, r and p.

Further, the solution $x(t)$ of Theorem 2.1 with $x(s) = \phi(0)$ is the unique continuous function satisfying

(i) $(x(t),y)$ is absolutely continuous on $[s,p]$ for $y \in \mathcal{D}(A^*)$

(ii) $$\frac{d}{dt}(x(t),y) = (x(t),A^*y) + \left(\int_{I_r} d_\theta B(t,\theta)x(t+\theta) + f(t),y\right)$$

a.e. on $I(y \in \mathcal{D}(A^*))$

(2.6) $$x_s = \phi$$

and it is given by

$$(2.7) \qquad x(t) = T(t,s)\phi + \int_s^t U(t,\tau)f(\tau)d\tau$$

Proof. For continuity, we need only show that the functions $T^n(t,s)\phi$ defined by (2.4) (with $f \equiv 0$, $\xi = \phi(s)$) are uniformly convergent in (t,s) $(0 \le s \le t \le p)$. Since for $t' \ge s' \ge s$ and $t \ge s$ we have

$$T^n(t',s')\phi - T^n(t,s)\phi =$$
$$= (T^n(t',s') - T^n(t',s))\phi + (T^n(t',s) - T^n(t,s))\phi,$$

we need only show that $\{T^n(t,s)\phi\}$ is equicontinuous in n and s, and also in n and t. Then the subsequence which converges uniformly in Theorem 2.1, and which converges for each (t,s) $(0 \le s \le t \le p)$ by Proposition 1.5, will converge to a continuous function. Uniformity in n, s follows from Proposition 1.5, while uniformity in n, t follows easily from the definition in (2.4) since s only appears as a lower limit of integration. A similar argument holds for $U(t,s)$.

Boundedness follows from the boundedness of $x(t)$ in Theorem 2.1. We show compactness of T_t - that of $T(t,s)$ is similar but simpler. Note firstly that $T(t,s)\phi$ satisfies (2.3) with $f \equiv 0$ and hence

$$(2.8) \qquad T(t,s)\phi = \Phi(t,s)\phi(s) + \int_s^t \Phi(t,\tau)\left(\int_{I_r} d_\theta B(\tau,\theta)T(\tau+\theta,s)\phi\right)d\tau.$$

For ϕ lying in a bounded set with bound C in $C(I_r,H)$, the set of values of

$$x(t) = \int_{I_r} d_\theta B(t,\theta)T(t+\theta,s)\phi$$

lies in the set $\{x \mid |x(t)| \le k_1(t)\}$ for some $k_1(t) \in L^1(I,R)$. This follows from the boundedness of T and the assumptions on B. By Proposition 1.5, the integrals on the righthand side of (2.8), for varying ϕ, have a uniformly convergent subsequence on $[t-r,t]$ (remembering that $t-r > s$).

Since $\phi(t,s)$ (for fixed s) is continuous in operator norm on

$[t-r,t]$ by (A_2), it follows that

$$|\Phi(t',s)\phi(s) - \Phi(t",s)\phi(s)| \le |\Phi(t',s) - \Phi(t",s)|C$$

and equicontinuity of $\{\Phi(t,s)\phi\}$ on $[t-r,t]$ follows from the uniform continuity of $\Phi(t,s)$ on $[t-r,t]$.

Since $\Phi(t,s)$ is compact for $t > s$, we may apply the Arzela-Ascoli theorem to get a uniformly convergent subsequence on $[t-r,t]$ from $\{\Phi(t,s)\phi\}$. Hence by (2.8), $\{T(t,s)\phi\}$ has a uniformly convergent subsequence on $[t-r,t]$ and this implies the compactness of $T_t(\cdot,s)$.

To prove the uniqueness of solutions of (i) and (ii):

If x,y are solutions, then $z = x-y$ is a solution of

(i)' $(z(t),y)$ is absolutely continuous for $y \in \mathcal{D}(A^*)$

(ii)' $\frac{d}{dt}(z(t),y) = (z(t),A^*y) + (\int_{I_r} d_\theta B(t,\theta)z(t+\theta),y)$

$$z_s = 0$$

By Proposition 1.6, $z(t)$ satisfies

$$z(t) = \int_s^t \Phi(t,\tau)(\int_{I_r} d_\theta B(\tau,\theta)z(\tau+\theta))d\tau$$

$$z_s = 0$$

and this has unique solution 0 by Theorem 2.1.

That the solution $x(t)$ of Theorem 2.1 satisfies (i) and (ii) follows from Proposition 1.6. It remains to prove that (2.7) holds. We prove this by showing that the function $x(t)$ defined by the right-hand side of (2.8) satisfies (i) and (ii) and then using the uniqueness just proved to show that $x(t)$ is the solution of Theorem 2.1.

The function $w(t) = T(t,s)\phi$ satisfies (i) and also (ii) with $f \equiv 0$. It suffices then to show that the function

$$v(t) = \int_s^t U(t,\tau)f(\tau)d\tau$$

satisfies (i) and also (ii) with $\phi = 0$. Clearly, $v_s = 0$ since $U(t,s) = 0$ for $s > t$. For $t \geq s$, we have

$$(v(t),y) = \int_s^t (U(t,\tau)f(\tau),y)d\tau$$

$$= \int_s^t (\Phi(t,\tau)f(\tau),y)d\tau + \int_s^t (\int_\tau^t \Phi(t,\sigma)g(\tau,\sigma)d\sigma,y)d\tau$$

where $g(\tau,\sigma) = \int_{I_r} d_\theta B(\sigma,\theta)U(\sigma+\theta,\tau)f(\tau)$.

Hence

$$(v(t),y) = \int_s^t (f(\tau),\Phi^*(t,\tau)y)d\tau + \int_s^t\int_\tau^t (g(\tau,\sigma),\Phi^*(t,\sigma)y)d\sigma d\tau.$$

Differentiating with respect to t then gives, for $y \in \mathcal{D}(A^*)$,

(2.9) $$\frac{d}{dt}(v(t),y) = (f(t),y) + \int_s^t (f(\tau),\Phi^*(t,\tau)A^*y)d\tau$$

$$+ \int_s^t (g(\tau,t),\Phi^*(t,t)y)d\tau + \int_s^t\int_\tau^t (g(\tau,\sigma),\Phi^*(t,\sigma)A^*y)d\sigma d\tau$$

(Justification for the differentiation is similar to that of Balakrishnan [1], Theorem 4.8.3, and we omit the lengthy but standard argument.)

$$= (f(t),y) + (\int_s^t \Phi(t,\tau)f(\tau)d\tau,A^*y) + \int_s^t (g(\tau,t),y)d\tau$$

$$+ \int_s^t (\int_\tau^t \Phi(t,\sigma)g(\tau,\sigma)d\sigma,A^*y)d\tau$$

$$= (f(t),y) + v(t),A^*y) + (\int_s^t (\int_{I_r} d_\theta B(t,\theta)U(t+\theta,\tau)f(\tau))d\tau,y)$$

$$= (f(t),y) + (v(t),A^*y) + (\int_{I_r} d_\theta B(t,\theta)\int_s^t U(t+\theta,\tau)f(\tau)d\tau,y)$$

But $\int_s^t U(t+\theta,\tau)f(\tau)d\tau = \int_s^{t+\theta} U(t+\theta,\tau)f(\tau)d\tau$ (since $U(t+\theta,\tau) = 0$ for $\tau > t+\theta$)

$$= v(t+\theta).$$

Hence by (2.9) $(x(t),y)$ satisfies the desired equation.

Theorem 2.3.

Let $(B^n) \subseteq \mathcal{B}$ be such that $B^n \to B$ weakly in $\mathcal{B}$ (see discussion after Proposition 1.2 for definitions) and let $|B_i^n| \le K'$ a.e. $(i = 0, \ldots, k)$. Denote by $T^n(t,s)$ and $U^n(t,s)$ the fundamental operator of (2.1) with coefficient B^n, and $T(t,s)$ and $U(t,s)$ those with coefficient B. Then

$$T^n(t,s)\phi \to T(t,s)\phi, \quad U^n(t,s)\xi \to U(t,s)\xi$$

uniformly in t on $s \le t \le p$ for each $\phi \in C(I_r,H)$, $\xi \in H$, and boundedly on $0 \le s \le t \le p$.

Proof.

$$(2.10) \qquad T^n(t,s)\phi = \Phi(t,s)\phi(s) + \int_s^t \Phi(t,\tau)\left(\int_{I_r} d_\theta B^n(\tau,\theta)T^n(\tau+\theta,s)\phi\right)d\tau$$

by Theorem 2.2, and also $|T^n(t,s)| \le K$ for K depending on K', M, r, p. Hence $w^n = \int_{I_r} d_\theta B^n(t,\theta)T^n(t+\theta,s)\phi$ is bounded a.e. By Proposition 1.5 and the boundedness of (w^n), it follows from (2.10) that there is a subsequence with indices n_k, say, of $(T^n(t,s)\phi)$ converging uniformly on $[s,p]$, with limit $T_1(t,s)\phi$, say. By Corollary 1.3, (w^n) converges weakly in $L^2([s,p],H)$ to w say. By Proposition 1.4, $\int_s^t \Phi\cdot$ is compact from L^2 to L^2, and so $\int_s^t \Phi w^n$ converges strongly in L^2 to $\int_s^t \Phi w$. Since $T^{n_k}(t,s)\phi$ converges uniformly to T_1, Corollary 1.3 implies that $w(t) = \int_{I_r} d_\theta B(t,\theta)T_1(t+s)\phi$. Hence

$$T_1(t,s)\phi = \Phi(t,s)\phi + \int_s^t \Phi(t,\tau)\left(\int_{I_r} d_\theta B(\tau,\theta)T_1(\tau+\theta,s)\phi\right)d\tau$$

and by uniqueness, $T_1(t,s)\phi = T(t,s)\phi$. Since every subsequence has a subsequence convergent to $T(t,s)\phi$, the whole sequence converges. A similar argument holds for U^n.

Corollary 2.4.

Let (f_n) satisfy $|f_n(t)| \le C$ a.e. on I and converge to f in $L^2(I,H)$; let B^n, B be as in the statement of the theorem and let

$\phi^n \to \phi$ in $C(I_r,H)$. Then the solution (x^n) (in the sense of (ii) in Theorem 2.2) of

$$\frac{dx^n}{dt} = Ax^n + \int_{I_r} d_\theta B^n(t,\theta)x^n(t+\theta) + f^n(t)$$

$$x_s^n = \phi^n$$

converges uniformly on $[s-r,p]$ to the solution $x(t)$ of

$$\frac{dx}{dt} = Ax + \int_{I_r} d_\theta B(t,\theta)x(t+\theta) + f(t)$$

$$x_s = \phi$$

Proof. Similar to that of the Theorem.

Corollary 2.5.

Let B^n, T^n be as in the theorem. Let (ϕ^n) be a bounded sequence in $C(I_r,H)$. Then for each $t > s+r$, the sequence $\{T_t^n(\cdot,s)\phi^n\}$ has a subsequence convergent in $C(I_r,H)$.

Proof. We have by Theorem 2.2, that

$$T^n(t,s)\phi^n = \Phi(t,s)\phi^n(s) + \int_s^t \Phi(t,\tau)\left(\int_{I_r} d_\theta B^n(\tau,\theta)T^n(\tau+\theta,s)\phi^n\right)d\tau.$$

By Theorem 2.2, $(T^n(t,s))$ is bounded, and this together with the boundedness of (ϕ^n) and the hypotheses on B^n imply that $w^n = \int_{I_r} d_\theta B^n(t,\theta)T^n(t+\theta,s)\phi^n$ is bounded a.e. on $[s,p]$. By Proposition 1.5, $\int_s^t \Phi(t,\tau)w^n(\tau)d\tau$ has a uniformly convergent subsequence.

As in the proof of Theorem 2.2, $(\Phi(t,s)\phi^n(0))$ has a subsequence converging uniformly on $[t-r,t]$ for each t satisfying $t-r > s$. Hence the result.

Corollary 2.6.

Let T^n, T be as in the theorem. Let λ belong to the resolvent set of $T_p^n(\cdot,0)$ for each n and to that of $T_p(\cdot,0)$. Then if $p > s+r$, we have

$$(\lambda - T_p^n(\cdot,0))^{-1} \to (\lambda - T_p(\cdot,0))^{-1}$$

strongly.

Proof. Follows from Corollary 2.5 and from Corollary 2.5 in Becker [2].

3. Periodic solutions.

For $x \in C(I_r^p,H)$ we define

$$\overline{x}(t) = (x(t+\nu_1), \ldots, x(t+\nu_m), \int_{I_r} C(t,\theta)x(t+\theta)d\theta)$$

where $-r \le \nu_1 < \ldots < \nu_m \le 0$ are fixed throughout the sequel, as is $C(t,\theta) \in L^1(I \times I_r, L(H))$, $|C(t,\theta)|$ uniformly bounded a.e. in (t,θ).

We consider equations of the form

$$(3.1)' \quad \frac{dx}{dt} = Ax + \sum_{i=1}^{k} B_i(t,\overline{x}(t))x(t+\omega_i) + \int_{I_r} B_o(t,\overline{x}(t),\theta)x(t+\theta)d\theta + f(t,x_t)$$

which we abbreviate in the usual way to

$$(3.1) \quad \frac{dx}{dt} = Ax + \int_{I_r} d_\theta B(t,\overline{x}(t),\theta)x(t+\theta) + f(t,x_t).$$

A *mild solution* of (3.1) is a continuous function satisfying for some $\phi \in C(I_r,H)$ the equations

$$x(t) = \Phi(t,0)\phi(0) + \int_0^t \Phi(t,\tau)(\int_{I_r} d_\theta B(\tau,\overline{x}(\tau),\theta)x(\tau+\theta) + f(\tau,x_\tau))d\tau \quad (t \in I)$$

$$x_o = \phi$$

A *periodic solution* of (3.1) is a mild solution satisfying

$$(3.2) \quad x_o = x_p$$

Theorem 3.1.

Let A be of compact type, let $p > r$, and let ω_i, ν_i and $C(t,\theta)$ be as above. Let $B_i(t,x_1,\ldots,x_{m+1})$ $(i = 1, \ldots, k)$ and $B_o(t,x_1,\ldots,x_{m+1},\theta)$ be measurable for fixed $x_1, \ldots, x_{m+1} \in H$ and continuous for almost all fixed t,θ. As before, denote $(B_o,B_1,\ldots,B_k)$ by $B(t,x_1,\ldots,x_{m+1},\theta)$. Let $f : I \times C(I_r,H) \to H$ be continuous and bounded by M'' on $I \times C(I_r,H)$. Let $\tilde{B} = (\tilde{B}_o,\ldots,\tilde{B}_k)$ lie in a weakly closed subset S of $S'_{M'} \times S_{M'} \times \ldots \times S_{M'}$ (see Proposition 1.1).

Suppose that for all $\tilde{B} \in S$, the equation

$$\frac{dx}{dt} = Ax + \int_{I_r} d_\theta B(t,\theta)x(t+\theta)$$

has only the zero solution satistying (3.2) in the sense of Theorem 2.2, (ii). Then (3.1) has a mild periodic solution.

Proof. We apply the Schauder fixed point theorem. Firstly, we show that if $h(t) \in M(I,H)$ and $|h(t)| \leq M''$ a.e. for $t \in I$, then for $\tilde{B} \in S$,

$$\frac{dx}{dt} = Ax + \int_{I_r} d_\theta B(t,\theta)x(t+\theta) + h(t) \tag{3.3}$$

has unique periodic solution $y(t)$, and there exists a constant K depending only on M', M'', A, r and p such that

$$|y(t)| \leq K \qquad (t \in I_r^p).$$

Let $T(t,s)$, $U(t,s)$ be the fundamental operators associated with B. Then there exists a solution y of (3.3) in the sense of Theorem 2.2, (ii), satisfying (3.2) iff there is a $\phi \in C(I_r,H)$ such that

$$(3.4) \qquad \begin{aligned} y(t) &= T(t,0)\phi + \int_0^t U(t,\tau)h(\tau)d\tau \qquad (t \in I) \\ y_p &= \phi. \end{aligned}$$

We can write the last condition as

$$(1 - T_p(\cdot,0))\phi = \left(\int_0^p U(p,\tau)h(\tau)d\tau\right)_t. \tag{3.5}$$

By the uniqueness assumption of the theorem, there does not exist a $\phi \in C(I_r,H)$ such that

$$(1 - T_p(\cdot,0))\phi = 0$$

So, by compactness of $T_p(\cdot,0)$, 1 lies in the resolvent set of $T_p(\cdot,0)$. We will show that $(1 - T_p(\cdot,0))^{-1}$ is bounded by a constant depending on M', A, r and p. If not, there is a sequence $(B^n) \subseteq \mathcal{B}$ satisfying the uniqueness bypothesis of the theorem and such that $\tilde{B}^n \in S$

and such that $(1 - T_p^n(\cdot,0))^{-1}$ is unbounded, where T^n is the fundamental operator corresponding to B^n. S is weakly compact in

$$L(L^2(I \times I_r,H)) \times L(L^2(I,H)) \times \ldots \times L(L^2(I,H)),$$

being a closed subset of a compact set by Proposition 1.1. So there exists B^o and a subsequence n_k such that $\tilde{B}^o \in S$ and $B^{n_k} \to B^o$ weakly in $\mathcal{B}$. If T^o is the fundamental operator corresponding to B^o, then by uniqueness, 1 lies in the resolvent set of $T_p^o(\cdot,0)$. By Corollary 2.7, $(1 - T_p^{n_k}(\cdot,0))^{-1}$ is uniformly bounded. Since any subsequence contains a subsequence for which these inverses are bounded, it follows that the whole sequence $(1 - T_p^n(\cdot,0))^{-1}$ is bounded, contradicting the assumption of unboundedness. Hence $(1 - T_p(\cdot,0))^{-1}$ is uniformly bounded as stated. By Theorem 2.2, T and U are bounded by a constant with the same dependency and so (3.5) implies ϕ is bounded and (3.4) implies that y is bounded.

Given $x \in C(I_r^p,H)$, define the map G, $G : C(I_r^p,H) \to C(I_r^p,H)$, $y = Gx$ by the requirement that y be the unique mild periodic solution of

$$\frac{dy}{dx} = Ay + \int_{I_r} d_\theta B(t,\overline{x}(t),\theta)y(t+\theta) + f(t,x_t).$$

We will show that G is compact and continuous.

Firstly compactness. By Proposition 1.6 we have

$$(3.6)\qquad \begin{aligned} (Gx)(t) &= y(t) = \Phi(t,0)\phi(0) + \int_0^t \Phi(t,\tau)\Big(\int_{I_r} d_\theta B(\tau,\overline{x}(\tau),\theta)y(\tau+\theta) + f(\tau,x_\tau)\Big)d\tau \qquad (t \in I)\\ (Gx)_p &= \phi \qquad (t \in I_r)\end{aligned}$$

for some $\phi \in C(I_r,H)$. Let (x^n) be a bounded sequence in $C(I_r^p,H)$. Then $y^n = Gx^n$ is uniformly bounded by the above discussion. Using the hypotheses on B and f, it follows that

$$\Big(\int_{I_r} d_\theta B(\tau,\overline{x}^n(\tau),\theta)y^n(\tau+\theta) + f(\tau,x_\tau^n)\Big)$$

is bounded a.e. on I. Hence by Proposition 1.5, the integrals on the right of (3.6) with x^n, y^n in place of x, y have a uniformly convergent subsequence with indices n_k. We may also suppose that $B(\tau,\overline{x}^{n_k}(\tau),\theta)$ converges in $\mathcal{B}$, and that $f(\tau,\overline{x}^{n_k})$ converges in L^2. Also, for any $0 < \varepsilon < p$, $\Phi(t,0)\phi^n(0)$ has a subsequence converging uniformly on $[\varepsilon,p]$ (since $\phi^n(0)$ is bounded by the above discussion and since we can then use an argument similar to that in Theorem 2.2 for the same purpose). Hence for such an ε, (y^n) has a subsequence (y^{n_k}) converging uniformly on $[\varepsilon,p]$. But $y_p^{n_k} = \phi^{n_k}$, so since $p > r$ we have ϕ^{n_k} uniformly convergent. By Corollary 2.4, this implies that y^{n_k} converges uniformly on I_r^p. Hence G is compact.

To prove continuity of G, let $x^n \to x$ in (I_r^p,H), let $y^n = Gx^n$ and let $y_o^n = \phi^n$. Then by the boundedness and continuity hypothesis on B in the statement of this theorem, it follows that

$$B(t,\overline{x^n}(t),\theta) \to B(t,\overline{x}(t),\theta) \quad \text{in } \mathcal{B} \text{ see definition before Corollary 1.3}$$

Also, $f(t,x_t^n)$ converges boundedly to $f(t,x_t)$, hence it converges in $L^2(I,H)$. By the compactness of G there is a subsequence (y^{n_k}) converging uniformly on I_r^p. Hence (ϕ^{n_k}) converges uniformly since $\phi^{n_k} = y_o^{n_k}$, and by Corollary 2.4, (y^{n_k}) converges to $y = Gx$. Since any subsequence has a subsequence converging uniformly to $y = Gx$, it follows that (y^n) converges to this y.

Thus G takes values in a fixed ball and is compact and continuous. By Schauder's theorem, there is a fixed point which is the desired periodic solution.

Remarks.

1) It is easily seen that x is a periodic mild solution of (3.1) iff for all $y \in \mathcal{D}(A^*)$

(i) $(x(t),y)$ is absolutely continuous on I and

(ii) $\frac{d}{dt}(x(t),y) = (x(t),A^*y) + (\int_{I_r} d_\theta B(t,\overline{x}(t),\theta)x(t+\theta) + f(t,x_t),y)$ a.e. on I

$x_o = x_p$

2) The sort of weakly closed subsets in the statement that could be useful are those of the form

$$|\lambda - B_i| \le C_i \quad \text{a.e.} \quad (i = 0, \ldots, k).$$

This set is a translation of a weakly compact set, and is compact by continuity of translation.

4. Concluding remarks.

We have only used $\int_{I_r} d_\theta B(t,\theta)x(t+\theta)$ as an abbreviation. If we could introduce a vector integral having the properties of Corollary 1.3, and also having an analogous property to that out-lined in Proposition 1.1 (e.g. if $\int |d_\theta B^n| < K$ implies the existence of a *weakly convergent* subsequence B^{n_k} such that the map $x \to \int d_\theta B^{n_k}(t,\theta)x(t+\theta)$ converges in the weak operator topology on $L^2(I,H)$) then the rest of the argument would go through as before. We hope to discuss the measure theory involved at another time.

For the case $H = R^n$, we may take $A \equiv 0$ in the foregoing discussion, since $x' = 0$ generates the identity semigroup which is of compact type in R^n. The specialization induced in Theorem 3.1 seems to be a new result even in this case. Waltham and Wong have discussed the situation in R^n for which B is independent of x and $f(t,x)$ has asymptotically sublinear growth. The treatment given here could be extended to this type of f. On the other hand, Waltham and Wong assume the uniqueness of the Cauchy problem associated with (3.1). The above treatment does not require such an assumption.

In the case $H = R^n$, the above measure theory requirements could be proved using slight generalizations of Helly's theorem, etc. Thus Theorem 3.1 holds for the case in which the righthand side of (3.1) is in fact a Stieltjies integral. We will not go into this here, hoping to give a more general treatment with vector measures having infinite dimensional ranges at a later date.

REFERENCES

[1] - BALAKRISHNAN, A.V., Applied Functional Analysis, Springer-Verlag, New York, (1976).

[2] - BECKER, R.I., Periodic solutions of semilinear equations of evolution of compact type, Submitted - J. Diff. Equations.

[3] - DUNFORD, N. and SCHWARTZ, J.T., Linear Operators, Vol. I, Interscience, New York, (1958).

[4] - FITZGIBBON, W.E., Semilinear functional differential equations, J. Diff. Equations, 29 (1978), 1-14.

[5] - HALE, J.K., Functional Differential Equations, Springer-Verlag, New York, (1971).

[6] - LAPTEV, G.I., Eigenvalue problems for second-order differential equations in Banach and Hilbert spaces, Differential'nye Uravneniya, 2 (9), (1966), 1151-1160.

[7] - OPIAL, Z., Linear problems for systems of nonlinear differential equations, J. Diff. Equations, 3 (1967), 580-594.

[8] - PAZY, A., A class of semilinear equations of evolution, Israel J. Math., 20 (1), (1975), 23-36.

[9] - WALTHAM, P. and WONG, J.S.W., Two point boundary value problems for nonlinear functional differential equations, Trans. A.M.S. 164 (1972), 39-54.

[10] - WARD, J.R., Semilinear boundary value problems in Banach space, In Nonlinear Equations in Abstract Spaces, V. Lakshmikantham (ed.), Academic Press (1978), 469-477.

STABILITY OF NONCONSERVATIVE LINEAR SYSTEMS

by J. Carr and M.Z.M. Malhardeen

1. Introduction.

In this paper we study the stability of some linear nonconservative problems of the form

$$\ddot{u} + Ku = 0, \quad t > 0, \quad u(0) = u_o, \quad \dot{u}(0) = u_1, \tag{1.1}$$

where u is in a Hilbert space H and K is a perturbation of a positive self-adjoint operator. The following problem provided the motivation for this study: for what values of p is the zero solution of the following system stable ?

$$\begin{aligned} &u_{tt} + u_{xxxx} + pu_{xx} = 0, \quad t > 0, \quad 0 < x < 1, \\ &u(0,t) = u_x(0,t) = u_{xx}(1,t) = u_{xxx}(1,t) = 0, \quad t \geq 0, \end{aligned} \tag{1.2}$$

with given initial conditions $u(x,0)$, $u_t(x,0)$, $x \in (0,1)$. With appropriate normalizations $u(x,t)$ represents the displacement of a thin elastic rod, fixed at $x = 0$, and subjected at $x = 1$ to a compressive tangential load of magnitude p. For $p > 0$, the problem is nonconservative due to the follower nature of the load.

The above problem, known as Beck's Problem, has been studied in the engineering literature by means of an eigenvalue analysis. It is by no means obvious that the eigenvalue analysis is justified and in Section 2 we give an example where this procedure gives the wrong result.

In Section 2 we compare and contrast the cases in which (i) H is finite dimensional and (ii) H is infinite dimensional. In Section 3 we briefly discuss the situation in which K is a positive self-adjoint operator. In Section 4 we study the stability of solutions of (1.1). For the class of problems that we consider, our results show that the instability mechanisms associated with (1.1) are finite dimensional in character. In particular this reduces the stability analysis to an eigenvalue analysis.

In Section 5 we apply these results to Beck's Problem. We prove that there exists $p_1 > 0$ such that the zero solution of (1.2) is stable for $0 < p < p_1$ and unstable for $p \geq p_1$. This improves the results obtained in [1] and [2]. Finally, in Section 6 we study the stability of a two parameter problem.

2. Instability mechanisms.

To obtain some insight into the mechanisms by which instability can occur in (1.1) we first consider a finite dimensional problem. Let $u \in R^n$ satisfy the linear equation

(2.1) $\quad \ddot{u} + K(p)u = 0$

where $K(p)$ is a real $n \times n$ matrix depending continuously on the real parameter p. The following result follows immediately from the theory of matrices.

Lemma 2.1. The zero solution of (2.1) is stable if and only if $K(p)$ is similar to a positive self-adjoint matrix.

Suppose that $K(0)$ has eigenvalues $\{\lambda_j(0)\}$ with $0 < \lambda_1(0) <$ $< \lambda_2(0) < \ldots < \lambda_n(0)$, so that the zero solution of (2.1) is stable when $p = 0$. As p is increased, the eigenvalues of $K(p)$ vary continuously in the complex plane. The zero solution of (2.1) becomes unstable only when

(i) an eigenvalue of $K(p)$ crosses the origin,

or

(ii) two eigenvalues of $K(p)$ coincide and then go complex. [We use the convention that λ is complex if it has nonzero imaginary part].

Case (i) is called Divergence instability and case (ii) is called Flutter instability.

Divergence and Flutter are the only ways in which a finite dimensional system of the form (2.1) may become unstable. For infinite dimensional systems the situation is much more complicated. The following example shows that even if the eigenvalues of K are all distinct and positive then the zero solution of (1.1) may be unstable.

<u>Example</u>. Let H be the sequence space ℓ^2. For $u = (u_n) \in \ell^2$ define linear operators A and B by

$$(Au) = (a_n), \quad (Bu) = (b_n)$$

where

$$a_{2n-1} = n^2 u_{2n-1}, \quad a_{2n} = (n^2+n^{-2})u_{2n},$$
$$b_{2n-1} = n^{-1}u_{2n}, \quad b_{2n} = 0.$$

The domain of A, $D(A) = \{(u_n) \in \ell^2 : \sum_{n=1}^{\infty} n^4 u_n^2 < \infty\}$ and A is a positive self-adjoint operator with compact resolvent. The operator B is compact and the eigenvalues of $A+B$ are positive and distinct.

We study the stability of the zero solution of

(2.2) $\quad \ddot{u} + (A+B)u = 0$

Since A is positive we can define $A^{1/2}$ with $D(A^{1/2}) = \{(u_n) \in \ell^2 : \sum_{n=1}^{\infty} n^2 u_n^2 < \infty\}$. Let X be the Hilbert space $D(A^{1/2}) \times \ell^2$ with inner product

$$\langle x_1, x_2 \rangle = (A^{1/2}y_1, A^{1/2}y_2) + (z_1, z_2)$$

$$x_i = \begin{bmatrix} y_i \\ z_i \end{bmatrix}$$

where $(\cdot,\cdot)$ is the inner product in ℓ^2. We can now rewrite (2.2) as

(2.3) $\quad \dot{x} = Fx$

where

$$x = \begin{bmatrix} u \\ \dot{u} \end{bmatrix}, \quad F = \begin{bmatrix} 0 & I \\ -(A+B) & 0 \end{bmatrix}$$

It is straightforward to check that F generates a strongly continuous group $T(t)$. We show that for any constant C, there exists t (depending on C) such that $||T(t)||_X \geq C$. This shows that the zero solution of (2.3) is unstable.

Let m be a positive integer and let

$$x(0) = \begin{bmatrix} u \\ 0 \end{bmatrix} \tag{2.4}$$

where

$$(u)_n = \delta_{2m,n}\alpha_m^{-1}, \qquad \alpha_m = (m^2+m^{-2})^{1/2}$$

An easy computation shows that the solution of (2.3) with $x(0)$ given by (2.4) is

$$x(t) = \begin{bmatrix} u(t) \\ \dot{u}(t) \end{bmatrix}$$

where

$$\begin{aligned} (u(t))_{2m-1} &= m(u)_{2m}[\cos \alpha_m t - \cos mt] \\ (u(t))_{2m} &= (u)_{2m}\cos \alpha_m t \end{aligned} \tag{2.5}$$

and $\quad (u(t))_n = 0 \quad$ otherwise.

Let $t_m = 2m^3\pi$, then for large m

$$\cos \alpha_m t_m - \cos mt_m = -2 + O(m^{-8}) \tag{2.6}$$

Using (2.5) and (2.6),

$$||T(t_m)||_X^2 \geq ||A^{1/2}u(t_m)||_{\ell^2}^2 \geq m^2.$$

Thus the zero solution of (2.3) is unstable.

Remark 1. Using the same calculations as above it is easy to show that the zero solution of

$$\ddot{u} + (A+pB)u = 0$$

is unstable for all nonzero p.

Remark 2. The eigenvectors of A+B are e_{2n-1} and $e_{2n-1} + n^{-1}e_{2n}$ where e_n is the sequence with one in the nth place and zeros elsewhere. Thus the eigenvectors of A+B are complete. The above example shows that completeness is not a sufficient condition for stability, contrary to the view expressed in the engineering literature.

Remark 3. The eigenvectors of A+B do not form a basis for ℓ^2. It might be thought that if the eigenvalues of K are distinct and positive and the corresponding eigenvectors form a basis for H, then the zero solution of (1.1) is stable. This is false in general however, the eigenvectors must form a Riesz Basis (see Section 4).

In Section 4 we show that for a certain class of operators K, the stability of the zero solution of (1.1) depends upon the eigenvalues and generalized eigenspaces of K in the same way as the finite dimensional case. In particular, if the eigenvalues of K are all distinct and positive, the zero solution of (1.1) is stable. The theory is applicable to many linear nonconservative problems; in particular it applies to Beck's Problem.

Another problem associated with (1.1) occurs in the analysis of the eigenvalues of K. In the engineering literature it is usually assumed that it is only the first pair of eigenvalues of K which *cause* the instability. It is of course necessary to consider all the eigenvalues of K.

In Section 5 we give a complete analysis of Beck's Problem for all $p \geq 0$. The calculations involved there are rather complicated so we give a simple analysis now for p less that the lowest critical value [2].

Theorem 2.1. There is a p_1 with $0 < p_1 < 4\pi^2$ such that the zero solution of (1.2) is stable for $0 \leq p < p_1$ and unstable for $p = p_1$.

Proof. We assume the result claimed above, that is, if the eigenvalues of K are all positive and distinct then the zero solution of (1.1) is stable. This result is proved in Section 4.

Recall that for Beck's Problem,

$$K(p) = \frac{d^4}{dx^4} + p\frac{d^2}{dx^2}$$

with

$$D(K(p)) = \{u \in W_2^4 : u(0) = u'(0) = u''(1) = u'''(1) = 0\}.$$

Let $p \geq 0$ be fixed. An easy calculation shows that λ is an eigenvalue of $K(p)$ if and only if $p^2 \neq -4\lambda$ and $g(p,\lambda) = 0$ where

$$g(p,\lambda) = p^2 + 2\lambda + p\lambda^{1/2}\sin(a)\sinh(b) + 2\lambda\cos(a)\cosh(b)$$

(2.7)

$$2a^2 = p + (p^2+4\lambda)^{1/2}, \quad b^2 = a^2-p.$$

Define a sequence $\{\alpha_n(p)\}$ by $\alpha_n(p) = \pi^2n^2(\pi^2n^2-p)$. Then for $0 \leq p \leq 4\pi^2$ and $n \geq 1$, a simple calculation shows that $g(p,\alpha_{2n}(p)) > 0$, $g(p,\alpha_{2n+1}(p)) < 0$. Thus for $n \geq 3$ and $0 \leq p \leq 4\pi^2$, $g(p,\lambda)$ cannot have a zero on the boundary of $[\alpha_{n-1}(p),\alpha_n(p)]$.

Since $K(0)$ is self-adjoint, solutions of $g(0,\lambda) = 0$ are all real. If $\{\lambda_n(0)\}$ denotes the solutions of $g(0,\lambda) = 0$, $\lambda \neq 0$, arranged in an increasing sequence, then simple calculations show that each $\lambda_n(0)$ is simple and that $\alpha_{n-1}(0) < \lambda_n(0) < \alpha_n(0)$.

Standard results show that the eigenvalues of $K(p)$, $\lambda_n(p)$, are continuous functions of p. Since zeros of $g(p,\lambda)$ occur in conjugate pairs, if there is a zero of $g(p,\lambda)$ which has nonzero imaginary part, then for some $p^* < p$, $g(p^*,\lambda)$ has a real non-simple zero.

Since $\lambda = 0$ is never an eigenvalue of $K(p)$ and $\alpha_2(4\pi^2) = 0$, there exists $p_1 < 4\pi^2$ such that the eigenvalues of $K(p)$ are positive and simple for $0 < p < p_1$ and $\lambda_1(p_1) = \lambda_2(p_1)$.

A simple computation shows that there exist non-trivial ϕ, $\psi \in D(K(p))$ with $K(p_1)\phi = \lambda_1(p_1)\phi$ and $K(p_1)\psi = \lambda_1(p_1)\psi + \phi$. Thus when $p = p_1$,

$$u(x,t) = t \cos\alpha t \ \phi(x) + 2\alpha \sin\alpha t \ \psi(x),$$

where $\alpha^2 = \lambda_1(p)$, is an unbounded solution of (1.2). This completes the proof of the Theorem.

3. Self-adjoint problems.

Before turning to non-self-adjoint problems we briefly study the self-adjoint case. Consider the equation

$$(3.1) \qquad \ddot{u} + Au = 0$$

where A is a positive self-adjoint operator in a Hilbert space H. We also assume that A^{-1} is compact and the eigenvalues of A are $\{\lambda_n\}$ where

$$0 < \lambda_1 < \lambda_2 < \ldots$$

Define a new Hilbert space $X = D(A^{1/2}) \times H$ with inner product $\langle \cdot , \cdot \rangle$ defined in the same way as the example in the previous section. We can rewrite (3.1) as the first order system

$$\dot{x} = Fx$$

where F is the skew self-adjoint operator

$$\begin{bmatrix} 0 & I \\ -A & 0 \end{bmatrix}$$

with domain $D(F) = D(A) \times D(A^{1/2})$. By Stone's Theorem [3], F generates a unitary group $U(t)$ of bounded linear operators on X. In particular, the zero solution of (3.2) is stable.

By the Spectral Theorem we can write down the general solution of (3.2) as

$$x(t) = \Sigma e^{\mu_k t} a_k \phi_k$$

where ϕ_k are the eigenvectors of F, $\mu_k = \pm i\lambda_k$ the corresponding eigenvalues of F and a_k the Fourier coefficients of the initial data. We note in passing that $x(t)$ is an almost periodic function of t.

4. Non-self-adjoint problems.

In this section we study the stability of solutions of

$$(4.1) \qquad \ddot{u} + (A+B)u = 0$$

where A satisfies the same conditions as in Section 3 and $B(D(B) \supset D(A))$ is weak relative to A in a sense to be made more precise later. As in Section 3 we rewrite (4.1) as

$$(4.2) \qquad \dot{x} = Fx$$

Then conditions that we put on A and B ensures that F is similar to a skew self-adjoint operator plus a bounded operator. Thus F is the generator of a strongly continuous group of bounded linear operators on $X = D(A^{1/2}) \times H$.

Suppose that the eigenvalues of $K = A+B$ are all positive and distinct. Then an obvious way of trying to prove that the zero solution of (4.2) is stable would be to prove that

$$(4.3) \qquad K = S^{-1}QS$$

where S is a bounded invertible operator on H and Q is a positive self-adjoint operator. If the eigenvectors of K are $\{\phi_k\}$ then the eigenvectors of Q are $\{S\phi_k\}$. Since Q is self-adjoint, $\{S\phi_k\}$ forms an orthonormal sequence which is complete in H. By definition this means that the $\{\phi_k\}$ must form a Riesz Basis for H. Conversely, if the $\{\phi_k\}$ forms a Riesz Basis, then there exists a bounded invertible linear operator S on H such that $\{S\phi_k\}$ is a complete orthonormal sequence. It is then easy to check that if Q is defined by (4.3), then Q is a positive self-adjoint operator.

There are many equivalent conditions for a sequence in H to be a Riesz Basis [4]. We only give one condition here.

Proposition. a sequence $\{\phi_n\}$ in H is a Riesz Basis if and only if

(i) $\inf |\phi_n| > 0$, $\sup |\phi_n| < \infty$,

and

(ii) any permutation of $\{\phi_n\}$ is a basis for H.

In the example given in Section 2 the eigenvectors are complete but they do not form a basis.

In [5] it is proved that the generalized eigenvectors of certain differential operators form a Riesz Basis. This result could also be deduced from the theory of spectral operators [6] (see [2] for the sequence of results in spectral operator theory which gives the desired stability criteria).

We give a more general result than is needed for obtaining stability results. Let $K = A+B$. Under certain conditions on A and B we prove that there is a bounded invertible operator S such that

$$S^{-1}KS = Q+F \tag{4.4}$$

where Q is a positive self-adjoint operator and F has finite dimensional range. F corresponds to the eigenspaces (and generalized eigenspaces) of K which arise from the finite number of eigenvalues of K which are either non-positive or non-simple. In particular, if the eigenvalues of K are all positive and distinct then F is zero. The canonical form (4.4) implies that the only instability mechanisms for (4.1) are finite dimensional in nature, i.e, Divergence or Flutter.

The results which we now give are drawn from Clark [7]. Clark states his results in terms of spectral operators, but we give the results in terms of similarity of operators.

Theorem 4.1. Let A be a positive self-adjoint operator on H with compact resolvent. Assume that all but a finite number of the eigenvalues of A are simple and that as $n \to \infty$

$$\begin{aligned} \lambda_n &= an^{\alpha}(1 + o(1)) \\ \lambda_{n+1} - \lambda_n &= a(n)n^{\alpha-1}(1 + o(1)) \end{aligned} \tag{4.5}$$

for some constants $\alpha > 1$, $a > 0$ and where $0 < c_1 < a(n) < c_2$ for large enough n, where c_1 and c_2 are constants.

Let B be a closed operator on H with $D(B) \supset D(A)$ having the following property: for each $\varepsilon > 0$ there exists $C(\varepsilon) > 0$ such that

(4.6) $\quad |Bu| \le \varepsilon|Au| + C(\varepsilon)|u|, \quad u \in D(A),$

where $C(\varepsilon) = O(\varepsilon^{-\beta})$ as $\varepsilon \to 0^+$, $0 \le \beta < \alpha-1$.

Then there exists a bounded invertible operator S on H such that $S^{-1}(A+B)S = Q+F$ where Q is a positive self-adjoint operator and F has finite dimensional range.

The proof of the above theorem is given in [7], so we only give a brief outline of the method used. We first introduce some notation. A sequence $\{P_n\}$ of projections in a Hilbert space H satisfying the orthogonality conditions $P_nP_m = \delta_{nm}P_n$ is called a p-sequence. (Note that the P_n need not be self-adjoint). A p-sequence $\{E_n\}$ is self-adjoint if $E_n^* = E_n$ for all n. A self-adjoint p-sequence $\{E_n\}$ is complete if $\Sigma E_n = I$.

<u>Proposition</u>. (Kato [8]). Let $\{P_n\}$ be a p-sequence and $\{E_n\}$ a complete self-adjoint p-sequence. Furthermore, assume that

(4.7) $\quad \dim P_o = \dim E_o < \infty,$

(4.8) $\quad \sum_{n=1}^{\infty} |E_n(P_n - E_n)u|^2 \le c|u|^2 \quad$ for all $u \in H$

where c is a constant with $0 \le c < 1$. Then there exists a bounded invertible operator S such that

$$P_n = S^{-1}E_nS, \quad n = 0, 1, 2, \ldots$$

Let $\{\lambda_n\}$ be the eigenvalues of A and let $\{\mu_n\}$ be the eigenvalues of $A+B$. Let $m \ge 1$. Let E_o be the projection onto the eigenspaces corresponding to $\lambda_1, \lambda_2, \ldots, \lambda_m$ and let E_n denote the projection onto the eigenspace corresponding to λ_{n+m} for $n \ge 1$. Similarly we define P_o to be the projection corresponding to $\mu_1, \ldots, \mu_m$ and P_n the projection corresponding to μ_{n+m} for $n \ge 1$. To prove

Theorem 4.1 it is sufficient to check that the hypothesis of the above Proposition are satisfied for m large enough.

The proof that the above Proposition applies may be divided into the following steps:

1. Construct a sequence of distinct circles Γ_n centred at $\lambda_n (n \geq m+1)$ such that each Γ_n contains exactly one eigenvalue μ_n of $A+B$.

2. From step 1, for $n \geq 1$, we have that

$$(4.9) \qquad E_n - P_n = \frac{1}{2\pi i}\int_{\Gamma_n} [R_\lambda(A+B) - R_\lambda(A)]d\lambda$$

   where $R_\lambda(A)$ is the resolvent of A. Proving (4.8) relies on obtaining suitable estimates on the integral in (4.9) and choosing m large enough.

3. Construct a contour Γ containing $\lambda_1, \ldots, \lambda_m$ and no other eigenvalue of A such that the integral of $||R_\lambda(A+B) - R_\lambda(A)||$ around Γ is small. This proves that $||P_o - E_o|| < 1$ which implies (4.7).

We now apply Theorem 4.1 to a class of differential operators. Our discussion is again based on Clark [7]. Let $J = [a,b]$ be a finite closed interval and let $H^m(J)$ be the Sobolev space consisting of all $f \in L^2(J)$ having generalized derivatives $f^{(j)}$ in $L^2(J)$ for $j \leq m$. Define a formal operator A by

$$(4.10) \qquad Af = (-1)^m f^{(2m)}$$

We need some more notation to describe the boundary conditions.

Let $H_o^m(J)$ be the closure in $H^m(J)$ of C^∞ functions whose support is a compact subset of (a,b). Let W be a closed subspace such that

$$H_o^{2m}(J) \subset W \subset H^{2m}(J)$$

We now define A by (4.10) with $D(A) = W$. We require that A satisfies the conditions of Theorem 4.1 with $\alpha = 2m$. Under quite general conditions, when the boundary conditions which determine W consist of $2m$ linearly independent homogeneous conditions, it is known that A satisfies the conditions of Theorem 4.1 with $\alpha = 2m$. For specific examples, it is easy to check directly that the conditions of Theorem 4.1 hold with $\alpha = 2m$.

The perturbing operator B is defined as the closure of the operator B_o where

$$B_o f = \sum_{k=0}^{2m-2} T_k f^{(k)}, \qquad D(B_o) = W$$

where T_k are bounded linear operators from $L^2(J)$ to $L^2(J)$.

To show that B satisfies the conditions of Theorem 4.1 we need the following Lemma.

<u>Lemma</u>. Let j,k be non-negative integers with $j < k$ and $k \geq 2$. Then there exists a constant C (depending only on j,k and the interval J) such that for all $\varepsilon > 0$ and all $f \in H^k(J)$

$$[\int |f^{(j)}(x)|^2 dx]^{1/2} \leq \varepsilon [\int |f^{(k)}(x)|^2 dx^{1/2} + C\varepsilon^{-\tau}[\int |f(x)|^2 dx]^{1/2}$$

where $\tau = j \mid (k-j)$ and all integrals are over J.

For a proof of the above Lemma see [9].

It follows by straightforward applications of the above Lemma that B satisfies the conditions of Theorem 4.1.

5. <u>Beck's Problem</u>.

In Section 2 we proved that there exists $p_1 > 0$ such that the zero solution of (1.2) is stable for $0 < p < p_1$ and unstable for $p = p_1$. In this Section we prove that the zero solution of (1.2) is unstable for $p > p_1$.

To check that the results of Section 4 apply we must study the eigenvalues of $A = d^4/dx^4$ with $D(A) = \{y \subset W_2^4 : y(0) = y'(0) = y''(1) = y'''(1) = 0\}$. An easy computation shows that if λ is an eigenvalue of A then

(5.1) $\cos \lambda^{1/4} \cosh \lambda^{1/4} = -1.$

It is easy to show from (5.1) that the eigenvalues of A are all distinct and that (4.5) is satisfied with $\alpha = 4$. Thus the stability results of Section 4 apply and so the eigenvalues of $K(p) = d^4/dx^4 + pd^2/dx^2$, $D(K(p)) = D(A)$, determine the stability of the zero solution of (1.2).

Let $\{\lambda_n(p)\}$ be the eigenvalues of $K(p)$. In Section 2 we proved that there exists $p_1 > 0$ such that for $0 \leq p < p_1$,

$$\lambda_1(p) < \lambda_2(p) < \lambda_3(p) < \ldots$$

and that $\lambda_1(p_1) = \lambda_2(p_2)$. For $p > p_1$ with $p - p_1$ small it is easy to show that $\lambda_1(p)$ and $\lambda_2(p) = \overline{\lambda_1(p)}$ are complex. Numerical computations indicate that for $p > p_1$, $\lambda_1(p)$, $\lambda_2(p)$ are complex. A proof of this would show that the zero solution of (1.2) is unstable for $p > p_1$. Unfortunately we do not have a simple direct proof of this. Instead we prove a result, concerning the eigenvalues of $K(p)$, which is interesting in its own right. We prove that as p increases from 0 to ∞, the eigenvalues of $K(p)$ become complex in pairs. Once a pair of eigenvalues becomes complex, they remain complex for higher values of p. Moreover, if $n < m$ then $\lambda_{2n-1}(p)$, $\lambda_{2n}(p)$ become complex for a lower value of p than $\lambda_{2m-1}(p)$ and $\lambda_{2m}(p)$. More precisely:

<u>Theorem</u> 5.1. There exists a sequence $\{p_i\}$ with $p_i \in (4\pi^2(i-1)^2, 4\pi^2 i^2)$ such that

(i) if $0 \leq p < p_1$ then the eigenvalues $\lambda_n(p)$ of $K(p)$ are positive and distinct;

(ii) if $p_i < p < p_{i+1}$ then $\lambda_1(p), \ldots, \lambda_{2i}(p)$ are complex while $\lambda_{2i+1}(p)$, $\lambda_{2i+2}(p), \ldots,$ are real and distinct.

The calculations that we do in order to prove Theorem 5.1 are rather involved so we first give a discussion of some of the difficulties. Recall from Section 2 that the eigenvalues $\lambda_n(p)$ are the zeros of the transcendental equation $g(\lambda,p) = 0$. Let $q_i = 2\pi i$,

$$\Omega_i = \{(\lambda,p) \in R^2 : 0 \quad p \le q_i, q_{i-1}^2(q_{i-1}^2 - p) \le \lambda \le q_i^2(q_i^2 - p) \text{ and } \lambda \ge 0\}$$

$$V_i = \{(\lambda,p) \in \Omega_i : g(\lambda, p) = 0 \text{ and } \lambda \ne 0\}.$$

Simple calculations show that V_i does not intersect V_j if $i \ne j$. Theorem 5.1 will be proved by investigating the geometric properties of the curves V_i.

Geometrically, Theorem 5.1 means that each V_i must have the form shown in Figure 1. Some of the possible forms for V_i that we must eliminate are shown in Figures 2 and 3. In Figure 2 the eigenvalues $\lambda_{2i-1}(p)$, $\lambda_{2i}(p)$ are real and distinct for $0 \le p < a$ and for $b < p < c$. In Figure 3, a pair of eigenvalues enters the real line when $p = a$ and there are 4 eigenvalues in $[q_{i-1}^4, q_i^4]$ for $a \le p \le b$. Lemma 5.1 shows that Figure 2 cannot occur while Lemma 5.2 eliminates all other possibilities than Figure 1.

It is convenient to use the variables α,β rather than λ,p where $\alpha^2 - \beta^2 = p$, $\alpha^2\beta^2 = \lambda$. Let

$$h(\alpha,\beta) = \alpha^4+\beta^4 + 2\alpha^2\beta^2\cos\alpha\cos\beta + \alpha\beta(\alpha^2-\beta^2)\sin\alpha\sinh\beta \tag{5.2}$$

$$\Omega = \{(\alpha,\beta) \in R^2 : 0 \le \beta \le \alpha\}$$

$$C = \{(\alpha,\beta) \in \Omega : h(\alpha,\beta) = 0, (\alpha,\beta) \ne (0,0)\}$$

$$\ell(p) = \{(\alpha,\beta) \in \Omega : \alpha^2 - \beta^2 = p\}$$

$$S(p) = C \cap \ell(p)$$

Then it is easy to check that there is a one-one relationship between the positive eigenvalues of $K(p)$ and the set $S(p)$. Thus Theorem 5.1 can be stated in the equivalent form:

<u>Theorem</u> 5.1. Let $C_i = C \cap \Omega_i$ where

$$\Omega_i = \{(\alpha,\beta) \in \Omega : 2\pi(i - 1) \le \alpha \le 2\pi i\}.$$

Then $\ell(p)$ intersect C_i in

(a) two distinct points of $0 \le p < p_i$

(b) one point if $p = p_i$

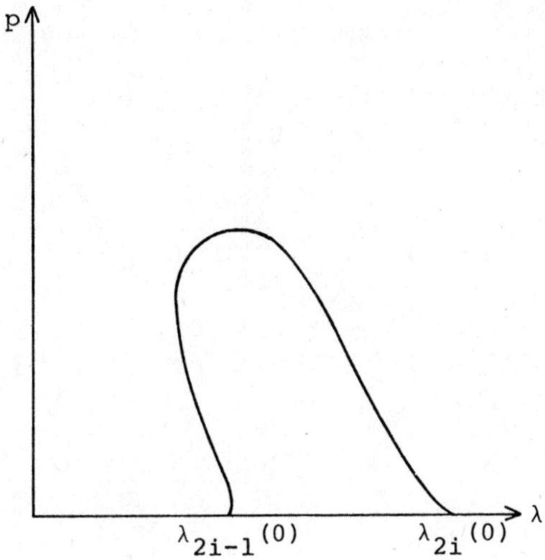

Fig. 1

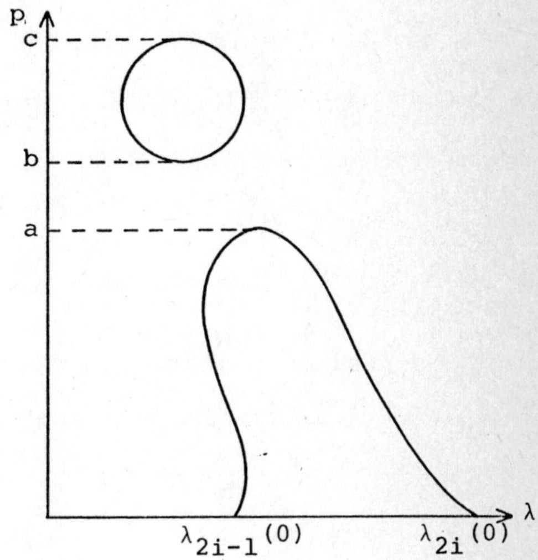

Fig. 2

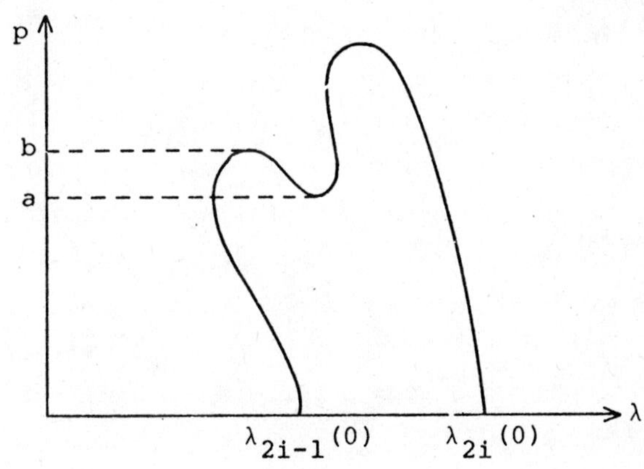

Fig. 3

(c) no points if $p > p_i$.

Using $h(2\pi i,\beta) > 0$ and $h(\alpha,0) > 0$ it follows that C_i does not intersect C_j if $i \neq j$ and C_i does not intersect the line $\beta = 0$. Using the same calculations as in the proof of Theorem 2.1, $\ell(0)$ intersect each C_i in two distinct points. Thus for each i, there is a least p_i such that $\ell(p) \cap C_i$ is empty for $p > p_i$. However this does not give any information about the set $\ell(p) \cap C_i$ for $0 \le p \le p_i$.

We first investigate intersection of C_i with straight line segments. For each $m \in [0,1]$ define a straight line segment r_m and a function H_m by

$$r_m = \{(\alpha,\beta) \in \Omega : \beta = m\alpha\},$$

$$H_m(\alpha) = \alpha^{-4}h(\alpha,m\alpha), \quad \alpha > 0.$$

Then the number of points in which r_m intersects C_i is equal to the number of roots of the equation $H_m(\alpha) = 0$ in the interval $[2\pi(i - 1),2\pi i]$.

<u>Lemma</u> 5.1. For each i there exists an $m_i \in (0,1)$ such that, in the interval $[2\pi(i -1),2\pi i]$ the equation $H_m(\alpha) = 0$ has

(i) two distinct roots if $m \in (m_i,1]$,

(ii) a double root if $m = m_i$,

(iii) no roots if $m \in [0,m_i]$.

The proof of Lemma 5.1 consists of simple but lengthy calculus arguments so we shall only record the highlights of the proof.

<u>Step A</u>. Prove that for each $m \in (0,1]$ there exists $\alpha_1(m)$ and $\alpha_2(m)$ with $2\pi(i - 1) < \alpha_1(m) < 2\pi i - \pi < \alpha_2(m) < 2\pi i$ such that $dH_m(\alpha)/d\alpha$ is

(i) positive if $\alpha \in [2\pi(i -1),\alpha_1(m))$,

(ii) negative if $\alpha \in (\alpha_1(m),\alpha_2(m))$,

(iii) positive if $\alpha \in (\alpha_2(m),2\pi i]$.

<u>Step B</u>. Prove that $f(m) = H_m(\alpha_2(m))$ is a decreasing function of

m with $f(1) < 0$, $\lim_{m\to 0^+} f(m) > 0$. The point m_i is the unique root of $f(m) = 0$.

Lemma 5.1 closely resembles but does not prove Theorem 5.1 since a $\ell(p)$ may intersect some C_i in more than one point. We complete the proof of Theorem 5.1 by means of a certain computation. Since the computation is straightforward but very long we only sketch the method used.

Let s denote the arc length of C_i measured from the point $a = (\alpha,\alpha)$ where $\alpha^4 = \lambda_{2i-1}(0)$. Let L_i be the total arc lenght of C_i so that $s(b) = L_i$ where $b = (\alpha,\alpha)$ with $\alpha^4 = \lambda_{2i}(0)$. At each point z on C_i there is a unique p such that $\ell(p)$ passes through z. Thus we can regard p as a function of s on C_i.

Lemma 5.2. For each i there exists s_i such that $p'(s) > 0$ if $s \in (0,s_i)$ and $p'(s) < 0$ if $s \in (s_i,L_i)$.

Theorem 5.1 follows easily from Lemma 5.2 with p_i the maximum value of $p(s)$, i.e., $p_i = p(s_i)$.

An easy computation shows that

$$\frac{dp(s)}{ds} = -2(\alpha^2 + \beta^2)^{1/2}\sin\phi_i(s)$$

where $\phi_i(s)$ is the angle between the tangent vector to C_i in the direction of s increasing and the tangent vector to $\ell(p)$ in the direction of α increasing. Thus to prove Lemma 5.2 we need to investigate the function $\phi_i(s)$.

We normalize $\phi_i(s)$ by fixing $\phi_i(0) \in (-\pi.0)$. Even though $\phi_i(s)$ could take any real value of $s \in (0,L_i)$, the fact that C_i forms part of the boundary of a simply connected region implies that $\phi_i(L_i) \in (0,\pi)$. Thus to prove Lemma 5.2 it is sufficient to prove that if $\phi_i(s) = m\pi$ then $\phi'(s) > 0$ for any integer m. A straightforward calculation gives the following result.

Lemma 5.3. Suppose that at some point on C_i, $\phi(s) = m\pi$, where m is an integer. Then at that point,

$$h = h^1 = 0,$$

$$h'' = -(\alpha + \beta^2)^{\beta-1}\frac{dh}{dn}\cdot\frac{d\phi}{ds}$$

where the prime denotes differentiation along $\ell(p)$ (i.e., keeping $\alpha^2 - \beta^2$ fixed) and dh/dn denotes differentiation along the normal to C_i in the direction pointing into the interior of the curve Q formed by C_i and the line segment joining a to b.

Since h is negative inside Q we have reduced the proof of Theorem 5.1 to the following.

Lemma 5.4. Suppose that $h = h' = 0$. Then $h'' > 0$.

The proof of Lemma 5.4 is straightforward but long so we omit the details. This completes the proof of Theorem 5.1. From Theorem 5.1 we deduce that the zero solution of (1.2) is unstable for $p \geq p_1$.

6. A two parameter problem.

In this Section we study the stability of the following system which depends on two non-negative real parameters p and q:

$$(6.1)\qquad \begin{aligned} &u_{tt} + u_{xxxx} + pu_{xx} = 0 \\ &u(0,t) = u_x(0,t) = u_{xx}(1,t) + qu_x(1,t) = u_{xxx}(1,t) = 0 \end{aligned}$$

The physical problem giving rise to (6.1) is the same as in Beck's Problem except that in addition the rod is restrained at $x = 1$ by a rotational spring, with q equal to the rotational spring constant.

It is easy to check that the above problem satisfies the conditions stated in Section 4 thus reducing the stability problem to an eigenvalue analysis.

Let $K(p,q)$ denote the operator associated with (6.1). It is easy to show that λ is an eigenvalue of $K(p,q)$ if and only if $4\lambda \neq -p^2$ and

(6.2) $\quad F(\alpha,\beta) = h(\alpha,\beta) + qf(\alpha,\beta) = 0,$

where α and β satisfy $\alpha^2 - \beta^2 = p$, $\alpha^2\beta^2 = \lambda$, h is defined by (5.2) and

$$f(\alpha,\beta) = (\alpha^2 + \beta^2)(\beta\sinh\beta\cos\alpha + \alpha\sin\alpha\cosh\beta).$$

In particular, $\lambda \geq 0$ is an eigenvalue of $K(p,q)$ if and only if there exist real α,β satisfying (6.2) with $\alpha^2 - \beta^2 = p$, $\alpha^2\beta^2 = \lambda$ and $(\alpha,\beta) \neq (0,0)$.

Let $\lambda_n(p,q)$ denote the eigenvalues of $K(p,q)$, ordered so that $\{\lambda_n(0,0)\}$ forms an increasing sequence. Since $K(0,q)$ is a positive self-adjoint operator all its eigenvalues are positive. An easy computation shows that the eigenvalues of $K(0,q)$ are simple and that $(n-1)^4\pi^4 < \lambda_n(0,q) < n^4\pi^4$.

The next result shows that the stability of the zero solution of (6.1) depends only on the first two eigenvalues $\lambda_1(p,q)$, $\lambda_2(p,q)$.

Lemma 6.1. Let $q \geq 0$ be fixed. Then $\lambda_1(p,q)$, $\lambda_2(p,q)$ cannot both be positive and distinct for all $0 \leq p \leq 4\pi^2$. Moreover, if $\lambda_1(p,q)$, $\lambda_2(p,q)$ are positive and distinct for $0 < p < p^*$ then so are all the other eigenvalues.

Lemma 6.1 is proved in exactly the same way as Theorem 2.1 so we omit the details.

Let $J(\beta) = F((p + \beta^2)^{1/2},\beta)$. The following result follows easily from Lemma 6.1.

Lemma 6.2. Let $p \in [0,4\pi^2]$ and $q \geq 0$.

(i) The equation

(6.3) $\quad J(\beta) = 0, \quad \beta \in (0,(4\pi^2 - p)^{1/2})$

has at most two roots.

(ii) If (6.3) has two roots, β_1, β_2 then $\lambda_1(p,q)$ and $\lambda_2(p,q)$ are both in $(0,4\pi^2(4\pi^2 - p))$ and are given by

$\lambda_i(p,q) = \beta_i^2(\beta_i^2 + p)$, $i = 1, 2$.

(iii) If (6.3) has only one root β, then $\lambda_1 \le 0$ and $\lambda_2(p,q) \in (0, 4\pi^2(4\pi^2 - p))$ with $\lambda_2(p,q) = \beta^2(\beta^2 + p)$.

In order to apply Lemma 6.2 to the study of the variation of the first two eigenvalues, we first prove some results concerning equation (6.3). We do this in four different cases depending upon the value of the paramenter p.

Lemma 6.3. Let $p \in [0,\pi^2]$. Then for each $q \ge 0$, (6.3) has two distinct solutions.

Proof. For each p and q in the given ranges, simple computations show that $J(0) > 0$ for $p > 0$, $J(((5\pi/4)^2 - p)^{1/2}) < 0$ and $J((4\pi^2 - p)^{1/2}) > 0$. The result now follows.

Let $p \in (\pi^2, 4\pi^2)$. Using the properties of the equation $h(\alpha,\beta) = 0$ proved in the previous section and the simplicity of the equation $f(\alpha,\beta) = 0$, it is easy to prove:

Lemma 6.4. For each $p \in (\pi^2, 4\pi^2)$ there exists $\beta(p)$ such that $h(\alpha(\beta),\beta) > 0$ if $\beta > \beta(p)$ and $g(\alpha(\beta),\beta)$ has the same sign as $\beta - \beta(p)$ where $\alpha(\beta) = (p + \beta^2)^{1/2}$.

Lemma 6.4 shows that β is a solution of (6.3) if and only if it is a solution of

(6.4) $\quad G(\beta) = q, \quad \beta \in (0, \beta(p))$

where $G(\beta) = -h(\alpha(\beta),\beta)/(f(\alpha(\beta),\beta))$. Hence if $p \in (\pi^2, 4\pi^2)$, Lemma 6.2 applies if equation (6.3) is replaced by equation (6.4).

Lemma 6.5. Let $p \in (\pi^2, p_1)$ where p_1 is the critical value in Beck's Problem. Then (6.4) has

(i) two solutions if $q \in [0, q_2(p))$,

(ii) one solution if $q \in [q_2(p), \infty)$

where

(6.5) $$q_2(p) = -p^{1/2}(\sin(p^{1/2}))^{-1}.$$

Proof. For each p and q in the given ranges it is easy to check that G has the following properties:

(a) G is continuous on $(0,\beta(p))$.

(b) $G(\beta) \to q_2(p)$ as $\beta \to 0$.

(c) $G(\beta) \to \infty$ as $\beta \to \beta(p)$.

(d) There exists $\beta_1 \in (0,\beta(p))$ such that $G(\beta_1) < 0$.

(e) Equation (6.4) has at most two solutions.

The result follows easily from these five properties.

To obtain more information about equation (6.4) it is important to study the behaviour of $G(\beta)$ for small β. Simple computations show that $G'(0) = 0$ and

$$G''(0) = \frac{1}{\sqrt{p}\,\sin^2\sqrt{p}}[p\sin\sqrt{p} - \sqrt{p}(2\sin^2\sqrt{p} - 3\cos\sqrt{p}) - (4\cos\sqrt{p} - 1)\sin\sqrt{p}]$$

Simple computations show that the equation $G''(0) = 0$ has a unique zero p^* in $(\pi^2,4\pi^2)$ and that $G''(0)$ has the same sign as $p - p^*$.

Lemma 6.6. Let $p \in [p_1,p^*)$. Then (6.4) has

(i) no solutions if $q \in [0,q_1(p))$,

(ii) two solutions if $q \in [q_1(p),q_2(p))$,

(iii) one solution if $q \in [q_2(p),\infty)$

where $q_1(p) = \min\{G(\beta) : \beta \in [0,\beta(p))\}$ and $q_2(p)$ is defined by (6.5).

Proof. It is easy to check that G has the following properties for p and q in given ranges:

(a) $G(0) = q_2(p)$, $G'(0) = 0$ and $G''(0) < 0$.

(b) $G(\beta) > 0$ for all $\beta \in [0,\beta(p))$.

(c) $G(\beta) \to \infty$ as $\beta \to \beta(p)$.

(d) Equation (6.4) has a most two solutions.

Lemma 6.6 follows easily from the above properties.

A similar argument to the one used above gives the following result.

Lemma 6.7. Let $p \in [p^*, 4\pi^2]$. Then (6.4) has

(i) no solution if $q \in (0, q_2(p)]$,

(ii) one solution if $q \in (q_2(p), \infty)$

Finally, simple computations show that the functions q_i have the following properties.

Lemma 6.8.

(i) $q_1(p_1) = 0$, $q_1(p^*) = q_2(p^*)$.

(ii) $q_1(p) < q_2(p)$ for all $p \in (p_1, p^*)$.

(iii) q_1 is a continuous increasing function of p.

Using the above results we can deduce the following result concerning the dependence of $\lambda_1(p,q)$ and $\lambda_2(p,q)$ on p and q. To state the result we require the following notation (see Figure 4).

$P = \{(p,q) : p \in [0, 4\pi^2], \ q \geq 0\}$.

$W_o = \{(p,q) : p \in [p_1, p^*], \ q \in [0, q_1(p)] \text{ or } p \in [p^*, 4\pi^2], \ q \in [0, q_2(p)]\}$.

$W_1 = \{(p,q) : p \in (\pi^2, 4\pi^2), \ q \in (q_2(p), \infty)\}$.

$W_2 = P - (W_o \cup \overline{W}_1)$.

Theorem 6.1.

(i) If $(p,q) \in W_2$ then $\lambda_1(p,q)$ and $\lambda_2(p,q)$ are positive and distinct.

(ii) If $(p,q) \in W_1$ then $\lambda_1 \leq 0$ and $\lambda_2 > 0$.

(iii) When (p,q) moves from W_2 into W_o, $\lambda_1(p,q)$ and $\lambda_2(p,q)$ coincide on the positive real axis and then become complex.

(iv) When (p,q) moves from W_2 into W_1, $\lambda_1(p,q)$ crosses the origin.

Since we have already shown that the stability of the zero solution of (6.1) depends only on the first two eigenvalues of $K(p,q)$, we get the following corollary.

<u>Corollary</u>.

The zero solution of (6.1) is stable for all $(p,q) \in W_2$. When (p,q) moves from W_2 into W_1 the system loses stability by Divergence. When (p,q) moves from W_2 into W_o it loses stability by Flutter (see Figure 5 for the stability diagram).

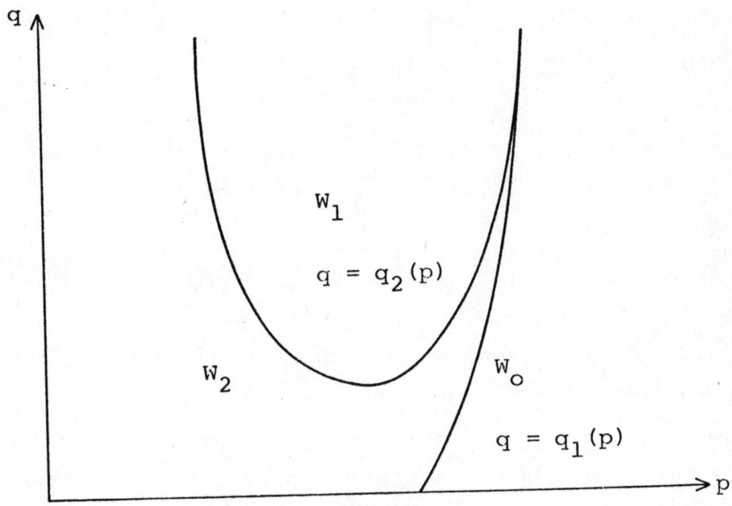

Fig. 4

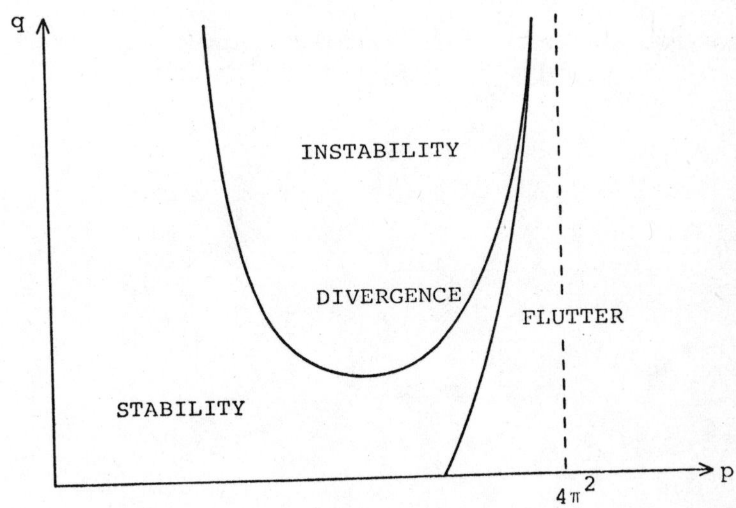

Fig. 5

Acknowledgement. The research of the first author was supported by the United States Army under AROD DAAG 29-76-GO294.

REFERENCES

[1] - WALKER, J.A. and INFANTE, E.F., A perturbation approach to the stability of undamped linear elastic systems subjected to follower forces, J. Math. Anal. Appl., 63(1978), 654-677.

[2] - CARR, J. and MALHARDEEN, M.Z.M., Beck's Problem, SIAM J. Appl. Math., 37(1979) (To appear).

[3] - REED, M. and SIMON, B., Methods of Modern Mathematical Physics, Vol. I: Functional Analysis, Academic Press, New York, (1972).

[4] - GOHBERG, I.C. and KREIN, M.G., Introduction to the Theory of Linear Non-self-adjoint Operators, Translations of Mathematical Monographs, Vol. 18, A.M.S., (1969).

[5] - MIHAILOV, V.P., Riesz bases in $L^2[0,1]$, Soviet Math., 3(1962), 851-855.

[6] - DUNFORD, N. and SCHWARTZ, J., Linear Operators, Part III, Wiley-Interscience, New York, (1971).

[7] - CLARK, C., On relatively bounded perturbations of ordinary differential operators, Pacific J. Math., 25(1968), 59-70.

[8] - KATO, T., Similarity for sequences of projections, Bull. Amer. Math. Soc., 73(1967), 904-905.

[9] - GOLDBERG, S., Unbounded Linear Operators, McGraw-Hill Book Company, New York, (1966).

AN ANALYSIS OF THE CHARACTERISTIC EQUATION OF THE SCALAR LINEAR DIFFERENCE EQUATION WITH TWO DELAYS

by L. A. V. Carvalho*

1. Introduction.

Let $R = (-\infty,\infty)$, $R^+ = [0,\infty)$ and consider the linear difference equation

(1.1) $x(t) = ax(t-r) + bx(t-s), \quad t \in R^+,$

where a, b, r and s are given real numbers with $a \neq 0$, $b \neq 0$ and $0 < r \leq s < \infty$. r and s are the delays of the equation.

If we attempt to obtain a solution of (1.1) of the form $x(t) = e^{\lambda t} (= \exp(\lambda t))$, λ complex, we obtain the following characteristic equation of (1.1):

(1.2) $ae^{-\lambda r} + be^{-\lambda s} = 1.$

It is our objective, in this work, to present conditions on a, b, r and s which determine the sign of the real part of λ. It is known [1,2,3,4] how these conditions affect the asymptotic behavior of the solutions of (1.1). As a particular result, the condition

$$|a| + |b| < 1$$

obtained by Melvin [4] as equivalent to the stability in the delays [1,2,3] of (1.1) will also be obtained here.

2. The analysis.

Let $\lambda = \alpha + \theta i$, i- the imaginary unit. Then, (1.2) can be split into the pair of equations in α and θ:

(2.1) $ae^{-\alpha r}\cos\theta r + be^{-\alpha s}\cos\theta s = 1$

(2.2) $ae^{-\alpha r}\sin\theta r + be^{-\alpha s}\sin\theta s = 0$

Then, we have:

---

* This research was partially supported by CAPES (Coordenação do Aperfeiçoamento de Pessoal de Nível Superior) - Brazil.

Lemma 2.1. To each $u > 0$ and to each pair of delays $r, s,$ fixed, but arbitrary, there corresponds a unique $\alpha = \alpha(u) \in R$ such that

$$|a|e^{-\alpha r} + |b|e^{-\alpha s} = 1/u.$$

The proof of this lemma easily follows from the ontoness of the function $f(\alpha) = |a|e^{-\alpha r} + |b|e^{-\alpha s}$.

In order to simplify the notation, let $a(u) = ae^{-\alpha(u)r}$, $b(u) = be^{-\alpha(u)s}$ and denote, for a given $u > 0$, the straight lines $a(u)x + b(u)y = 1$ and $a(u)x + b(u)y = 0$ by $L_1(u)$ and $L_2(u)$, respectively. Also, if x is a nonzero real number, put $\text{sign}(x) = 1$ if $x > 0$ and $\text{sign}(x) = -1$ if $x < 0$. Then, we note that $L_1(u)$ and $L_2(u)$ are parallel lines, $L_1(u)$ passes through $u(\text{sign}(a),\text{sign}(b))$, the slope of $L_1(u)$ is positive if $\text{sign}(a) = \text{sign}(b)$ and negative otherwise, $L_2(u)$ passes through the origin and $L_1(u)$ never collapses with $L_2(u)$ for $u > 0$.

Along with equations (2.1) and (2.2) consider also, for a given $u > 0$, the following pair of equations in $\beta, \gamma \in R$:

(2.3) $a(u)\cos\beta + b(u)\cos\gamma = 1$

(2.4) $a(u)\sin\beta + b(u)\sin\gamma = 0.$

One clearly notes that if the system (2.3), (2.4) does not have a solution (β,γ), neither does the system (2.1), (2.2).

Next, we have:

Lemma 2.2. Equation (2.3) has a solution if and only if $L_1(u) \cap Q \neq \emptyset$, where $Q = \{(x,y) \in R^2 : x \leq 1, \ y \leq 1\}$.

Proof. Suppose β, γ solve (2.3). Then, since $|\cos\beta| \leq 1$ and $|\cos\gamma| \leq 1$ it follows that $L_1(u) \cap Q \neq \emptyset$. And conversely, if $(x,y) \in L_1(u) \cap Q$, just let $\beta = \arccos x$ and $\gamma = \arccos y$ to obtain the result.

Corollary 2.1. The system (2.1), (2.2) has no solution if $u > 1$.

Proof. Since $L_1(u)$ passes through $u(\text{sign}(a), \text{sign}(b)) = (u', u")$, if $u > 1$ and there existed $(x,y) \in L_1(u) \cap Q$ we would have

$$a(u)(x-u') + b(u)(y-u") = 0,$$

i.e., $(x-u', y-u") \in L_2(u)$ with $\text{sign}(\frac{x-u'}{y-u"}) = -\text{sign}(\frac{b}{a})$, which is impossible because of the position of $L_2(u)$. Thus, if $u > 1$, (2.3) cannot have a solution. The result then follows.

The proof of the next lemma easily follows from the definition of $\alpha(u)$.

Lemma 2.3. $\alpha(u)$ is an increasing function of u.

We observe now that if $(\cos\beta, \cos\gamma)$ satisfies (2.3) then, in order that (2.3) be satisfied it is necessary and sufficient that the following equation be satisfied:

(2.5) $-\text{sign}(a).\text{sign}(b).a(u)|\sin\beta| + b(u)|\sin\gamma| = 0.$

In fact, if a and b have the same sign, the same is true of $a(u)$ and $b(u)$ and, since, in this case, (2.5) is equivalent to

$$-a(u)|\sin\beta| + b(u)|\sin\gamma| = 0,$$

the result follows from the fact that cos is an even function and sin is an odd function. If a and b have opposite sign then the same is true of $a(u)$ and $b(u)$ and, since, in this case, (2.5) is equivalent to

$$a(u)|\sin\beta| + b(u)|\sin\gamma| = 0,$$

the result again follows from the fact that cos is even and sin is odd.

As a consequence of this observation we see that there is no loss of generality if instead of working with (2.3), (2.4) we work with (2.3), (2.5) and thus, it follows that if (2.3), (2.5) does not have a solution for u in a certain range, so do (2.1), (2.2), that is to say, (1.2) does not have a solution λ such that $\text{Re}(\lambda) = \alpha(u)$.

Definition 2.1. Let $(\cos\beta, \cos\gamma) \in L_1(u)$. We put $C(\cos\beta, \cos\gamma) = (-\text{sign}(a).\text{sign}(b)|\sin\beta|, |\sin\gamma|)$ and call it the "companion" of $(\cos\beta, \cos\gamma)$.

It is easy to see that $C : L_1(u) \cap Q \to Q$ is a continuous map. $C(L_1(u) \cap Q)$ is called the "companion curve" of $L_1(u)$.

Definition 2.2. If there exists a point $(x, \mathrm{sign}(b)) \in L_1(u) \cap Q$ we put $P(u) = (x, \mathrm{sign}(b))$ and call it the "initial point of $L_1(u)$". If there exists a point $(\mathrm{sign}(a), y) \in L_1(u) \cap Q$ we put $R(u) = (\mathrm{sign}(a), y)$ and call it the "terminal point of $L_1(u)$".

One should observe that the initial point of $L_1(u)$ is never defined for $u > 1$ and that it also fails to exist for sufficiently small values of $u > 0$ due to the fact that $r \le s$. The terminal point of $L_1(u)$, nevertheless, is always defined for $0 < u \le 1$. The idea of the initial and terminal point of $L_1(u)$ is simple: if one accordingly parametrizes $L_1(u)$ then $P(u)$ is the first point of $L_1(u)$ that is in Q and $R(u)$ is the last.

We also introduce the following sets:

$$L_2^-(u) = \{(x,y) \in Q : a(u)x + b(u)y \le 0\}$$

$$L_2^+(u) = \{(x,y) \in Q : a(u)x + b(u)y \ge 0\}.$$

Then, we have:

Lemma 2.4. Suppose $u \le 1$. Then, $C(R(u)) \in L_2^{\mathrm{sign}(b)}(u)$ and if $P(u)$ is defined then $C(P(u)) \in L_2^{-\mathrm{sign}(b)}(u)$.

Proof. $C(R(u)) = (0, |\sin(\arccos y)|)$ and $C(P(u)) = (-\mathrm{sign}(a).\mathrm{sign}(b)|\sin(\arccos x)|, 0)$ substituted into $a(u)x + b(u)y$ lead to the result.

Lemma 2.5. The system (2.3), (2.5) has a solution if and only if $P(u)$ is defined.

Proof. If $P(u)$ is defined then Lemma 2.4 and the continuity of the companion curve imply the result.

In order to prove the converse, suppose first that $a > 0$, $b > 0$ and that $P(u)$ is not defined. Then, either $u > 1$ or there exists

$(-1,k) \in L_1(u) \cap Q$, $0 < k < 1$. If the first hypothesis holds then (2.3), (2.5) does not have a solution by Corollary 2.1. So, suppose that the second hypothesis holds and that (2.3), (2.5) has a solution. This means that there exists $(x,y) \in L_1(u) \cap Q$ with $C(x,y) \in L_2(u)$. From $(x,y) \in L_1(u) \cap Q$ we get

(i) $a(u)x + b(u)y = 1.$

From $C(x,y) \in L_2(u)$ we obtain that

$$a^2(u)(1-x^2) = b^2(u)(1-y^2),$$

i.e.,

$$\{a(u)x - b(u)y\}\{a(u)x + b(u)y\} = a^2(u) - b^2(u).$$

Using (i) we get from this last equation that

(ii) $a(u)x - b(u)y = a^2(u) - b^2(u).$

The hypothesis that $(-1,k) \in L_1(u)$ implies that $a(u) = b(u)k - 1$. This, together with (i) and (ii), yields

(iii) $\{b(u)k - 1\}x + b(u)y = 1$

$\{b(u)k - 1\}x - b(u)y = \{b(u)k - 1\}^2 - b^2(u),$

which give

(iv) $2\{b(u)k - 1\}x = 1 + \{b(u)k - 1\}^2 - b^2(u).$

But, (iii) implies that $x = \{1-b(u)y\}/\{b(u)k-1\}$ which, substituted into (iv) yields

$$2\{1 - b(u)y\} = 1 + \{b(u)k - 1\}^2 - b^2(u).$$

Hence,

$$b(u) = 2(y-k)/(1-k^2).$$

Now, since $b(u) > 0$ and $0 < k < 1$ we see from this last equation that we must have $y > k$. But then, since $(x,y) \in L_1(u)$, $y > k$ implies that $(x,y) \notin Q$, a contradiction. Hence, (2.3), (2.5) has no solution if $P(u)$ is not defined. This finishes the proof for the case $a > 0$,

$b > 0$. The proof for the cases when $a < 0$ or $b < 0$ is handled similary and we omit it.

Corollary 2.2. The range of α's for which (1.2) has a solution $\lambda = \alpha + i\theta$ is bounded.

Proof. Indeed, for $u > 1$, $P(u)$ is not defined. So, for $\alpha > \alpha(1)$ there exists no solution $\lambda = \alpha + i\theta$ of (1.2). Also, as $u \to 0$, $\alpha(u) \to -\infty$. But then, $1/a(u) \to 0$ and $1/b(u) \to 0$ in such a way that $L_1(u)$ tends to occupy a horizontal position, $L_1(u)$ and $L_2(u)$ tending to collapse with the x-axis. So for some u_o, $0 < u_o < 1$, $(-\text{sign}(a), \text{sign}(b)) \in L_1(u_o)$. For $u < u_o$, $P(u)$ is not defined. Hence, for $\alpha < \alpha(u_o)$, (1.2) has no solution.

Lemma 2.6. A necessary and sufficient condition for (1.2) to have a solution is the system (2.3), (2.5) to have a solution (β,γ) with $r/s = \beta/\gamma$.

Proof. Just note that $r/s = \beta/\gamma$ if and oly if there exists θ such that $\theta r = \beta$ and $\theta s = \gamma$.

Lemma 2.7. Suppose $a > 0$ and $b > 0$. Then (1.2) has a real solution for $u = 1$.

Proof. In fact, $u = 1$ implies that $L_1(u) \cap Q = (1,1)$. Since $C(1,1) = (0,0) \in L_2(1)$ we have just to pick $\theta = 0$ to get the solution $\lambda = \alpha(1)$ of (1.2). This solution is, of course, unique (modulo 2π).

Corollary 2.3. Suppose $a > 0$ and $b > 0$. Then, every solution of (1.2) has negative real part if and only if $a+b < 1$.

Proof. If $a+b < 1$ then, since $a(1) + b(1) = 1$ with $a(1) = ae^{-\alpha(1)r}$ and $b(1) = be^{-\alpha(1)s}$, it follows that either $e^{-\alpha(1)r} > 1$ or $e^{-\alpha(1)s} > 1$. Hence, we must have $\alpha(1) < 0$ and the result follows from Lemma 2.7 and Corollary 2.2.

On the other hand, if $\alpha(1) < 0$ then $a(1) > a$ and $b(1) > b$ imply that

$$a(1) + b(1) = 1 > a+b.$$

Remark. Nothing was said about the delays in the above lemma. It follows, as a corollary, that (if $a > 0$ and $b > 0$ then) every root of (1.2) has a negative real part for every fixed pair of delays r, s, $0 < r \leq s < \infty$ if and only if $a+b < 1$.

Let now u_o be as in the proof of Corollary 2.2, i.e, $0 < u_o < 1$ and

$$-a(u_o)\text{sign}(a) + b(u_o)\text{sign}(b) = 1.$$

Then, we have:

Lemma 2.8. Suppose $a > 0$ and $b > 0$. Then, there exists a root $\lambda = \alpha(u_o) + i\theta$ of (1.2) if and only if $r/s = p/q$ with p odd and q even.

Proof. Note that in case $a > 0$ and $b > 0$, $\beta = \pi$ and $\gamma = 2\pi$ yield a solution for (2.3), (2.5) when $u = u_o$. It thus follows from Lemma 2.6 that, in this case, (1.2) has a root $\lambda = \alpha(u_o) + i\theta$ if and only if there exists θ such that $\theta r = \pi (\text{mod } 2\pi)$ and $\theta s = 0(\text{mod } 2\pi)$, i.e, if an only if there exists integers k and n such that $\theta r = (2k+1)\pi$ and $\theta s = 2n\pi$, which gives the result.

Corollary 2.4. Suppose $a > 0$ and $b > 0$. Then, every root of (1.2) has positive real part for all delays r, s, $0 < r \leq s < \infty$ if and only if $b > a+1$.

Proof. Let u_o be as above and suppose that $b > a+1$ and that there exists v, $u_o \leq v \leq 1$ such that $\alpha(v) \leq 0$. Then, due to the monotonicity of $\alpha(u)$ it follows that $\alpha(u_o) \leq 0$. But, this signifies that

$$-a(u_o) + b(u_o) = -ae^{-\alpha(u_o)r} + be^{-\alpha(u_o)s} = 1$$

with

$$e^{-\alpha(u_o)s} \geq e^{-\alpha(u_o)r} \geq 1.$$

Hence, $1 \geq (-a+b)e^{-\alpha(u_o)s}$, i.e., $b \leq a+1$, a contradiction.

And conversely, if for any delays r, s, $0 < r \leq s < \infty$, every root of (1.2) has positive real part then, in particular, for r odd and s even the roots of (1.2) have positive real part. But then, Lemma 2.8 implies that $\alpha(u_o) > 0$ and so, $e^{-\alpha(u_o)s} < e^{-\alpha(u_o)r} < 1$, which implies that

$$1 = -a(u_o) + b(u_o) < (-a+b)e^{-\alpha(u_o)r},$$

and, as a consequence, we must have $b > a+1$.

From the proof of the above corollary, we immediately obtain the following

Corollary 2.5. Suppose $a > 0$ and $b > 0$. Then, if every root of (1.2) has positive real part for some delays r, s, where r/s = odd/even, it follows that $b > a+1$.

Now, it is clear that using techniques similar to the above, similar results can be obtained for the remaining cases, namely, $a < 0$ and $b < 0$, $a > 0$ and $b < 0$, and $a < 0$ and $b > 0$. Below, we list a few and prove some of these results, before stating the main theorems.

Lemma 2.9. Suppose $a < 0$ and $b < 0$. Then, there exists a root $\lambda = \alpha(1) + i\theta$ of (1.2) if and only if r/s = odd/odd.

The proof goes as in Lemma 2.8, upon nothing that, when $a < 0$, $b < 0$ and $u = 1$, $\beta = (2k+1)\pi$ and $\gamma = (2n+1)\pi$ with k and n being integers, yield the only solutions (2.3), (2.5) can have.

Corollary 2.6. Suppose $a < 0$ and $b < 0$. Then, the largest real part of the solutions of (1.2) is less than or equal to $\alpha(1)$.

Corollary 2.7. Suppose $a < 0$ and $b < 0$. Then, the real part of every root of (1.2) is negative for all delays r, s, $0 < r \leq s < \infty$, if and only if $|a| + |b| < 1$.

Proof. Since $L_1(u)$ passes through $u(\text{sign}(a), \text{sign}(b))$ it follows, in this case, that

$$|a|e^{-\alpha(1)r} + |b|e^{-\alpha(1)s} = 1.$$

Thus, $|a| + |b| < 1$ if and only if $\alpha(1) < 0$ and the result follows from the above corollary.

Corollary 2.8. Suppose that $a < 0$, $b < 0$, $|a| + |b| = 1$ and $r/s \neq$ odd/odd. Then, every root of (1.2) has negative real part.

Corollary 2.9. Suppose $a < 0$ and $b < 0$. Then, if every root of (1.2) has negative real part for some pair of delays r, s such that $r/s =$ odd/odd, it follows that $|a| + |b| < 1$.

Lemma 2.10. Suppose $a < 0$ and $b < 0$. Then, every root of (1.2) has positive real part for all delays r, s, $0 < r \leq s < \infty$ if and only if $|b| \geq |a| + 1$.

Proof. Let u_o satisfy, as before, the equation

(*) $\quad -a(u_o)\text{sign}(a) + b(u_o)\text{sign}(b) = 1.$

It follows that for $u < u_o$, $P(u)$ is not defined and thus, there exists no root $\lambda = \alpha + i\theta$ of (1.2) with $\alpha < \alpha(u_o)$. Suppose then that there exists v, $u_o \leq v \leq 1$ such that $\alpha(v) \leq 0$. Due to the monotonocity of $\alpha(u)$ it follows that $\alpha(u_o) \leq 0$. But, this signifies (via (*)) that

$$-|a|e^{-\alpha(u_o)r} + |b|e^{-\alpha(u_o)s} = 1$$

with $e^{-\alpha(u_o)s} \geq e^{-\alpha(u_o)r} \geq 1$. Hence,

$$1 = -|a|e^{-\alpha(u_o)r} + |b|e^{-\alpha(u_o)s} \geq (-|a| + |b|),$$

a contradiction.

And conversely, if for all delays r, s, $0 < r \leq s < \infty$, every root of (1.2) has positive real part, then, in particular, for r even, p odd, we get $\lambda = \alpha(u_o) + i\theta$ as a root of (1.2). Hence, $\alpha(u_o) > 0$ and so, $e^{-\alpha(u_o)s} < e^{-\alpha(u_o)r} < 1$, which implies that

$$1 = -|a|e^{-\alpha(u_o)r} + |b|e^{-\alpha(u_o)s} < -|a| + |b|.$$

Lemma 2.11. Suppose $a < 0$ and $b > 0$. Then, there exists a root $\lambda = \alpha + i\theta$ of (1.2) with $\alpha = \alpha(1)$ if and only if $r/s =$ odd/even.

Corollary 2.10. Suppose $a < 0$ and $b > 0$. Then, every root of (1.2) has negative real part for every pair of delays r, s, $0 < r \leq s < \infty$, if and only if $|a| + b < 1$.

Corollary 2.11. Suppose $a < 0$, $b > 0$, $|a| + b = 1$ and $r/s \neq$ odd/even. Then, every root of (1.2) has negative real part.

Corollary 2.12. Suppose $a < 0$ and $b > 0$. Then, if every root of (1.2) has negative real part for some delays r, s, $0 < r \leq s < \infty$ such that $r/s =$ odd/even it follows that $|a| + b < 1$.

Lemma 2.12. Suppose $a > 0$ and $b < 0$. Then, (1.2) has a root $\lambda = \alpha + i\theta$ with $\alpha = \alpha(1)$ if and only if $r/s =$ even/odd.

Corollary 2.13. Suppose $a > 0$ and $b < 0$. Then, every root of (1.2) has negative real part for all delays r, s, $0 < r \leq s < \infty$ if and only if $a + |b| < 1$.

Corollary 2.14. Suppose $a > 0$, $b < 0$, $a + |b| = 1$ and $r/s \neq$ even/odd. Then, every root of (1.2) has negative real part.

Lemma 2.13. Suppose $a > 0$ and $b < 0$. Then, every root of (1.2) has positive real part for every pair of delays r, s, $0 < r \leq s < \infty$ if and only if $|b| > a+1$.

Lemma 2.14. Suppose $a < 0$ and $b > 0$. Then, every root of (1.2) has positive real part for all delays r, s, $0 < r \leq s < \infty$ if and only if $b > |a| + 1$.

3. Summary of results.

Note that if $0 < s \leq r < \infty$ then the roles of a and b in the above analysis are interchanged. But then, due to the apparent symmetry that there exists between the cases "$a > 0$, $b > 0$" and "$a < 0$, $b < 0$" and between the cases "$a > 0$, $b < 0$" and "$a < 0$,

Corollary 2.3. In order to see that, let u' be such that $\alpha(u') = 0$, i.e., $u' = 1/(a+b)$ in case $a > 0$ and $b > 0$. Then, we have:

Lemma 4.1. Suppose $a > 0$ and $b > 0$. Then, every root of (1.2) has negative real part for every pair of delays r, s, $0 < r,s < \infty$ if and only if $u' > 1$.

Proof. Since the roots $\lambda = \alpha(u) + i\theta$ of (1.2) occur only if $u \leq 1$, the hypothesis that $u' > 1$ implies (in view of the monotonicity of $\alpha(u)$) that $\alpha(u) < \alpha(u') = 0$ for all $u \leq 1$.

On the other hand, if every root of (1.2) has negative real part for all delays then, in view of Lemma 2.7, $\alpha(1) < 0$ which implies that $u' > 1$.

5. Example.

In [3] Silkowski considered the following example:

(4.1) $x(t) = -(1/2)x(t-1) - (1/2)x(t-2), \quad t \geq 0,$

with initial condition satisfying $|x(t)| \leq 1$ for $t \in [-2,0]$, and showed through the method of Hurwitz that (4.1) is asymptotically stable, i.e., every root of its characteristic equation has negative real part. He also found that $y(t) = \sin\{(n+3/2)\pi t\}$ is a solution of

$$y(t) = -(1/2)y(t-1+1/(2n+3)) - (1/2)y(t-2),$$

$t \geq 0$, which does not tend to zero as $t \to \infty$. Thus, he gave an example of a scalar difference equation with two delays which is asymptotically stable for a fixed pair of delays $(r = 1$ and $s = 2)$ and fails to be asymptotically stable for arbitrarily small perturbations of these delays $(r = 1 - 1/(2n+3)$ and $s = 2)$. We can get even sharper results for this equation by using the techniques of the analysis given in section 2 above. In fact, note that in this example, $a = b = -(1/2) < 0$ and $|a| + |b| = 1$ and, thus, the situation is clearly described by Lemma 2.9 and Corollaries 2.6 to 2.9.

$b > 0$", the analysis of (1.2) that is obtained when $0 < s \le r < \infty$ can be coupled with the one we just made, in order to obtain the following theorems:

Theorem 3.1. All roots of (1.2) have negative real part for every pair of delays r, s, $0 < r,s < \infty$ if and only if $|a| + |b| < 1$. If $|a| + |b| = 1$ and either $a < 0$ or $b < 0$ then there exists one and only one category of delays r, s among the following three categories:

$$C_1 = \{(r,s) : 0 < r,s < \infty \text{ and } r/s = \text{odd/odd}\},$$

$$C_2 = \{(r,s) : 0 < r,s < \infty \text{ and } r/s = \text{odd/even}\},$$

$$C_3 = \{(r,s) : 0 < r,s < \infty \text{ and } r/s = \text{even/odd}\},$$

which yield a root of (1.2) with real part equal to zero; this class of delays is uniquely determined by the signs of a and b and, moreover, all the remaining classes of delays yield only roots with negative real part.

Theorem 3.2. All roots of (1.2) have positive real part for $(a,b,r,s) \in A_i \times B_i$, $i = 1, 2$, where

$$A_1 = \{(a,b) \in R^2 : |b| > |a| + 1\}$$

$$A_2 = \{(a,b) \in R^2 : |a| > |b| + 1\}$$

$$B_1 = \{(r,s) \in R^2 : 0 < r \le s < \infty\}$$

$$B_2 = \{(r,s) \in R^2 : 0 < s \le r < \infty\}$$

Remark. Observe that

$$\{(r,s) : 0 < r,s < \infty \text{ and } r/s \text{ is irrational}\}$$

is a *remaining class of delays* in Theorem 3.1.

4. Alternative procedure.

There exists, in the spirit of this analysis, an alternative procedure to obtain results like those stated, for instance, by

Note, incidentally, that every root of the characteristic equation of

$$x(t) = -(1/2)x(t-r) - (1/2)x(t-s), \quad t \geq 0,$$

namely, the equation

$$(4.2) \quad 1 = -(1/2)e^{-\lambda r} - (1/2)e^{-\lambda s},$$

has negative real part for every fixed pair of rationally independent delays. We check this fact. First, we observe that there is no loss of generality if we choose $r = 1$ and pick s irrational. Then, we observe that since, in this case, $\alpha(1) = 0$, the most one can expect is to get a root of (4.2) (with $r = 1$ and s irrational) with real part equal to zero, i.e., to get a real θ such that

$$(4.3) \quad -(1/2)\cos\theta - (1/2)\cos\theta s = 1,$$

$$(4.4) \quad -(1/2)\sin\theta - (1/2)\sin\theta s = 0.$$

But, (4.4) implies that $\sin\theta = -\sin\theta s$. Hence, either $\theta = -\theta s+2k\pi$ or $\theta = \theta s + (2k+1)\pi$, $k = 0, 1, 2, \ldots,$ which implies that either $\theta = 2k\pi/(1+s)$ or $\theta = (2k+1)\pi/(1-s)$. Therefore, either $\theta = 0$ or $\theta \neq n\pi$ for any integer n. But, $\theta = 0$ is not a solution of (4.3) and $\theta \neq n\pi$ implies that $\cos\theta > -1$. This, together with (4.3), implies that $\cos\theta s < -1$, an impossibility.

REFERENCES

[1] - CARVALHO, L.A.V., On Lyapunov Functionals for Linear Difference Equations, Ph.D. Thesis, Brown University, Providence, R.I., June, 1979.

[2] - HALE, J.K., Parametric stability in difference equations, Boll. Un. Mat. It. (4), 10(1974).

[3] - SILKOWSKI, R.A., Star Shaped Regions of Stability in Hereditary Systems, Ph.D. Thesis, Brown University, Providence, R.I., June, 1976.

[4] - MELVIN, W.R., Stability properties of functional differential equations, J. Math. Anal. Appl., 48(1974), 749-763.

A LIAPUNOV FUNCTIONAL FOR A MATRIX RETARDED DIFFERENCE-DIFFERENTIAL EQUATION WITH SEVERAL DELAY

by Walter de Bona Castelan

1. Introduction.

In this paper we construct a Liapunov functional that characterizes the asymptotic behavior of the solutions of the linear autonomous matrix retarded difference-differential equation with several delays

$$(1.1)\quad \dot{X}(t) = AX(t) + \sum_{k=1}^{m} B_k X(t-\tau_k), \quad t > 0$$

where $X(t)$ is an n-vector function of time, A, B_k, $k = 1,\ldots,m$, are constant $n\times n$ matrices, and $0 < \tau_1 < \tau_2 <\ldots< \tau_m$.

For linear autonomous retarded difference-differential equations with one delay this, same problem was considered recently in [7] for the scalar case, and in [6] for the matrix case. Here, we extend the results obtained in [7] and [6] for equation (1.1).

As in [7], we first construct, by well-known methods, a Liapunov function that gives necessary and sufficient conditions for the asymptotic stability of the solutions of a difference equation approximation of the scalar retarded difrerence-differential equation with two delays rationally related

$$(1.2)\quad \dot{x}(t) = ax(t) + b_1x(t-\tau_1) + b_2x(t-\tau_2), \quad t > 0.$$

Then taking appropriate limits on this Liapunov function we obtain the desired Liapunov functional for equation (1.2). The Liapunov functional for equation (1.1) is then obtained as a generalization of the one for equation (1.2).

The Liapunov functional for equation (1.1) depends critically on a matrix function which must satisfy a special functional differential equation. For this equation we prove existence and uniqueness, as well as we give an algebraic representation of the solutions, in the case the delays are rationally related. The results obtained for this equation are extensions of those described in [2].

As in [6], we analyse the structure of the Liapunov functional for equation (1.1).

This Liapunov functional gives necessary and sufficient conditions for the asymptotic stability of the solutions of equation (1.1); moreover it gives the best possible estimate for the rates of growth or decay of the solutions.

2. <u>The retarded difference-differential equations with several delays.</u>

Denote by $L_2([a,b],R^n)$ the space of all Lebesque square integrable functions defined on $[a,b]$ with values in R^n. With $\tau \geq 0$ fixed, consider the Hilbert space $H = R^n \times L_2([-\tau,0],R^n)$ with the inner product

$$<u_1,u_2> = v_1^T v_2 + \int_{-\tau}^{0} \phi_1^T(\theta)\phi_2(\theta)\, d\theta$$

where $u_i = (v_i,\phi_i) \in H$, and the induced norm

$$\|(v,\phi)\|_H = v^T v + \int_{-\tau}^{0} \phi^T(\theta)\phi(\theta)\, d\theta.$$

Here, the superscript T denotes the transpose of a matrix. Let $X:[-\tau,\infty) \to R^n$; then for $t \geq 0$ we define the function $X_t:[-\tau,0]\to R^n$ by $X_t(\theta) = X(t+\theta)$.

Consider the matrix retarded difference-differential equation with several delays

$$(2.1) \quad \dot{X}(t) = AX(t) + \sum_{k=1}^{m} B_k X(t-\tau_k), \quad t > 0,$$

where A, B_k, $k = 1,\ldots,m$, are $n\times n$ constant matrices, $X(t)$ is an n-vector, and $0 < \tau_1 < \tau_2 <\ldots < \tau_m = \tau$. Let

$$X_o(0) = \xi, \quad X_o = \phi$$

be a given initial condition, with

(2.2) $(\xi,\phi) \in H$.

A solution of this initial value problem is for each $t > 0$, a function $X \in L_2([- ,t],R^n)$ such that x is absolutely continuous for $t \geq 0$, satisfies (2.1) a.e. on $[0,t]$ and $X(0) = \xi, X(\theta) = \phi(\theta)$ a.e. for $\theta \in [-\tau, 0]$. It is known [1, 4] that the initial value problem (2.1) - (2.2) has a unique solution, defined on $[-\tau,\infty)$, which depends continuously on the initial data in the norm of H.

The initial value problem (2.1) - (2.2) can be rewritten as

$$(2.3)\quad \frac{d}{dt}\begin{bmatrix} X_t(0) \\ X_t \end{bmatrix} = A \begin{bmatrix} X_t(0) \\ X_t \end{bmatrix},$$

$$(2.4)\quad (X_o(0),X_o) = (\xi,\phi) \in H,$$

where

$$(2.5)\quad A\begin{bmatrix} X_t(0) \\ X_t \end{bmatrix} = \begin{bmatrix} AX_t(0) + \sum_{k=1}^{m} B_k X_t(-\tau_k) \\ \dfrac{\partial X_t(\theta)}{\partial\theta}, \quad -\tau \leq \theta \leq 0 \end{bmatrix}$$

The above operator A has a domain $\mathcal{D}(A)$, dense in H, defined by

$$\mathcal{D}(A) = \{(\xi,\phi) \in H \mid \phi \text{ is A.C. in } [-\tau,0],$$
$$\phi' \in L_2[-\tau,0],\ \phi(0) = \xi\}.$$

The operator A is the generator of a C_o-semigroup $I(t):H \to H$ given by

$$I(t)(\xi,\phi) = (X(t),X_t)$$

the solution pair of (2.3), (2.4).

Let $G(A)$ denote the spectrum of A, i.e.,

$$G(A) = \{\lambda \mid \det[\lambda I - A - \sum_{k=1}^{m} B_k e^{-\lambda\tau_k}] = 0\}.$$

Then, [1,4], there exists a constant γ such that $Re(\lambda)\le\gamma$ for all $\lambda \in G(A)$. Also, for every $\varepsilon > 0$, there exists a constant $K \ge 1$ such that

$$(2.6) \quad \|T(t)\|_{(H,H)} \le Ke^{(\gamma+\varepsilon)t}$$

Finally [4], a useful representation of the solutions of (2.1) is given for every $t,u \ge 0$ by the formula

$$(2.7) \quad x_{t+u}(0) = S(u)x_t(0) + \sum_{k=1}^{m} \int_{-\tau_k}^{0} S(u-\alpha-\tau_k)B_k x_t(\alpha)\,d\alpha,$$

where the matrix S is the solution of the matrix initial value problem

$$(2.8) \quad \begin{aligned} &\frac{d}{dt}S(t) = S(t)A + \sum_{k=1}^{m} S(t-\tau_k)B_k \\ &S(0) = I, \; S(t) = 0 \quad \text{for} \quad t < 0 \end{aligned}$$

3. <u>A Liapunov function for a difference equation approximation of a scalar retarded difference-differential equation with two delays.</u>

In this section we develop a Liapunov function for a difference equation approximation of the scalar retarded difference - differential equation with two delays

$$(3.1) \quad x(t) = ax(t) + bx(t-\tau_1) + cx(t-\tau_2), \quad t \ge 0,$$

where we assume $\tau_1 = \frac{p}{q}\tau_2$, p and q positive integers, $p < q$.

Consider N fixed, and the intervals $[0,\infty)$, $[-\tau_2,0]$ subdivided in subintervals of equal length $\frac{\tau_2}{Npq}$. The values of the function $x_t(\theta)$ at the mesh points $x_{k\frac{\tau_2}{Npq}}(-J\frac{\tau_2}{Npq})$ will be denoted x_k^J, $k = 0,1,\ldots,$ $J = 0,\ldots,Npq$. A difference equation approximation of (3.1) is then given by

$$x_{k+1}^{o} = \left(1+\frac{\tau_2}{Npq}a\right)x_k^{o} + \frac{\tau_2}{Npq}\left(bx_k^{Np^2}\right) + \frac{\tau_2}{Npq}\left(cx_k^{Npq}\right), \quad k = 0,1,\ldots,$$

$$x_{k+1}^{J} = x_k^{J-1}, \quad J = 1,\ldots,Npq.$$

This difference equation can be rewritten in the form

$$(3.2) \quad y_{k+1} = \hat{A}y_k,$$

where y_k denotes the $(Np^2+Npq+1)$-dimensional vector

$$(3.3) \quad y_k = [x_k^{o}, bx_k^{1}, \ldots, bx_k^{Np^2}, cx_k^{1}, \ldots, cx_k^{Npq}]^T,$$

$$(3.4) \quad \hat{A} = \begin{bmatrix} \left(1+\frac{\tau_2}{Npq}a\right) & 0 & \ldots & 0 & \frac{\tau_2}{Npq} & 0 & \ldots & 0 & \frac{\tau_2}{Npq} \\ b & 0 & \ldots & 0 & 0 & 0 & \ldots & 0 & 0 \\ 0 & 1 & \ldots & 0 & 0 & 0 & \ldots & 0 & 0 \\ \vdots & \vdots & \vdots & \vdots & \vdots & \vdots & \vdots & \vdots & \vdots \\ 0 & 0 & \ldots & 1 & 0 & 0 & \ldots & 0 & 0 \\ c & 0 & \ldots & 0 & 0 & 0 & \ldots & 0 & 0 \\ 0 & 0 & \ldots & 0 & 0 & 1 & \ldots & 0 & 0 \\ \vdots & \vdots & \vdots & \vdots & \vdots & \vdots & \vdots & \vdots & \vdots \\ 0 & 0 & & 0 & 0 & 0 & & 1 & 0 \end{bmatrix}$$

The method of construction of a Liapunov function for the difference equation (3.2) in order to obtain necessary and sufficient conditions for the asymptotic stability of its solutions is well-known [9, 10]. Such a Liapunov function is given by $\hat{V}(y_k) = y^T D y_k$, where D is a positive definite matrix. Consider the forward difference $\Delta\hat{V}(y_k) = \hat{V}(y_{k+1}) - \hat{V}(y_k)$, which is given by $\Delta\hat{V}(y_k) = -y_k^T E y_k$, where $-E = \hat{A}^T D \hat{A} - D$. Now, if E is a positive definite matrix, then all the solutions of (3.2) are asymptotically stable. On the other hand, if we assume asymptotic stability, i.e., all the eigenvalues of $\hat{A}$

have modulus strictly less than one, then given any positive definite matrix E, the equation $\hat{A}^T DA - D = -E$ has a unique solution D which is also a positive definite matrix. Now, assume that for some real number $\hat{\delta}$, $0 \leq \hat{\delta} < 1$ all the eigenvalues of the matrix $\frac{1}{\sqrt{1-\hat{\delta}}}\hat{A}$ have modulus strictly less than one.

Then given any positive definite matrix E, there is a unique positive matrix D which satisfies the equation

(3.5) $\hat{A}^T D\hat{A} - (1-\hat{\delta})D = -E$

Note that if $\hat{V}(y_k) = y_k^T Dy_k$, then $\Delta\hat{V}(y_k) = -y_k^T Ey_k \;\; -\hat{\delta}y_k Dy_k \;\; -\hat{\delta}\hat{V}(y_k)$.

The special matrix $\hat{A}$ is seen to be equivalent to a matrix in companion form. In this case, it is sufficient to take the matrix E semidefinite and not identically zero in order to have the existence, uniqueness and positive definiteness of the matrix D, the solution of the equation (3.5), [10]. Therefore we can choose particularly simple matrices E. Given the special form of the matrix $\hat{A}$, we restrict ourselves to certain choices of E, in order to obtain as simple a form as possible for the matrix D. The unique solution D of the equation (3.5) will be represented in the form

$$(3.6)\quad D = \begin{bmatrix} \alpha & \tilde{r}^T & r^T \\ \tilde{r} & \tilde{Q} & \approx\!\!Q \\ r & \approx\!\!Q & Q \end{bmatrix},$$

where α is a scalar, $r^T = (r_1,\ldots,r_{Npq})$, $\tilde{r}^T = (\tilde{r}_1,\ldots,\tilde{r}_{Np^2})$ are Npq-, Np^2-dimensional vectors, $Q = (q_{ij})$, $\tilde{Q} = (\tilde{q}_{ij})$ are $Npq \times Npq$, $Np^2 \times Np^2$ symetric matrices, and $\approx\!\!Q = (\approx\!\!q_{ij})$, $\approx\!\!Q = (\approx\!\!q_{ij})$ are $Np^2 \times Npq$, $Npq \times Np^2$ matrices respectively.

Now, consider substitution of (3.4) and (3.6) into the equation (3.5). Then a particularly simple form for the matrix D can be obtained if we proceed in the following manner. The matrix $E = E^T = (e_{ij})$ is chosen to have zero entries everywhere except the elements

e_{11}, $e_{1,Np^2+1} = e_{Np^2+1,1}$, $e_{1,Npq+1} = e_{Npq+1,1}$, e_{Np^2+1,Np^2+1}, $e_{Np^2+1,Npq+1} = e_{Npq+1,Np^2+1}$, and $e_{Npq+1,Npq+1}$. Also, given the simplicity of the structure of the matrix $\hat{A}$, the vector $\tilde{r}^T$ can be chosen as $\tilde{r}^T = (r_{Npq-Np^2+1},\ldots,r_{Npq})$, and the matrices Q, $\tilde{Q}$, $\tilde{\tilde{Q}}$ and $\tilde{\tilde{\tilde{Q}}}$ to be related by

$$\tilde{q}_{i,j} = q_{Npq-Np^2+i,Npq-Np^2+j} \qquad \text{for } i \neq j,\ i,j = 1,\ldots,Np^2,$$

$$\tilde{\tilde{q}}_{i,j} = q_{Npq-Np^2+i,j} \qquad \text{for } j \neq Npq-Np^2+i,\ i = 1,\ldots,Np^2,\ j = 1,\ldots,Npq,$$

$$\tilde{\tilde{q}}_{i,j} = 0 \qquad \text{for } j = Npq-Np^2+i,\ i = 1,\ldots,Np^2,$$

$$\tilde{\tilde{\tilde{q}}}_{i,j} = q_{i,Npq-Np^2+j} \qquad \text{for } i \neq Npq-Np^2+j,\ i = 1,\ldots,Npq,\ j = 1,\ldots,Np^2,$$

$$\tilde{\tilde{\tilde{q}}}_{i,j} = 0 \qquad \text{for } i = Npq-Np^2+j,\ j = 1,\ldots,Np^2.$$

The above choices imply that the following equations are satisfied

(3.7.a) $$q_{i,j} = (1-\hat{\delta})q_{i-1,j-1}, \quad i = 2,\ldots,Npq, \quad j = 2,\ldots,Npq$$

(3.7.b) $$(1+\frac{\tau_2}{Npq}a)r_i + bq_{Npq-Np^2+1,i} + cq_{1,i} - (1-\hat{\delta})r_{i-1} = 0, \quad i \neq Npq-Np^2+1, \quad i = 2,\ldots,Npq,$$

(3.7.c) $$\frac{\tau_2}{Npq}r_i - (1-\hat{\delta})q_{i-1,Npq} = 0, \quad i = 2,\ldots,Npq$$

and that the nonzero elements of the matrix E are given by

(3.7.d) $$e_{1,1} = -[(1+\frac{\tau_2}{Npq}a)[(1+\frac{\tau_2}{Npq}a)\alpha + br_{Npq-Np^2+1} + cr_1] + [(1+\frac{\tau_2}{Npq}a)r_{Npq-Np^2+1} + b\tilde{q}_{1,1} + c\tilde{\tilde{q}}_{1,1}]b$$

$$+ [(1+\frac{\tau_2}{Npq}a)r_1 + b\tilde{q}_{1,1} + cq_{1,1}]c - (1-\hat{\delta})\alpha]],$$

(3.7.e) $$e_{1,Np^2+1} = e_{Np^2+1,1} = e_{1,Npq+1} = e_{Npq+1,1}$$
$$= - [(1+\frac{\tau_2}{Npq}a)\alpha\frac{\tau_2}{Npq} + br_{Npq-Np^2+1}\frac{\tau_2}{Npq}$$
$$+ cr_1\frac{\tau_2}{Npq} - (1-\hat{\delta})r_{Npq}],$$

(3.7.f) $$e_{Np^2+1,Np^2+1} = (\frac{\tau_2}{Npq})^2\alpha - (1-\hat{\delta})\tilde{q}_{Np^2,Np^2},$$

(3.7.g) $$e_{Npq+1,Npq+1} = (\frac{\tau_2}{Npq})^2\alpha - (1-\hat{\delta})q_{Npq,Npq},$$

(3.7.h) $$e_{Np^2+1,Npq+1} = e_{Npq+1,Np^2+1} = (\frac{\tau_2}{Npq})^2\alpha .$$

From equation (3.7.a) it follows that

(3.8.a) $$q_{Npq-Np^2+i,Npq-Np^2+j} = (1-\hat{\delta})^{-Np^2+i}q_{Npq,Npq-(i-j)},$$
$$i = 2,\ldots,Np^2, \quad j = 1,\ldots,i-1,$$

(3.8.b) $$q_{i,j} = (1-\hat{\delta})^{-Npq+i}q_{Npq,Npq-(i-j)},$$
$$i = 2,\ldots,Npq, \quad j = 1,\ldots,i-1,$$

(3.8.c) $$q_{Npq-Np^2+i,j} = (1-\hat{\delta})^{-Np^2+i}q_{Npq,Np^2-(i-j)},$$
$$i = 1,\ldots,Np^2, \quad j = 1,\ldots,Npq-Np^2+i-1,$$

(3.8.d) $$q_{Npq-Np^2+i,j} = (1-\hat{\delta})^{-Npq+j}q_{2Npq-Np^2-(j-i),Npq},$$
$$i = 1,\ldots,Np^2-1, \quad j = Npq-Np^2+i+1,\ldots,Npq,$$

(3.8.e) $$\tilde{q}_{i,i} = (1-\hat{\delta})^{i-1}\tilde{q}_{1,1}, \quad i = 1,\ldots,Np^2$$

(3.8.f) $$q_{i,i} = (1-\hat{\delta})^{i-1}q_{1,1}, \quad i = 1,\ldots,Npq$$

Using the equations (3.7.b-c) and (3.8.b-d), we obtain

$$(3.9.a)\quad (1+\frac{\tau_2}{Npq}a)r_i + b(1-\hat{\delta})^{-Np^2}\frac{\tau_2}{Npq}r_{Np^2+i}$$
$$+ c(1-\hat{\delta})^{-Npq+i-1}\frac{\tau_2}{Npq}r_{Npq-i+2} - (1-\hat{\delta})r_{i-1} = 0,\quad i = 2,\ldots,Npq-Np^2$$

and

$$(3.9.b)\quad (1+\frac{\tau_2}{Npq}a)r_i + b(1-\hat{\delta})^{-Npq+i-1}\frac{\tau_2}{Npq}r_{2Npq-Np^2-i+2}$$
$$+ c(1-\hat{\delta})^{-Npq+i-1}\frac{\tau_2}{Npq}r_{Npq-i+2} - (1-\hat{\delta})r_{i-1} = 0,$$
$$i = Npq-Np^2+2,\ldots,Npq$$

Let $\tilde{V}(x_k) = \frac{\tau_2}{Npq}\hat{V}(y_k)$, where $x_k = (x_k^o, x_k^1,\ldots,x_k^{Npq})$, $\tilde{\alpha} = \frac{\tau_2}{Npq}\alpha$, $\tilde{\beta} = b^2\tilde{q}_{1,1}$, $\tilde{\gamma} = c^2 q_{1,1}$, $\delta = \frac{1}{2}\frac{Npq}{\tau_2}\hat{\delta}$. Then using the equations (3.7.c), (3.8.a-f) and (3.9.a-b), we obtain the desired Liapunov function in the form

$$(3.10)\quad \tilde{V}(x_k) = \alpha(x_k^o)^2 + 2x_k^o\sum_{i=1}^{Np^2} r_{Npq-Np^2+i}bx_k^i\frac{\tau_2}{Npq}$$
$$+ 2x_k^o\sum_{i=1}^{Npq} r_i cx_k^i\frac{\tau_2}{Npq}$$
$$+ 2\sum_{i=2}^{Np^2}\sum_{j=1}^{i-1}(1-2\frac{\tau_2}{Npq}\delta)^{i-Np^2-1}r_{Npq-(i-j)+1}b^2x_k^ix_k^j(\frac{\tau_2}{Npq})^2$$
$$+ 2\sum_{i=2}^{Npq}\sum_{j=1}^{i-1}(1-2\frac{\tau_2}{Npq}\delta)^{i-Npq-1}r_{Npq-(i-j)+1} + c^2x_k^ix_k^j(\frac{\tau_2}{Npq})^2$$
$$+ 2\sum_{i=1}^{Np^2}\sum_{j=1}^{Npq-Np^2+i-1}(1-2\frac{\tau_2}{Npq}\delta)^{-Np^2+i-1}r_{Np^2-(i-j)+1}bcx_k^ix_k^j(\frac{\tau_2}{Npq})^2$$
$$+ 2\sum_{i=1}^{Np^2-1}\sum_{j=Npq-Np^2+i+1}^{Npq}(1-2\frac{\tau_2}{Npq}\delta)^{j-Npq+1}r_{2Npq-Np^2-(j-i)+1}bcx_k^ix_k^j(\frac{\tau_2}{Npq})^2$$
$$+ \sum_{i=1}^{Np^2}(1-2\frac{\tau_2}{Npq}\delta)^{i-1}\tilde{\beta}(x_k^i)^2\frac{\tau_2}{Npq} + \sum_{i=1}^{Npq}(1-2\frac{\tau_2}{Npq}\delta)^{i-1}\tilde{\gamma}(x_k^i)^2\frac{\tau_2}{Npq}$$

and its forward difference, divided by $\frac{\tau_2}{Npq}$,

$$(3.11)\quad \frac{\Delta\tilde{V}(x_k)}{\tau_2/Npq} = [2\tilde{\alpha}(a+\delta) + \frac{\tau_2}{Npq}a^2\tilde{\alpha} + 2br_{Npq-Np^2+1}$$

$$+ 2cr_1 + \frac{\tau_2}{Npq}abr_{Npq-Np^2+1} + \frac{\tau_2}{Npq}acr_1 + \tilde{\beta}$$

$$+2bcr_{Npq-Np^2+1,1}(1-2\frac{\tau_2}{Npq}\delta)^{-Np^2+1}\frac{\tau_2}{Npq} + \tilde{\gamma}](x_k^o)^2$$

$$+ [\tilde{\alpha} - r_{Npq} + \frac{\tau_2}{Npq}a\tilde{\alpha} + br_{Npq-Np^2+1}\frac{\tau_2}{Npq} + cr_1\frac{\tau_2}{Npq}$$

$$+ 2\frac{\tau_2}{Npq}r_{Npq}].[2bx_k^o x_k^{Np^2} + 2cx_k^o x_k^{Npq}]$$

$$+ [\frac{\tau_2}{Npq}\tilde{\alpha}b^2-(1-2\frac{\tau_2}{Npq}\delta)^{Np^2-1}\tilde{\beta}+2\frac{\tau_2}{Npq}(1-2\frac{\tau_2}{Npq}\delta)^{Np^2-1}\tilde{\beta}].(x_k^{Np^2})^2$$

$$+ [\frac{\tau_2}{Npq}\tilde{\alpha}c^2-(1-2\frac{\tau_2}{Npq}\delta)^{Npq-1}\tilde{\gamma}+2\frac{\tau_2}{Npq}(1-2\frac{\tau_2}{Npq}\delta)^{Npq-1}\tilde{\gamma}].(x_k^{Npq})^2$$

$$+ 2\frac{\tau_2}{Npq}\tilde{\alpha}bcx_k^{Npq}x_k^{Np^2} - 2\delta\tilde{V}(x_k),$$

where the r_i satisfy

$$(3.12.a)\quad \frac{r_i-r_{i-1}}{\tau_2/Npq} = -\frac{a+2\delta}{1-2\frac{\tau_2}{Npq}\delta}r_i - b(1-2\frac{\tau_2}{Npq}\delta)^{-Np^2-1}r_{Np^2+i}$$

$$- c(1-2\frac{\tau_2}{Npq}\delta)^{-Npq+i-2}r_{Npq-i+2}, \quad i = 2,\dots,Npq-Np^2,$$

and

$$(3.12.b)\quad \frac{r_i-r_{i-1}}{\tau_2/Npq} = -\frac{a+2\delta}{1-2\frac{\tau_2}{Npq}\delta}r_i - b(1-2\frac{\tau_2}{Npq}\delta)^{-Npq+i-2}r_{2Npq-Np^2-i+2}$$

$$- c(1-2\frac{\tau_2}{Npq}\delta)^{-Npq+i-2}r_{Npq-i+2}, \quad i = Npq-Np^2+2,\dots,Npq$$

4. A Liapunov functional for a scalar retarded difference-differential equation with two delays.

Using the results of the Section 3, we can obtain a Liapunov functional in a explicit form for the scalar retarded difference--differential equation (3.1). We consider the limiting process described by

$$x_k^J \xrightarrow[N\to\infty]{} x_t(\theta), \quad -\tau_2 \le \theta \le 0,$$

$$r_i \xrightarrow[N\to\infty]{} r(-\theta), \quad -\tau_2 \le \theta \le 0,$$

$$\tilde{V}(x_k) \xrightarrow[N\to\infty]{} V(x_t(0),x_t),$$

$$\frac{\Delta\tilde{V}(x_k)}{\tau_2/Npq} \xrightarrow[N\to\infty]{} \dot{V}(x_t(0),x(t),$$

$$\lim_{N\to\infty} (1-2\frac{\tau_2}{Npq}\delta)^{i-Npq+1} \longrightarrow e^{2\delta\theta}, \quad -\tau_2 \le \theta \le 0.$$

The substitution of the above formal limits into the equations (3.10), (3.11) and (3.12.a-b) yield the equation

$$(4.1) \qquad V(x_t(0),x_t) = \tilde{\alpha}x_t^2(0) + 2x_t(0)\int_{-\tau_1}^{0} r(\tau_2-\tau_1-\theta)bx_t(\theta)d\theta$$

$$+ 2x_t(0)\int_{-r_2}^{0} r(-\theta)cx_t(\theta)d\theta$$

$$+ 2\int_{-\tau_1}^{0}\int_{\theta}^{0} e^{2\delta(\theta+\tau_1)} r(\tau_2+\theta-\beta)b^2x_t(\theta)\ x_t(\beta)d\beta d\theta$$

$$+ 2\int_{-\tau_2}^{0}\int_{\theta}^{0} e^{2\delta(\theta+\tau_2)} r(\tau_2+\theta-\beta)c^2x_t(\theta)\ x_t(\beta)d\beta d\theta$$

$$+ 2\int_{-\tau_1}^{0}\int_{\theta-(\tau_2-\tau_1)}^{0} e^{2\delta(\theta+\tau_1)} r(\tau_1+\theta-\beta)bcx_t(\theta)x_t(\beta)d\beta d\theta$$

$$+ 2\int_{-\tau_1}^{0}\int_{-\tau_2}^{\theta-(\tau_2-\tau_1)} e^{2\delta(\theta+\tau_2)} r(2\tau_2-\tau_1+\beta-\theta)bcx_t(\theta)x_t(\beta)d\beta d\theta$$

$$+ \int_{-\tau_1}^{0} e^{2\delta\theta}\tilde{\beta}x_t^2(\theta)d\theta + \int_{-\tau_2}^{0} e^{2\delta\theta}\tilde{\gamma}x_t^2(\theta)d\theta,$$

(4.2) $$\dot{V}(x_t(0),x_t) = [2\tilde{\alpha}(a+\delta) + 2br(\tau_2-\tau_1) + 2cr(0)$$

$$+ \tilde{\beta} + \tilde{\gamma}]x_t^2(0) + [\tilde{\alpha}-r(\tau_2)]. [2bx_t(0)x_t(-\tau_1)+2cx_t(0)x_t(-\tau_2)]$$

$$- e^{-2\delta\tau_1}\tilde{\beta}x_t^2(-\tau_1) - e^{-2\delta\tau_1}\tilde{\gamma}x_t^2(-\tau_2)$$

$$- 2\delta V(x_t(0),x_t),$$

(4.3.a) $$\dot{r}(-\theta) = -(a+2\delta)r(-\theta)-be^{2\delta\tau_1}r(\tau_1-\theta)$$

$$- ce^{2\delta(\theta+\tau_2)}r(\tau_2+\theta), \qquad -\tau_2+\tau_1 \le \theta \le 0,$$

and

(4.3.b) $$\dot{r}(-\theta) = -(a+2\delta)r(-\theta) - be^{2\delta(\theta+\tau_2)}r(2\tau_2-\tau_1+\theta)$$

$$-ce^{2\delta(\theta+\tau_2)}r(\tau_2+\theta), \qquad -\tau_2 \le \theta \le -\tau_2+\tau_1.$$

Now, we introduce the notation defined by the equations

$$r(\theta) = w(\tau_2-\theta)e^{\delta(\tau_2-\theta)}$$

and

$$\tilde{\alpha} = \tilde{\tilde{\alpha}} + r(\tau_1),$$

$$\tilde{\beta} = e^{\delta\tau_1}\tilde{\tilde{\beta}},$$

$$\tilde{\gamma} = e^{\delta\tau_2}\tilde{\tilde{\gamma}}.$$

The substitution of the above notation into the equations (4.1), (4.2) and (4.3.a-b) yields the equations

(4.4) $$V(x_t(0),x_t) = \tilde{\tilde{\alpha}}x_t^2(0) + \int_{-\tau_1}^{0} e^{\delta(\tau_1+2\theta)}\tilde{\tilde{\beta}}x_t^2(\theta)d\theta$$

$$+ \int_{-\tau_2}^{0} e^{\delta(\tau_2+2\theta)} \tilde{\tilde{\gamma}} x_t^2(\theta)\,d\theta + w(0)x_t^2(0)$$

$$+ 2x_t(0) \int_{-\tau_1}^{0} w(\tau_1+\theta) e^{\delta(\tau_1+\theta)} b x_t(\theta)\,d\theta$$

$$+ 2x_t(0) \int_{-\tau_2}^{0} w(\tau_2+\theta) e^{\delta(\tau_2+\theta)} c x_t(\theta)\,d\theta$$

$$+ 2 \int_{-\tau_1}^{0} \int_{\theta}^{0} w(\beta-\theta) e^{\delta(2\tau_1+\theta+\beta)} b^2 x_t(\theta) x_t(\beta)\,d\beta\,d\theta$$

$$+ 2 \int_{-\tau_2}^{0} \int_{\theta}^{0} w(\beta-\theta) e^{\delta(2\tau_2+\theta+\beta)} c^2 x_t(\theta) x_t(\beta)\,d\beta\,d\theta$$

$$+ 2 \int_{-\tau_1}^{0} \int_{\theta-(\tau_2-\tau_1)}^{0} w(\tau_2-\tau_1+\beta-\theta) e^{\delta(\tau_1+\tau_2+\theta+\beta)} bc x_t(\theta) x_t(\beta)\,d\theta\,d\beta$$

$$+ 2 \int_{-\tau_1}^{0} \int_{-\tau_2}^{\theta-(\tau_2-\tau_1)} w(-\tau_2+\tau_1+\theta-\beta) e^{\delta(\tau_1+\tau_2+\theta+\beta)} bc x_t(\theta) x_t(\beta)\,d\theta\,d\beta,$$

(4.5) $$\dot{V}(x_t(0),x_t) = [2\tilde{\tilde{\alpha}}(a+\delta)+2w(0)(a+\delta)+2bw(\tau_1)e^{\delta\tau_1}$$

$$+ 2cw(\tau_2)e^{\delta\tau_2} + \tilde{\tilde{\beta}}e^{\delta\tau_1} + \tilde{\tilde{\gamma}}e^{\delta\tau_2}]x_t^2(0)$$

$$+ 2\tilde{\tilde{\alpha}} b x_t(0)x_t(-\tau_1) + 2\tilde{\tilde{\alpha}} c x_t(0) x_t(-\tau_2)$$

$$- e^{-\delta\tau_1}\tilde{\tilde{\beta}} x_t^2(-\tau_1) - e^{-\delta\tau_2}\tilde{\tilde{\gamma}} x_t^2(-\tau_2) - 2\delta V(x_t(0),x_t),$$

(4.6.a) $$\dot{w}(\alpha) = (a+\delta)w(\alpha) + be^{\delta\tau_1}w(\tau_1-\alpha) + ce^{\delta\tau_2}w(\tau_2-\alpha), \quad 0 \le \alpha \le \tau_1,$$

and

(4.6.b) $$\dot{w}(\alpha) = (a+\delta)w(\alpha) + be^{\delta\tau_1}w(\alpha-\tau_1) + ce^{\delta\tau_2}w(\tau_2-\alpha), \quad \tau_1 \le \alpha \le \tau_2.$$

Equations (4.4), (4.5) and (4.6.a-b) have been obtained by

formally taking limits on the Liapunov for the difference equation approximation of the functional differential equation (3.1). It is easy to see that (4.4) is a well-defined functional on H; moreover, a straightforward although laborius computation shows that, if (4.6.a-b) is satisfied, (4.5) represents the rate of change of the functional (4.4) along the solutions of the scalar retarded functional differential equation (3.1) with initial conditions in the domain of the generator. We postpone such an analysis, and use the above results as motivation for the method of analysis presented in the next section.

5. A Liapunov functional for the matrix retarded difference-differential equation with several delays.

In this section we consider a Liapunov functional for the matrix retarded difference-differential equation with several delays (2.1)-(2.2). The form of this functional is suggested by the results of the previous section. On the space H consider the real symmetric quadratic form

$$(5.1) \quad V(\xi,\phi) = \xi^T M\xi + \sum_{k=1}^{m} e^{\delta\tau_k}\int_{-\tau_k}^{0}\phi^T(\theta)R_k e^{2\delta\theta}\phi(\theta)d\theta$$

$$+ \xi^T Q(0)\xi + 2\xi^T\sum_{k=1}^{m}\int_{-\tau_k}^{0} Q(\tau_k+\theta)e^{\delta(\tau_k+\theta)}B_k\phi(\theta)d\theta$$

$$+ 2\sum_{k=1}^{m}\int_{-\tau_k}^{0}\int_{\theta}^{0}\phi^T(\theta)B_k^T Q(\beta-\theta)e^{\delta(\theta+\beta+2\tau_k)}B_k\phi(\beta)d\beta d\theta$$

$$+ 2\sum_{k=2}^{m}\sum_{j=1}^{k-1}\int_{-\tau_j}^{0}\int_{\theta-(\tau_k-\tau_j)}^{0}\phi^T(\theta)B_j^T Q(\beta-\theta+\tau_k-\tau_j)e^{\delta(\theta+\beta+\tau_j+\tau_k)}B_k\phi(\beta)d\beta d\theta$$

$$+ 2\sum_{k=2}^{m}\sum_{j=1}^{k-1}\int_{-\tau_j}^{0}\int_{-\tau_k}^{\theta-(\tau_k-\tau_j)}\phi^T(\theta)B_j^T Q^T(\theta-\beta+\tau_j-\tau_k)e^{\delta(\theta+\beta+\tau_j+\tau_k)}B_k\phi(\beta)d\beta d\theta,$$

where δ is a real number, M, R_k, $k = 1,\dots,m$, are constant $n\times n$ real positive definite matrices, and $Q(\alpha)$, $0 \le \theta \le \tau_m$, is a continuously differentiable matrix; here we assume that $Q(\alpha)$ is a solution of the initial value problem for the functional differential equation

$$(5.2.a_1)\quad Q'(\alpha) = (A^T+\delta I)Q(\alpha) + \sum_{k=1}^{m} e^{\delta\tau_k} B_k^T Q^T(\tau_k-\alpha) \quad \text{for } 0 \le \alpha \le \tau_1,$$

$$(5.2.a_2)\quad Q'(\alpha) = (A^T+\delta I)Q(\alpha) + e^{\delta\tau_1} B_1^T Q(\alpha-\tau_1)$$

$$+ \sum_{k=2}^{m} e^{\delta\tau_k} B_k^T Q^T(\tau_k-\alpha) \quad \text{for } \tau_1 \le \alpha \le \tau_2,$$

...

$$(5.2.a_m)\quad Q'(\alpha) = (A^T+\delta I)Q(\alpha) + \sum_{k=1}^{m-1} e^{\delta\tau_k} B_k^T Q(\alpha-\tau_k)$$

$$+ e^{\delta\tau_m} B_m^T Q^T(\tau_m-\alpha) \quad \text{for } \tau_{m-1} \le \alpha \le \tau_m,$$

$$(5.3)\quad Q(0) = Q^T(0) = Q_o,$$

where Q_o is an arbitrary symmetric matrix.

If the Fréchet differentiable functional (5.1) is evaluated along a solution of (2.1) - (2.2) with initial conditions in $\mathcal{D}(A)$, then this yields a function of time, which we denote by $V(t) = V(X_t(0),X_t)$. This function of time is differentiable along such solution. After a straightforward but laborious computation we obtain that

$$(5.4)\quad \dot V(t) = \frac{d}{dt}V(X_t(0),X_t) = -2\delta V(X_t(0),X_t)$$

$$+ X_t^T(0)[(A^T+\delta I)Q(0)+Q(0)(A+\delta I)+(A^T+\delta I)M+M(A+\delta I)$$

$$+ \sum_{k=1}^{m} e^{\delta\tau_k} B_k Q^T(\tau_k) + \sum_{k=1}^{m} e^{\delta\tau_k} Q(\tau_k)B_k + \sum_{k=1}^{m} e^{\delta\tau_k} R_k]X_t^T(0)$$

$$+ 2X_t^T(0)M \sum_{k=1}^{m} B_k X_t(-\tau_k) - \sum_{k=1}^{m} e^{-\delta\tau_k} X_t(-\tau_k) R_k X_t(-\tau_k)$$

$$+ 2X_t^T(0)\int_{-\tau_1}^{0} [(A^T+\delta I)Q(\tau_1+\theta)-\dot{Q}(\tau_1+\theta)+B_1^T e^{\delta\tau_1} Q^T(-\theta)$$

$$+ \sum_{j=2}^{m} e^{\delta\tau_j} B_j^T Q^T(-\theta+\tau_j-\tau_1)]e^{\delta(\theta+\tau_1)} B_1 X_t(\theta)d\theta$$

$$+ 2X_t^T(0) \sum_{k=2}^{m}\int_{-(\tau_k-\tau_{k-1})}^{0} [(A^T+\delta I)Q(\tau_k+\theta)-\dot{Q}(\tau_k+\theta)$$

$$+ \sum_{j=1}^{k-1} e^{\delta\tau_j} B_j^T Q(\theta+\tau_k-\tau_j)+e^{\delta\tau_k} B_k^T Q^T(-\theta)$$

$$+ \sum_{j=k+1}^{m+1} e^{\delta\tau_j} B_j^T Q^T(-\theta+\tau_j-\tau_k)]e^{\delta(\theta+\tau_k)} B_k X_t(\theta)d\theta$$

$$+ 2X_t^T(0) \sum_{k=2}^{m} \int_{-(\tau_k-\tau_{k-2})}^{-(\tau_k-\tau_{k-1})} [(A^T+\delta I)Q(\tau_k+\theta)-\dot{Q}(\tau_k+\theta)$$

$$+ \sum_{j=1}^{k-2} e^{\delta\tau_j} B_j^T Q(\theta+\tau_k-\tau_j)+e^{\delta\tau_{k-1}} B_{k-1}^T Q^T(-\theta+\tau_{k-1}-\tau_k)$$

$$+ e^{\delta\tau_k} B_k^T Q^T(-\theta) + \sum_{j=k+1}^{m+1} e^{\delta\tau_j} B_t^T Q^T(-\theta+\tau_j-\tau_k)]e^{\delta(\theta+\tau_k)} B_k X_t(\theta)d\theta$$

$$+ \dots$$

$$+ 2X_t^T(0) \sum_{k=2}^{m} \int_{-(\tau_k-\tau_1)}^{-(\tau_k-\tau_2)} [(A^T+\delta I)Q(\tau_k+\theta)-\dot{Q}(\tau_k+\theta)+e^{\delta\tau_1} B_1^T Q(\theta+\tau_k-\tau_1)$$

$$+ \sum_{j=2}^{k-1} e^{\delta\tau_j} B_j^T Q^T(-\theta+\tau_j-\tau_k) + e^{\delta\tau_k} B_k^T Q^T(-\theta)$$

$$+ \sum_{j=k+1}^{m+1} e^{\delta\tau_j} B_j^T Q^T(-\theta+\tau_j-\tau_k)]e^{\delta(\theta+\tau_k)} B_k X_t(\theta)d\theta$$

$$+ 2X_t^T(0) \sum_{k=2}^{m} \int_{-\tau_k}^{-(\tau_k-\tau_1)} [(A^T+\delta I)Q(\tau_k+\theta)-\dot{Q}(\tau_k+\theta)$$

$$+ \sum_{j=1}^{k-1} e^{\delta\tau_j} B_j^T Q^T(-\theta+\tau_j-\tau_k) + e^{\delta\tau_k} B_k^T(-\theta)$$

$$+ \sum_{j=k+1}^{m+1} e^{\delta\tau_j} B_j^T Q^T(-\theta+\tau_j-\tau_k)] e^{\delta(\theta+\tau_k)} B_k X_t(\theta)d\theta,$$

where $B_{m+1} = 0$.

Note that equation (5.2.a_1) implies

$$Q'(\theta+\tau_1) = (A^T+\delta I)Q(\theta+\tau_1) + B_1^T e^{\delta\tau_1} Q^T(-\theta)$$

$$+ \sum_{j=2}^{m} e^{\delta\tau_j} B_j^T Q^T(-\theta+\tau_j-\tau_1) \qquad \text{for} \quad -\tau_1 \le \theta \le 0$$

and

$$Q'(\theta+\tau_k) = (A^T+\delta I)Q(\theta+\tau_k) + \sum_{j=1}^{k-1} e^{\delta\tau_j} B_j^T Q^T(-\theta+\tau_j-\tau_k)$$

$$+ e^{\delta\tau_k} B_k^T Q^T(-\theta) + \sum_{j=k+1}^{m+1} e^{\delta\tau_j} B_j^T Q^T(-\theta+\tau_j-\tau_k)$$

$$\text{for} \quad -\tau_k \le \theta \le -(\tau_k-\tau_1),\ k = 2,\dots,m.$$

Also equations (5.a_2) to (5.a_m) imply

$$Q'(\theta+\tau_k) = (A^T+\delta I)Q(\theta+\tau_k) + \sum_{i=1}^{\ell} e^{\delta\tau_i} B_i^T Q(\theta+\tau_k-\tau_i)$$

$$+ \sum_{i=\ell+1}^{m} e^{\delta\tau_i} B_i^T Q^T(-\theta+\tau_i-\tau_k) \quad \text{for} \quad -(\tau_k-\tau_\ell) \le \theta \le -(\tau_k-\tau_{+1}),$$

$$\ell = 1,\dots,k-1$$

$$k = 2,\dots,m.$$

The substitution of the above equations into (5.4) shows that

(5.5) $$\dot V(t) = \frac{d}{dt}V(X_t(0),X_t) = -2\delta V(X_t(0),X_t)$$

$$+ X_t^T(0)[(A^T+\delta I)Q(0)+Q(0)(A+\delta I)+(A^T+\delta I)M+M(A+\delta I)$$

$$+ \sum_{k=1}^{m} e^{\delta\tau_k} B_k^T Q^T(\tau_k) + \sum_{k=1}^{m} e^{\delta\tau_k} Q(\tau_k) B_k + 2 \sum_{k=1}^{m} e^{\delta\tau_k} R_k] X_t(0)$$

$$- h(X_t(0), X_t(-\tau_1), \ldots, X_t(-\tau_m)) \equiv U(X_t(0), X_t),$$

where

$$h(X_t(0), X_t(-\tau_1), \ldots, X_t(-\tau_m))$$
$$= [X_t^T(0), -e^{-\delta\tau_1} X_t^T(-\tau_1), \ldots, e^{-\delta\tau_m} X_t^T(-\tau_m)].$$

$$(5.6)\quad \begin{bmatrix} \sum_{k=1}^{m} R_k e^{\delta\tau_k} & MB_1 e^{\delta\tau_1} & MB_2 e^{\delta\tau_2} & \cdots & MB_m e^{\delta\tau_m} \\ B_1^T M e^{\delta\tau_1} & R_1 e^{\delta\tau_1} & 0 & \cdots & 0 \\ B_2^T M e^{\delta\tau_2} & 0 & R_2 e^{\delta\tau_2} & \cdots & 0 \\ \vdots & \vdots & \vdots & \vdots & \vdots \\ B_m^T M e^{\delta\tau_m} & 0 & 0 & \cdots & R_m e^{\delta\tau_m} \end{bmatrix} \begin{bmatrix} X_t(0) \\ -e^{-\delta\tau_1} X_t(-\tau_1) \\ -e^{-\delta\tau_2} X_t(-\tau_2) \\ \vdots \\ -e^{-\delta\tau_m} X_t(-\tau_m) \end{bmatrix}$$

The above computations to obtain $\dot{V}(t)$ is valid only on solutions with initial data on $\mathcal{D}(A)$. A direct application of Theorem 3.9 of [11] shows that for any solution with initial condition on H we have

$$\dot{V}(t) \equiv \overline{\lim_{\Delta \to 0}} \frac{V(X_{t+\Delta}) - V(X_t)}{\Delta} \le U(X_t(0), X_t).$$

We wish to show that it is possible to give an estimate of the rate of growth or decay of the solutions of the matrix retarded difference-differential equation with several delays (2.1)-(2.2), through the use of a functional of the form (5.1) and its derivative (5.5).

Let $\gamma = \max\{\operatorname{Re}\lambda \mid \lambda \in \sigma(A)\}$. Given $\varepsilon > 0$ and $-\delta = \gamma + 2\varepsilon$ it is our purpose to show there exist positive definite matrices M, R_k, $k = 1, \ldots, m$, and a differentiable matrix $Q(\alpha)$, $0 \le \alpha \le \tau_m$, satisfying

(5.2.a_1-a_m) - (5.3) so that for the functionals $V(\xi,\phi)$ and $U(\xi,\phi)$ given by (5.1) and (5.5) with these matrices we have that

(5.7) $$c_1 \|(\xi,\phi)\|_H^2 \le V(\xi,\phi) \le c_2 \|(\xi,\phi)\|_H^2$$

and

(5.8) $$\dot{V}(\xi,\phi) \le -2\delta V(\xi,\phi)$$

for some positive constants c_1, c_2. In this case, a norm is induced by the square root of the Liapunov functional (5.1), which will be denoted by $\tilde{\|}(\xi,\phi)\tilde{\|}_H = \{V(\xi,\phi)\}^{1/2}$. Using the relationships (5.7) and (5.8), we obtain that the norms $\|\cdot\|_H$ and $\tilde{\|}\cdot\tilde{\|}_H$ are equivalent on H, and that

(5.9) $$\tilde{\|}(x_t(0),x_t)\tilde{\|}_H \le \tilde{\|}(x_t(0),x_t)\tilde{\|}_H e^{-\delta t},$$

or

(5.10) $$\tilde{\|}(x_t(0),x_t)\tilde{\|}_H \le \left[\frac{c_2}{c_1}\right]^{1/2} \|(x_t(0),x_t)\|_H e^{-\delta t}$$

The above estimates are precisely those stated in (2.6), and the norm $\tilde{\|}\cdot\tilde{\|}_H$ is the best possible one in the sense that it yields (2.6) with $K = 1$. Also, if $\gamma < 0$ the relationship (5.9) or (5.10) shows that the Liapunov functional (5.1) proves uniform exponential asymptotic stability for the solutions of (2.1) - (2.2).

In this manner we have the following

<u>Theorem</u> 1: Consider the matrix retarded difference-differential equation with several delays

$$\dot{X}(t) = AX(t) + \sum_{k=1}^{m} B_k X(t-\tau_k), \quad t > 0$$

where $0 < \tau_1 < \ldots < \tau_m$, and the Liapunov functional given by equation (5.1). Let $\gamma = \max\{\mathrm{Re}\,\lambda \mid \det[\lambda I - A - \sum_{k=1}^{m} B_k e^{-\lambda\tau_k}] = 0\}$ and $\varepsilon > 0$. Assume there exist constant positive matrices M, R_k, $k = 1,\ldots,m$, and a differentiable matrix $Q(\alpha)$, $0 \le \alpha \le \tau_m$, that satisfy

(5.2.a_1-a_m) with $Q(0) = Q(0)^T$, such that the inequalities (5.7) and (5.8) hold with c_1, c_2 positive and with $-\delta = \gamma+2\varepsilon$. Then the solution of the difference-differential equation satisfy the exponential bound (5.10) and the equation is exponentially asymptotically stable if $\delta > 0$.

In the next two sections we show that an appropriate differentiable matrix $Q(\alpha)$ as required by Theorem 1 can always be chosen, in the case the delays are rationally related; also we analyse its structure.

In a further section we determine appropriate matrices M, R_k, $k = 1,\dots,m$, so that the conditions for Theorem 1 are always satisfied.

In this manner we show the existence of a Liapunov functional of the form (5.1) for the matrix retarded difference-differential equation with several delays, in the case the delays are rationally related.

6. <u>A special system of functional differential equations</u>.

Consider the system of functional differential equations

(6.1.a_1) $$Q_1'(\alpha) = CQ_1(\alpha) + \sum_{k=1}^{m} D_k Q_k^T(k\tau-\alpha),$$

(6.1.a_2) $$Q_2'(\tau+\alpha) = CQ_2(\tau+\alpha) + D_1Q_1(\alpha) + \sum_{k=2}^{m} D_k Q_{k-1}^T((k-1)\tau-\alpha)$$

...

(6.1.a_m) $$Q_m'((m-1)\tau+\alpha) = CQ_m((m-1)\tau+\alpha) + \sum_{k=1}^{m-1} D_k Q_{m-k}((m-k-1)\tau+\alpha) + D_m Q_1^T(\tau-\alpha),$$

where C, D_k, $k = 1,\dots,m$, are $n\times n$ constant matrices, $Q_k((k-1)\tau+\alpha)$, $0 \le \alpha \le \tau$, $k = 1,\dots,m$, are differentiable matrices and $\tau > 0$.

Let

(6.2) $$Q_k(((k-1) + \tfrac{1}{2})\tau) = K_k, \quad k = 1,\dots,m$$

be a given initial condition, where the K_k are arbitrary $n \times n$ matrices.

It is our purpose to show that the solutions of system (6.1.a_1-a_m) with initial conditions given by (6.2) exist and are unique and that the linear vector space of all solutions of such a system has dimension $m \times n^2$; moreover, we give an algebraic representation of these solutions. The results we obtain are extensions of those described in [2].

If we define the matrices

$$(6.3) \quad \begin{aligned} F_k(\alpha) &= Q_k(k\tau-\alpha) \\ G_k(\alpha) &= Q_k((k-1)\tau+\alpha) \end{aligned} \quad , \ 0 \le \alpha \le \tau \ , \quad k = 1,\dots,m$$

then the system of functional differential equations (6.1.a_1-a_m) with initial condition (6.2) reduces to the system of ordinary differential equations

$$(6.4) \quad G_k'(\alpha) = \sum_{\ell=1}^{k-1} D_\ell G_{k-\ell}(\alpha) + CG_k(\alpha) + \sum_{\ell=k}^{m} D_\ell F_{\ell-k+1}(\alpha),$$

$$F_k'(\alpha) = -\sum_{\ell=1}^{k-1} F_{k-\ell}(\alpha) D_\ell^T - F_k(\alpha)C^T - \sum_{\ell=k}^{m} G_{\ell-k+1}(\alpha) D_\ell^T,$$

$$k = 1,\dots,m,$$

with the initial conditions

$$(6.5) \quad F_k(\tfrac{\tau}{2}) = K_k, \quad G_k(\tfrac{\tau}{2}) = K_k^T, \quad k = 1,\dots,m.$$

In the above equations, $G_o(\alpha) \equiv 0 \equiv F_o(\alpha)$.

We introduce the notation

$$F_k(\alpha) = (f_{j\ell}^k(\alpha)) = \begin{bmatrix} f_{1*}^k(\alpha) \\ \vdots \\ f_{n*}^k(\alpha) \end{bmatrix} = [f_{*1}^k(\alpha),\dots,f_{*n}^k(\alpha)]$$

where f^k_{j*} and $f^k_{*\ell}$ are, respectively, the jth row and the ℓth column of $F_k(\alpha)$. Defining the n^2-vector

$$f_k(\alpha) = (f^k_{1x}(\alpha),\dots,f^k_{nx}(\alpha))^T$$

and making use of the Kronecker (or direct) product of two matrices [8], then equations (6.3) and (6.4) can be rewritten as the $2\,m\,n^2$ system of ordinary differential equations

$$\text{(6.6)}\qquad \begin{bmatrix} g_1(\alpha) \\ f_1(\alpha) \\ g_2(\alpha) \\ f_2(\alpha) \\ \vdots \\ g_m(\alpha) \\ f_m(\alpha) \end{bmatrix}' = \begin{bmatrix} C\otimes I & D_1\otimes I & 0 & D_2\otimes I & \cdots & 0 & D_m\otimes I \\ -I\otimes D_1 & -I\otimes C & -I\otimes D_2 & 0 & \cdots & -I\otimes D_m & 0 \\ D_1\otimes I & D_2\otimes I & C\otimes I & D_3\otimes I & \cdots & 0 & 0 \\ -I\otimes D_2 & -I\otimes D_1 & -I\otimes D_3 & -I\otimes C & \cdots & 0 & 0 \\ \vdots & \vdots & \vdots & \vdots & \ddots & \vdots & \vdots \\ D_{m-1}\otimes I & D_m\otimes I & D_{m-2}\otimes I & 0 & \cdots & C\otimes I & 0 \\ -I\otimes D_m & -I\otimes D_{m-1} & 0 & -I\otimes D_{m-2} & \cdots & 0 & -I\otimes C \end{bmatrix} \cdot \begin{bmatrix} g_1(\alpha) \\ f_1(\alpha) \\ g_2(\alpha) \\ f_2(\alpha) \\ \vdots \\ g_m(\alpha) \\ f_m(\alpha) \end{bmatrix}$$

with the initial conditions

$$\text{(6.7)}\qquad \begin{aligned} f_i(\tfrac{\tau}{2}) &= [k^i_{1*},\dots,k^i_{n*}]^T \\ g_i(\tfrac{\tau}{2}) &= [k^{i^T}_{*1},\dots,k^{i^T}_{*n}]^T \end{aligned}\quad,\quad i = 1,\dots,m.$$

Now, we consider the uniqueness of solutions of $(6.1.a_1-a_m)-(6.2)$. Indeed, if the initial value problem $(6.1.a_1-a_m)$ - (6.2) has a differentiable solution $(Q_1(\alpha),\dots,Q_m((m-1)\tau+\alpha))$ defined on $0 \le \alpha \le \tau$, then the matrices $F_k(\alpha)$, $G_k(\alpha)$, $k = 1,\dots,m$, defined in (6.3) will satisfy equations (6.4) - (6.5); it follows the vectors $f_k(\alpha)$, $g_k(\alpha)$, $k = 1,\dots,m$ as defined above, will satisfy equations (6.6) - (6.7). But the linearity of all the equations, and the uniqueness of solutions of (6.6) - (6.7), imply that the solution $(Q_1(\alpha),\ldots,Q_m((m-1)\tau+\alpha))$ of $(6.1.a_1-a_m)$ - (6.2) is unique.

Now, we consider the existence of solutions of $(6.1.a_1-a_m)$-(6.2). The initial value problem (6.6) - (6.7) has a unique solution $f_k(\alpha)$, $g_k(\alpha)$, $k = 1,\ldots,m$, defined on $0 \le \alpha \le \tau$,and this implies the existence of unique 2m-tuple of differentiable matrices $(F_1(\alpha), G_1(\alpha),\ldots,F_m(\alpha), G_m(\alpha))$ defined on $0 \le \alpha \le \tau$ and satisfying (6.4) - (6.5). These equations can be rewritten as

$$\frac{d}{d\alpha}G_k(\alpha) = \sum_{\ell=1}^{k-1} D_\ell G_{k-\ell}(\alpha) + CG_k(\alpha) + \sum_{\ell=k}^{m} D_\ell F_{\ell-k+1}(\alpha)$$

$$\frac{d}{d\alpha}F_k^T(\tau-\alpha) = \sum_{\ell=1}^{k-1} D_\ell F_{k-\alpha}^T(\tau-\alpha) + CF_k(\tau-\alpha) + \sum_{\ell=k}^{m} D_\ell G_{\ell-k+1}(\tau-\alpha),$$

$$k = 1,\ldots,m,$$

with the initial condition

$$G_k(\tfrac{\tau}{2}) = K^k = F_k^T(\tfrac{\tau}{2}), \quad k = 1,\ldots,m.$$

Then it follows, from the uniqueness of the solutions, that

$$G_k^T(\tau-\alpha) = F_k(\alpha) = Q_k(k\tau-\alpha), \quad 0 \le \alpha \le \tau, \; k = 1,\ldots,m.$$

The structure of the solutions of system (6.5) it is wellknown, [3,5]. Let $\lambda_1,\ldots,\lambda_p$, $p < 2\,m\,n^2$, be the distinct eigenvalues of the matrix in (6.6), that is, solutions of the determinentel equation

$$(6.8)\quad \det\begin{bmatrix} (\lambda I-C)\otimes I & -D_1\otimes I & 0 & -D_2\otimes I & \cdots & 0 & -D_m\otimes I \\ I\otimes D_1 & I\otimes(\lambda I+C) & I\otimes D_2 & 0 & \cdots & I\otimes D_m & 0 \\ -D_1\otimes I & -D_2\otimes I & (\lambda I-C)\otimes I & -D_3\otimes I & \cdots & 0 & 0 \\ I\otimes D_2 & I\otimes D_2 & I\otimes D_3 & I\otimes(\lambda I+C) & \cdots & 0 & 0 \\ \vdots & \vdots & \vdots & \vdots & \vdots & \vdots & \vdots \\ -D_{m-1}\otimes I & -D_m\otimes I & -D_{m-2}\otimes I & 0 & \cdots & (\lambda I-C)\otimes I & 0 \\ I\otimes B_m & I\otimes B_{m-1} & 0 & I\otimes B_{m-2} & \cdots & 0 & I\otimes(\lambda I+C) \end{bmatrix} = 0,$$

each λ_j, $j = 1,\ldots,p$, with algebraic multiplicity m_j and geometric multiplicities n_j^r, $\sum_{r=1}^{s} n_j^r = m_j$, $\sum_j m_j = 2\,m\,n^2$. Then $2\,m\,n^2$ linearly independents solutions of (6.5) are given by

$$(6.9)\qquad \phi^q_{j,r}(\alpha) = e^{\lambda_j(\alpha-\frac{\tau}{2})} \sum_{i=1}^{q} \frac{(\alpha-\frac{\tau}{2})^{q-i}}{(q-i)!}\, e^i_{j,r}\,,$$

where $q = 1,\ldots,n_j^r$, $r = 1,\ldots,s$, $j = 1,\ldots,p$, $\sum_{r=1}^{s} n_j^r = m_j$, $\sum_j m_j = 2\,m\,n^2$, and the $2\,m\,n^2$ linearly independent eigenvectors and generalized eigenvectors are given by

$$[\lambda_j I - H]e^i_{j,r} = -\,e^{i-1}_{j,r}\,, \quad e^o_{j,s} = 0,$$

where H is the $2\,m\,n^2 \times 2\,m\,n^2$ matrix in (6.6). Now, we change the notation and return from the vector to the matrix form. Then we see that $2\,m\,n^2$ linearly independent solutions of (6.4) are given by

$$(6.10)\quad \begin{bmatrix} \Phi^{1,q}_{j,r}(\alpha) \\ \Psi^{1,q}_{j,r}(\alpha) \\ \vdots \\ \Phi^{m,q}_{j,r}(\alpha) \\ \Psi^{m,q}_{j,r}(\alpha) \end{bmatrix} = e^{\lambda_j(\alpha-\frac{\tau}{2})} \sum_{i=1}^{q} \left[\frac{(\alpha-\frac{\tau}{2})^{q-i}}{(q-i)!} \begin{bmatrix} L^{1,q}_{j,r} \\ M^{1,q}_{j,r} \\ \vdots \\ L^{m,q}_{j,r} \\ \Psi^{m,q}_{j,r} \end{bmatrix} \right],$$

for $q = 1,\ldots,n^r_j$, $r = 1,\ldots,s$, $j = 1,\ldots,p$, $\sum_{r=1}^{s} n^r_j = m_j$, $\sum_j m_j = 2\,m\,n^2$, and where the generalized eigenmatrix 2m-tuple $(L^{1,i}_{j,r}, M^{1,i}_{j,r},\ldots,L^{m,i}_{j,r}, M^{m,i}_{j,r})$ associated with the eigenvalues λ_j satisfy the equations

$$(6.11)\quad -\sum_{\ell=1}^{k-1} D_\ell L^{k-\ell,i}_{j,r} + (\lambda_j I-C)L^{k,i}_{j,r} - \sum_{\ell=k}^{m} D_\ell M^{\ell-k+1,i}_{j,r} = -L^{k,i-1}_{j,r}$$

$$\sum_{\ell=1}^{k-1} M^{k-\ell,i}_{j,r} D^T_\ell + M^{k,i}_{j,r}(\lambda_j I-C^T) + \sum_{\ell=k}^{m} L^{\ell-k+1,i}_{j,r} D^T_\ell = -M^{k,i-1}_{j,r}$$

$$k = 1,\ldots,m,$$

for $i = 1,\ldots,n^r_j$, $r = 1,\ldots,s$ $L^{k,o}_{j,s} = 0$, $M^{k,o}_{j,s} = 0$. These equations have a very special structure; indeed, if they are multiplied by -I, transposed, and written in reverse order, we obtain

$$(6.12)\quad -\sum_{\ell=1}^{k-1} D_\ell M^{k-\ell,i^T}_{j,r} + (-\lambda_j I-C)M^{k,i^T}_{j,r} - \sum_{\ell=k}^{m} D_\ell L^{\ell-k+1,i^T}_{j,r} = M^{k,i-1^T}_{j,r}$$

$$\sum_{\ell=1}^{k-1} L^{k-\ell,i^T}_{j,r} D^T_\ell + L^{k,i^T}_{j,r}(-\lambda_j I-C^T) - \sum_{\ell=k}^{m} M^{\ell-k+1,i^T}_{j,r} D^T_\ell = L^{k,i-1^T}_{j,r}$$

$$k = 1,\ldots,m,$$

for $i = 1,\ldots,n^r_j$, $r = 1,\ldots,s$, $L^{k,o}_{j,s} = 0$, $M^{k,o}_{j,s} = 0$. This result show that if λ_j is a solution of (6.8), $-\lambda_j$ will also be a solution; moreover, λ_j and $-\lambda_j$ have the same geometric multiplicities and the same algebraic multiplicity. Hence, the distinct eigenvalues always appear in pairs $(\lambda_j, -\lambda_j)$. An examination of equations (6.11) and (6.12) shows that if the generalized eigenmatrix 2m-tuple corresponding to λ_j is $(L^{1,i}_{j,r}, M^{1,i}_{j,r}, \ldots$ $\ldots, L^{m,i}_{j,r}, M^{m,i}_{j,r})$, then the generalized eigenmatrix 2m-tuple corresponding to $-\lambda_j$ will be $((-1)^{i+1} L^{1,i^T}_{j,r}, (-1)^{i+1} M^{1,i^T}_{j,r}, \ldots$ $\ldots, (-1)^{i+1} L^{m,i^T}_{j,r}, (-1)^{i+1} M^{m,i^T}_{j,r})$.

The above remarks show that if the solution (6.10) corresponding to λ_j is added to the solution (6.10) corresponding to $-\lambda_j$ multiplied by $(-1)^{q+1}$, we obtain $m\,n^2$ linearly independent solutions of (6.4) given

$$\begin{bmatrix} \Xi^{1,q}_{j,s}(\alpha) \\ \Pi^{1,q}_{j,s}(\alpha) \\ \vdots \\ \Xi^{m,q}_{j,s}(\alpha) \\ \Pi^{m,q}_{j,s}(\alpha) \end{bmatrix} = e^{\lambda_j(\alpha-\frac{\tau}{2})} \sum_{i=1}^{q} \frac{(\alpha-\frac{\tau}{2})}{(q-i)!} \begin{bmatrix} L^{1,i}_{j,r} \\ M^{1,i}_{j,r} \\ \vdots \\ L^{m,i}_{j,r} \\ M^{m,i}_{j,r} \end{bmatrix}$$

$$+ e^{-\lambda_j(\alpha-\frac{\tau}{2})} \sum_{i=1}^{q} \frac{(\alpha-\frac{\tau}{2})}{(q-i)!} (-1)^{q+1} \begin{bmatrix} M^{1,i^T}_{j,r} \\ L^{1,i^T}_{j,r} \\ \vdots \\ M^{m,i^T}_{j,r} \\ L^{m,i^T}_{j,r} \end{bmatrix}$$

which satisfy the conditions $\Xi^{k,q}_{j,s}(\frac{\tau}{2}) = \Pi^{k,q^T}_{j,s}(\frac{\tau}{2})$, $k = 1,\ldots,m$.
These conditions are precisely conditions (6.5). It therefore follows that the m-tuples

$$(6.13) \quad (Q^q_{1,j,r}(\alpha),\ldots,Q^q_{k,j,r}((k-1)\tau+\alpha),\ldots,Q^q_{m,j,r}((m-1)\tau+\alpha)),$$

where

$$(6.14) \quad Q^q_{k,j,r}((k-1)\tau+\alpha) = \sum_{i=1}^{q} \frac{(\alpha-\frac{\tau}{2})^{q-i}}{(q-i)!}$$

$$[e^{\lambda_j(\alpha-\frac{\tau}{2})} L^{k,i}_{j,r} + (-1)^{q+1} e^{-\lambda_j(\alpha-\frac{\tau}{2})} M^{k,i^T}_{j,r}] ,$$

for $q = 1,\ldots,n^r_j$, $\sum_{r=1}^{s} n^r_j = m_j$, $\sum_{2j} m_j = 2\, m\, n^2$ are $m\, n^2$ linearly independent solutions of $(6.1.a_1-a_m)$. In this manner, we have that system $(6.1.a_1-a_m)$ has $m\, n^2$ linearly independent solutions given by (6.13), (6.14), where the generalized eigenmatrix 2m-tuples $(L^{1,i}_{j,r}, M^{1,i}_{j,r},\ldots,L^{m,i}_{j,r}, M^{m,i}_{j,r})$ satisfy equation (6.11) for one of the elements of the pair $(\lambda_j,-\lambda_j)$, each of which is a solution of equation (6.8).

Making use of the above results, it is clear that the initial value problem for equations $(6.1.a_1-a_m)$ with initial condition

$$(6.15) \quad Q_1(0) = Q_1^T(0) = Q_o$$

$$Q_{k+1}(k\tau) = Q_k(k\tau), \quad k = 1,\ldots,m-1,$$

where Q_o is an arbitrary symmetric $n\times n$ matrix has a unique solution

Now, consider the functional differential equations

$$(6.16.a_1) \quad Q'(\alpha) = CQ(\alpha) + \sum_{k=1}^{m} D_k Q^T(k\tau-\alpha), \quad 0 \le \alpha \le \tau$$

$$(6.16.a_2)\quad Q'(\alpha) = CQ(\alpha) + D_1Q(\alpha-\tau) + \sum_{k=2}^{m} D_kQ^T(k\tau-\alpha), \quad \tau \le \alpha \le 2\tau,$$

...

$$(6.16.a_m)\quad Q'(\alpha) = CQ(\alpha) + \sum_{k=1}^{m-1} D_kQ(\alpha-k\tau) + D_mQ^T(m\tau-\alpha), \quad (m-1)\tau \le \alpha \le m\tau,$$

with the initial condition

$$(6.17)\quad Q(0) = Q^T(0) = Q_o,$$

where Q_o is an arbitrary symmetric $n\times n$ matrix.

Equations $(6.16.a_1-a_m)$ can be rewritten in the form

$$(6.16'.a_1)\quad Q'(\alpha) = CQ'(\alpha) + \sum_{k=1}^{m} D_kQ^T(k\tau-\alpha),$$

$$(6.16'.a_2)\quad Q'(\tau+\alpha) = CQ(\tau+\alpha) + D_1Q(\alpha) + \sum_{k=2}^{m} D_kQ^T((k-1)\tau-\alpha),$$

...

$$(6.16'.a_m)\quad Q'((m-1)\tau+\alpha) = CQ((m-1)\tau+\alpha) + \sum_{k=1}^{m-1} D_kQ((m-k-1)\tau+\alpha) + D_mQ(\tau-\alpha)$$

where $0 \le \alpha \le \tau$.

It is clear that the initial value problem $(6.16'a_1-a_m)-(6.17)$ has a unique differentiable solution $Q(\alpha)$, $0 \le \alpha \le m\tau$ if and only if the initial value problem $(6.1.a_1-a_m)$ - (6.15) has a unique differentiable solution $(Q_1(\alpha), Q_2(\tau+\alpha),\ldots,Q_m((m-1)\tau+\alpha))$, $0 \le \alpha \le \tau$. In this case, we have

$$(6.18)\quad Q(\alpha) = Q_k(\alpha), \qquad (k-1)\tau \le \alpha \le k\tau, \quad k = 1,\ldots,m$$

We note that the conditions (6.15) and the form of equations $(6.16.a_1-a_m)$ make clear the differentiability of solution $Q(\alpha)$ given by (6.18) at the points $\alpha = k\tau$, $k = 1,\ldots,m-1$.

The above results shows that the initial value problem $(6.16.a_1-a_m)$ - (6.17) has always an unique solution, which is obtained in an explicit form using equations (6.13), (6.14) and (6.15).

Finally we note that the initial value - problem $(5.2.a_1-a_m)$ - - (5.3) with the delays $\tau_1,\dots,\tau_m$ rationally related is a special case of the initial value - problem $(6.16.a_1-a_m)$ - (6.17).

7. The matrix function $\tilde{Q}(\alpha)$ and its properties.

In this section we consider an integral representation for the solution $Q(\alpha)$ of the functional differential equation $(5.2.a_1-a_m)$. Let the matrix function $\tilde{Q}(\alpha)$ be defined by

$$(7.1)\qquad \tilde{Q}(\alpha) = \int_0^\infty S^T(\beta)e^{\delta\beta}WS(\beta-\alpha)e^{\delta(\beta-\alpha)}d\beta, \qquad -\infty < \alpha < \infty,$$

where W is an arbitrary symmetric matrix, $S(\beta)$ is the solution of the equation (2.8) and $-\delta = -\gamma - 2\varepsilon,\ \varepsilon > 0$. Note that this integral always converges since, for every $\varepsilon > 0$ we have $\|S(t)\| \le \tilde{K}e^{(j+\varepsilon)t}$ for some $\tilde{K} \ge 1$.

Immediately follows from (7.1) that

$$(7.2)\qquad \tilde{Q}(0) = \tilde{Q}^T(0) = \int_0^\infty S^T(\beta)e^{\delta\beta}WS(\beta)e^{\delta\beta}d\beta$$

and

$$(7.3)\qquad \tilde{Q}(\alpha) = \int_\alpha^\infty S^T(\beta)e^{\delta\beta}WS(\beta-\alpha)e^{\delta(\beta-\alpha)}d\beta$$

$$= \int_0^\infty S^T(\alpha+\xi)e^{\delta(\alpha+\xi)}WS(\xi)e^{\delta\xi}d\xi = Q^T(-\alpha), \qquad \alpha \ge 0.$$

Moreover, $\tilde{Q}(\alpha)$ satisfies the functional differential equation $(5.2.a_1-a_m)$. To prove this, we use the equation (2.8) in the definition of $\tilde{Q}(\alpha)$; then it follows that

$$\tilde{Q}'(\alpha) = -\tilde{Q}(\alpha)[A+\delta I] - \sum_{k=1}^{m}\tilde{Q}(\alpha+\tau_k)B_k e^{\delta\tau_k} - S^T(\alpha)e^{\delta\alpha}W$$

The property (7.3) implies that

$$\tilde{Q}'(\alpha) = \frac{d}{d\alpha}[\tilde{Q}^T(-\alpha)];$$

then we obtain

$$\tilde{Q}'(\alpha) = (A^T+\delta I)\tilde{Q}(\alpha) + \sum_{k=1}^{m} e^{\delta\tau_k} B_k^T \tilde{Q}^T(\tau_k-\alpha) + S^T(-\alpha)e^{-\delta\alpha}W.$$

Since for $\alpha > 0$, $S(-\alpha) = 0$, it follows that $\tilde{Q}(\alpha)$ satisfies

$$\tilde{Q}'(\alpha) = (A^T+\delta I)\tilde{Q}(\alpha) + \sum_{k=1}^{m} e^{\delta\tau_k} B_k^T \tilde{Q}^T(\tau_k-\alpha), \quad \alpha \geq 0.$$

Using the property (7.3) in the equation above, we obtain the desired result, i.e., $\tilde{Q}(\alpha)$ satisfies

$$\tilde{Q}'(\alpha) = (A^T+\delta I)\tilde{Q}(\alpha) + \sum_{k=1}^{m} e^{\delta\tau_k} B_k^T \tilde{Q}^T(\tau_k-\alpha) \quad \text{for } 0 \leq \alpha \leq \tau_1,$$

$$\tilde{Q}'(\alpha) = (A^T+\delta I)\tilde{Q}(\alpha) + e^{\delta\tau_1} B_1^T \tilde{Q}(\alpha-\tau_1)$$

$$+ \sum_{k=2}^{m} e^{\delta\tau_k} B_k^T \tilde{Q}^T(\tau_k-\alpha) \qquad \text{for } \tau_1 \leq \alpha \leq \tau_2,$$

...

$$\tilde{Q}'(\alpha) = (A^T+\delta I)\tilde{Q}(\alpha) + \sum_{k=1}^{m-1} e^{\delta\tau_k} B_k^T \tilde{Q}(\alpha-\tau_k) + e^{\delta\tau_m} B_m^T Q^T(\tau_m-\alpha)$$

$$\text{for } \tau_{m-1} \leq \alpha \leq \tau_m.$$

Now, we wish to show that the following relationship is satisfied:

$$(7.4) \quad Y \equiv \tilde{Q}'(0) + \tilde{Q}'^T(0) = (A^T+\delta I)\tilde{Q}(0) + \sum_{k=1}^{m} e^{\delta\tau_k} B_k^T Q^T(\tau_k)$$

$$+ \tilde{Q}(0)(A+\delta I) + \sum_{k=1}^{m} \tilde{Q}(\tau_k) B_k e^{\delta\tau_k} = -W.$$

To prove this, we use $\tilde{Q}(0)$ and $\tilde{Q}(\tau)$ given by (7.1), and also the property (7.2); then it follows that

$$Y = \int_0^\infty [A^T S^T(\beta) + \sum_{k=1}^m B_k S^T(\beta-\tau_k)] e^{\delta\beta} W e^{\delta\beta} S(\beta) d\beta$$

$$+ \int_0^\infty S^T(\beta) e^{\delta\beta} W e^{\delta\beta} [S(\beta)A + \sum_{k=1}^m S(\beta-\tau_k)B_k] d\beta$$

$$+ 2\delta \int_0^\infty S^T(\beta) e^{\delta\beta} W e^{\delta\beta} S(\beta) d\beta.$$

Making use of (2.8), we obtain the desired result,

$$Y = \int_0^\infty \dot{S}^T(\beta) e^{\delta\beta} W e^{\delta\beta} S(\beta) d\beta + \int_0^\infty S^T(\beta) e^{\delta\beta} W e^{\delta\beta} \dot{S}(\beta) d\beta$$

$$+ 2\delta \int_0^\infty S^T(\beta) e^{\delta\beta} W e^{\delta\beta} S(\beta) d\beta$$

$$= \int_0^\infty \frac{d}{d\beta} [S^T(\beta) e^{\delta\beta} W e^{\delta\beta} S(\beta)] d\beta = -S^T(0) W S(0) = -W.$$

We note that to each constant symmetric positive definite matrix W there corresponds a matrix function $\tilde{Q}(\alpha)$ given by (7.1), and that this matrix $\tilde{Q}(\alpha)$ is the unique solution of $(5.2.a_1-a_m)-(5.3)$ with the initial condition prescribed by

$$\tilde{Q}(0) = \tilde{Q}^T(0) = \int_0^\infty S^T(\beta) e^{\delta\beta} W e^{\delta\beta} S(\beta) d\beta,$$

which is necessarily positive definite. Conversely, to each $Q(0) = Q^T(0) = Q_o$, the equations $(5.2.a_1-a_m) - (5.3)$ have a unique differentiable solution $Q(\alpha)$ which, through (7.4) defines a unique symmetric matrix W. For this W (7.1) gives the unique solution for $(5.a_1-a_m) - (5.3)$ with the prescribed initial condition. Then it follows that the map $W \to \tilde{Q}(0)$ defined by

$$\tilde{Q}(0) = \int_0^\infty S^T(\beta) e^{\delta\beta} W e^{\delta\beta} S(\beta) d\beta,$$

as a map on the space of $n \times n$ symmetric matrices is one - to - one, onto, and it maps positive definite matrices W into positive definite matrices $\tilde{Q}(0)$.

We note that (7.1) is an integral representation for the unique solution of the initial-value problem (5.2.a_1-a_m) - (5.3) in the case the delays are rationally related.

8. Structure of the Liapunov functional and its derivative.

We can put into evidence the particular structure of the Liapunov functional (5.1) and its derivative (5.5), using the characterization given in the previous section for the matrix function $\tilde{Q}(\alpha)$. Indeed, substitution of (7.1) for $Q(\alpha)$ into the Liapunov functional (5.1), yields

$$V(\xi,\phi) = \xi^T M\xi + \sum_{k=1}^{m} e^{\delta\tau_k} \int_{-\tau_k}^{0} \phi^T(\theta) R_k e^{2\delta\theta} \phi(\theta) d\theta + \xi^T \tilde{Q}(0)\xi$$

$$+ 2\xi^T \sum_{k=1}^{m} \int_{-\tau_k}^{0} \tilde{Q}(\tau_k+\theta) e^{\delta(\tau_k+\theta)} B_k \phi(\theta) d\theta$$

$$+ \sum_{k=1}^{m} \sum_{j=1}^{m} \int_{-\tau_j}^{0} \int_{-\tau_k}^{0} \phi^T(\theta) B_j^T \tilde{Q}(\beta-\theta+\tau_k-\tau_j) e^{\delta(\theta+\beta+\tau_j+\tau_k)} B_k \phi(\beta) d\beta d\theta$$

$$= \xi^T M\xi + \sum_{k=1}^{m} e^{\delta\tau_k} \int_{-\tau_k}^{0} \phi^T(\theta) R_k e^{2\delta\theta} \phi(\theta) d\theta$$

$$+ \int_0^{\infty} [S(\beta)\xi + \sum_{k=1}^{m} \int_{-\tau_k}^{0} S(\beta-\alpha-\tau_k) B_k \phi(\alpha) d\alpha]^T e^{\delta\beta} W e^{\delta\beta} [S(\beta)\xi$$

$$+ \sum_{k=1}^{m} \int_{-\tau_k}^{0} S(\beta-\alpha-\tau_k) B_k \phi(\alpha) d\alpha] d\beta.$$

If the above functional is evaluated along the solutions of the equation (5.1), then, using (2.7), we obtain

$$(8.1) \quad V(x_t(0),x_t) = x_t^T(0) M x_t(0) + \sum_{k=1}^{m} e^{\delta\tau_k} \int_{-\tau_k}^{0} x_t^T(\theta) R_k e^{2\delta\theta} x_t(\theta) d\theta$$

$$+ \int_0^{\infty} x_{t+\beta}(0) e^{\delta\beta} W e^{\delta\beta} x_{t+\beta}(0) d\beta,$$

where $x_{t+\beta}(0)$ is the solution of (2.1) - (2.2) for $\beta \geq 0$ with initial condition $x_t(\theta)$, $-\tau_m \leq \theta \leq 0$.

Similarly, using the above notation, equation (5.5) can be rewritten as

$$U(x_t(0),x_t) = -2\delta V(x_t(0),x_t)$$

$$+ x_t^T(0)[-W + (A^T+\delta I)M + M(A+\delta I) + 2\sum_{k=1}^{m} e^{\delta\tau_k}R_k]x_t(0)$$

$$- [x_t^T(0), -e^{-\delta\tau_1}x_t^T(-\tau_1),\ldots, -e^{-\delta\tau_m}x_t^T(-\tau_m)].$$

(8.2)
$$\begin{bmatrix} \sum_{k=1}^{m} R_k e^{\delta\tau_k} & MB_1 e^{\delta\tau_1} & MB_2 e^{\delta\tau_2} & \cdots & MB_m e^{\delta\tau_m} \\ B_1^T M e^{\delta\tau_1} & R_1 e^{\delta\tau_1} & 0 & \cdots & 0 \\ B_2^T M e^{\delta\tau_2} & 0 & R_2 e^{\delta\tau_2} & \cdots & 0 \\ \vdots & \vdots & \vdots & \vdots & \vdots \\ B_m^T M e^{\delta\tau_m} & 0 & 0 & \cdots & R_m e^{\delta\tau_m} \end{bmatrix} \begin{bmatrix} x_t(0) \\ -e^{-\delta\tau_1}x_t(-\tau_1) \\ -e^{-\delta\tau_2}x_t(-\tau_2) \\ \vdots \\ -e^{-\delta\tau_m}x_t(-\tau_m) \end{bmatrix}$$

9. Bounds for the Liapunov functional and its derivative.

In this section we show that the structure of the Liapunov functional (5.1) and its derivative (5.5) developed in the previous section can be used to choose positive definite matrices M, R_k, $k = 1,\ldots,m$, and W (and therefore $Q(\alpha)$) so that bounds of the form (5.7) and (5.8) always hold.

First we show there exist positive constant c_1 and c_2 such that

$$c_1\|(\xi,\phi)\|_H^2 \leq V(\xi,\phi) \leq c_2\|(\xi,\phi)\|_H^2$$

for all $(\xi,\phi) \in \mathcal{D}(A)$.

Let $\lambda_{min}(M)$ and $\lambda_{max}(M)$ be the minimum and maximum eigenvalues of the positive definite matrix M; the corresponding notation will be used for the positive definite matrices R_k, $k = 1,\dots,m$, and W. From (8.1) we obtain

$$\min(\lambda_{min}(M),\ e^{-|\delta|\tau_m}\lambda_{min}(R_m))\,\|(\xi,\phi)\|_H^2 \le V(\xi,\phi).$$

Now, using (2.6), we obtain

$$\|x_\beta(0)\|_{R^n} \le \|(x_\beta(0),x_\beta)\|_H \le Ke^{(\gamma+\varepsilon)}\|(x_o(0),x_o)\|_H.$$

Therefore, from (8.1) it follows that

$$V(\xi,\phi) \le [\max(\lambda_{max}(M),\ e^{|\delta|\tau_1}\lambda_{max}(R_1),\dots$$

$$\dots,\ e^{|\delta|\tau_m}\lambda_{max}(R_m)) + \frac{\lambda_{max}(W)}{2\varepsilon}K^2]\,\|(\xi,\phi)\|_H^2.$$

Thus, the desired values of C_1 and C_2 are

$$C_1 = \min(\lambda_{min}(M),\ e^{-|\delta|\tau_m}\lambda_{min}(R_m))$$

and

$$C_2 = \max(\lambda_{max}(M),\ e^{|\delta|\tau_1}\lambda_{max}(R_1),\dots$$

$$\dots,\ e^{|\delta|\tau_m}\lambda_{max}(R_m)) + \frac{\lambda_{max}(W)}{2\varepsilon}K^2.$$

Secondly, we show that the equation (5.8) holds. Equation (8.2) shows that we only need to choose the positive definite matrices M, R_k, $k = 1,\dots,m$, and W so that the function

$$h_1(x_t(0),x_t(-\tau_1),\dots,x_t(-\tau_m))$$

$$= [x_t^T(0),\ -e^{-\delta\tau_1}x_t^T(-\tau_1),\dots,\ -e^{-\delta\tau_m}x_t^T(-\tau_m)].$$

$$\begin{bmatrix} W-(A^T+\delta I)M-M(A+\delta I)-\sum_{k=1}^{m} R_k e^{\delta\tau_k} & MB_1 e^{\delta\tau_1} & MB_2 e^{\delta\tau_2} & \cdots & MB_m e^{\delta\tau_m} \\ B_1^T M e^{\delta\tau_1} & R_1 e^{\delta\tau_1} & 0 & \cdots & 0 \\ B_2^T M e^{\delta\tau_2} & 0 & R_2 e^{\delta\tau_2} & \cdots & 0 \\ \vdots & \vdots & \vdots & \vdots & \vdots \\ B_m^T M e^{\delta\tau_m} & 0 & 0 & \cdots & R_m e^{\delta\tau_m} \end{bmatrix} \begin{bmatrix} X_t(0) \\ -e^{-\delta\tau_1} X_t(-\tau_1) \\ -e^{-\delta\tau_2} X_t(-\tau_2) \\ \vdots \\ -e^{-\delta\tau_m} X_t(-\tau_m) \end{bmatrix}$$

is nonnegative. This form can be rewritten as

(9.1) $$h_1(X_t(0),X_t(-\tau_1),\ldots,X_t(-\tau_m)) =$$

$$= X_t^T(0)[W - (A^T+\delta I)M - M(A+\delta I) - 2\sum_{k=1}^{m} R_k e^{\delta\tau_k}]X_t(0)$$

$$+ \sum_{k=1}^{m} e^{-\delta\tau_k} X_t^T(-\tau_k)[R_k - B_k^T M R_k^{-1^T} MB_k]X_t(-\tau_k)$$

$$+ \sum_{k=1}^{m} [-X_t^T(-\tau)B_k^T M R_k^{-1} e^{-\delta\tau_1} + X_t^T(0)]R_k e^{\delta\tau_k}$$

$$\cdot [-e^{-\delta\tau_k} R_k^{-1^T} MB_k X_t(-\tau_k) + X_t(0)].$$

Now, taking $M = I$, $R_k = d_k I$, $d_k > 0$ sufficiently large, $k = 1,\ldots,m$, then the matrices

$$R_k - B_k^T M R_k^{-1} MB_k \ , \quad k = 1,\ldots,m,$$

will certainly be positive definite. Moreover, choosing $W = d_W I$, $d_W > 0$ sufficiently large, also the matrix

$$W - (A^T+\delta I)M - M(A+\delta I) - 2\sum_{k=1}^{m} R_k e^{\delta\tau_k}$$

will certainly be positive definite.

The last term in (9.1) is always nonnegative, by the above choice of the R_k, $k = 1,\dots,m$.

Then, it is seen that the form $h_1(X_t(0), X_t(-\tau_1),\dots,X_t(-\tau_m))$ can be always made nonnegative.

In this manner, we have the following

Theorem 2. Consider the matrix retarded difference-differential equation with several delays

$$\dot{X}(t) = AX(t) + \sum_{k=1}^{m} B_k(t-\tau_k), \quad t > 0,$$

where the delays are rationally related and $0 < \tau_1 < \dots < \tau_m$, and the Liapunov functional V given by equation (5.1). Let

$$\gamma = \max\{\mathrm{Re}\,\lambda \mid [\det \lambda I - A - \sum_{k=1}^{m} B_k e^{-\lambda\tau_k}] = 0\}$$

and $\varepsilon > 0$. Then there exist constant positive definite matrices M, R_k, $k = 1,\dots,m$ and a differentiable matrix $Q(\alpha)$, $0 \le \alpha \le \tau_m$ with $Q(0) = Q(0)^T$ such that the functional V is positive definite, bounded above, and

$$\dot{V} \le 2(\gamma+\varepsilon)V.$$

Of course, if $\gamma < 0$, then the above result implies exponential asymptotic stability.

The author wishes to thank Prof. Ettore F. Infante for several conversations and suggestions.

REFERENCES

[1] - BORISOVIC, J.G. and TURBABIN, A.S., On the Cauchy problem for linear non-homogeneous differential equations with retarded argument, Soviet Math. Dokl. 10(1969), 401-405.

[2] - CASTELAN, W.B. and INFANTE, E.F., On a functional equation arising in the stability theory of difference-differential equations, Quart. Appl. Math. 35(1977), 311-319.

[3] - CODDINGTON, E.A. and LEVINSON, N., Theory of Ordinary Differential Equations, McGraw-Hill, New York, (1955).

[4] - HALE, J.K., Theory of Functional Differential Equations, Appl. Math. Science Series, Springer-Verlag, New York/ Berlin, (1977).

[5] - HALE, J.K., Ordinary Differential Equations, Interscience, New York, (1967).

[6] - INFANTE, E.F. and CASTELAN, W.B., A Liapunov functional for a matrix differential-difference equation, J. Diff. Equations 29(1978). 439-451.

[7] - INFANTE, E.F. and WALKER, J.A., A Liapunov functional for a scalar differential-difference equation, Proc. Roy. Soc. Edinburgh, 79A(1977), 307-316.

[8] - LANCASTER, P., Theory of Matrices, Academic Press, New York, (1969).

[9] - LASALLE, J.P., Stability theory for difference equations, in "A Study of Ordinary Differential Equations" (J.K. Hale), Studies in Mathematics Series, American Mathematical Association, (to appear).

[10] - LASALLE, J.P., The Stability and Control of Discrete Processes, Appl. Math. Science Series, Springer-Verlag , New York/ Berlin, (to appear).

[11] - WALKER, J.A., On the application of Liapunov's direct method to linear dynamical systems, J. Math. Anal. Appl. 53(1976), 187-220.

A COMPACTNESS THEOREM FOR INTEGRAL OPERATORS AND APPLICATIONS

by M. Cecchi, M. Marini and P. L. Zezza

§ 1.

Let $BC(A)$ the space of bounded continuous function from $J = [a,b)$ to A, where A is an open set of R^n, for $-\infty < a < b \le +\infty$, $(BC(R^n) = BC)$, and let S be a closed set, $S \subset A$, with a non empty interior. We define S_1 by:

$$S_1 = \{c \in R_o^+ \quad \text{such that} \quad \exists\, d \in S : ||d|| = c\};$$

then we have the following theorem:

<u>Theorem</u> 1. Let $h(t,u) \in c[[a,b) \times A, R^n]$ such that

(1) $\quad ||h(t,u)|| \le g(t,||u||), \quad t \in [a,b), \quad ||u|| \in S_1$

where $g : [a,b) \times S_1 \to R_o^+$ satisfies the conditions:

(i) $g(t,v)$ is continuous in v for $t \in [a,b)$;

(ii) if $g_\eta(t) = \max\limits_{v \in [0,\eta] \cap S_1} g(t,v)$, then

$g_\eta(t)$ is summable in $[a,b)$ and

$$\int_a^b g_\eta(t)dt < +\infty, \qquad \forall\, \eta \in R^+$$

Let $L : BC \to BC$ a bounded linear operator, then the operator

$$K : x \to L\int_a^t h(s,x(s))ds$$

$$K : \text{dom } K \subset BC \to BC$$

with $\text{dom } K = \{x \in BC \text{ such that } x(t) \in S, \ \forall\, t \in [a,b)\}$

is completely continuous.

<u>Proof</u>.

Let $\tilde{K} : \text{dom } K \to BC$ be defined by

$$\tilde{K} : x \to \int_a^t h(s,x(s))ds.$$

Then we have

$$\operatorname{Im}\tilde{K} = BC_\ell = \{x \in BC \text{ such that } \exists \lim_{t\to b} x(t) = \ell_x ||\ell_x|| < +\infty\}$$

That is, let $x \in \operatorname{dom} K$, and let us show that there exists the limit for $t \to b$ of

(2) $\tilde{K}x = \int_a^t h(s,x(s))ds.$

Since $x \in \operatorname{dom} K$, we have for $\forall\, t \in [a,b)$

$$||x(t)|| \le \nu;$$

then from (1) we have

$$\int_a^t ||h(s,x(s))||ds \le \int_a^t g(s,||x(s)||)ds \le \int_a^t g_2(s)ds < +\infty,$$

then (2) is absolutely convergent and then convergent.

Let us prove now that the operator $\tilde{K}$ is completely continuous. To show this it is enough to show that $\tilde{K}$ is continuous in dom K and takes Δ sets, $\Delta \subset \operatorname{dom} K$, into $\tilde{K}(\Delta)$ of relatively compact sets that is [1]:

(a) uniformly bounded

(b) equicontinuous

(c) uniformly convergent in the sense that

$\forall\, \varepsilon > 0$, $\exists\, \delta(\varepsilon) > 0$ such that $\forall\, t > \delta(\varepsilon)$, $\forall\, y \in \tilde{K}(\Delta)$,

$$||y(t) - \lim_{t\to b} y(t)|| < \varepsilon.$$

Continuity of $\tilde{K}$.

Let $\{x_n\}$ a sequence in dom K convergent for $x \in \operatorname{dom} K$ and let us show that $\{\tilde{K}x_n\} \to \tilde{K}x$.

Since h is continuous we have

(3) $||h(t,x_n(t)) - h(t,x(t))|| \to 0$

and

$$(4)\quad \begin{aligned} &||h(t,x_n(t)) - h(t,x(t))|| \leq ||h(t,x_n(t))|| + \\ &+ ||h(t,x(t))|| \leq g(t,||x_n(t)||) + g(t,||x(t)||). \end{aligned}$$

We have also that for n big enough and $t \in [a,b)$

$$||x_n(t)|| \leq ||x_n|| \leq ||x|| + \varepsilon$$

then

$$g(t,||x_n(t)||) \leq g_{||x||+\varepsilon}(t);$$

and from (4) we have that

$$||h(t,x_n(t))-h(t,x(t))|| \leq g_{||x||+\varepsilon}(t) + g_{||x||}(t).$$

From (1), (3) and Lebesgue dominated convergence theorem we see that $\tilde{K}$ is continuous.

Let $\Delta = \{x \in \text{dom } K$ such that $||x|| < \gamma\}$ and let us show that $\tilde{K}(\Delta)$ is relatively compact, that is, conditions (a), (b), (c) are satisfied.

Equiboundedness.

$$||\tilde{K}x|| = ||\int_a^t h(s,x(s))ds|| \leq \int_a^t ||h(s,x(s))||ds \leq$$

$$\leq \int_a^t g(s,||x(s)||)ds \leq \int_a^t g_\gamma(s)ds = \Gamma < +\infty, \qquad (x \in \Delta).$$

Equicontinuity.

Let $t_1,t_2 \in [a,b)$ and $t_2 > t_1$. If $x \in \Delta$ we have

$$||(\tilde{K}x)(t_2)-(\tilde{K}x)(t_1)|| = ||\int_{t_1}^{t_2} h(s,x(s))ds|| \leq$$

$$\leq \int_{t_1}^{t_2} g(s,||x(s)||)ds \leq \int_{t_1}^{t_2} g_\gamma(s)ds.$$

From the integrability of $g_\gamma(s)$ follows the equicontinuity.

Uniform convergence.

Let $x \in \Delta$. We have

$$||\{\lim_{t\to b}(\tilde{K}x)(t)\} - (\tilde{K}x)(t)|| = ||\int_a^b h(s,x(s))ds - \int_a^t h(s,x(s))ds|| =$$

$$= ||\int_t^b h(s,x(s))ds|| \le \int_t^b ||h(s,x(s))||ds \le$$

$$\le \int_t^b g(s,||x(s)||)ds \le \int_t^b g_\gamma(s)ds$$

Since from (1) $\lim_{t\to b}\int_t^b g_\gamma(s)ds = 0$ we have uniform convergence.

It is easy now to show that $K = L\tilde{K}$ is completely continuous on dom K. If Δ is bounded from the continuity of L and from the compactness of $\overline{\tilde{K}(\Delta)}$ follows the compactness of $\overline{L(\tilde{K}(\Delta))} = \overline{K(\Delta)}$.

Remark. Condition (ii) of Theorem 1 is satisfied if

$$g(t,v) = p(t)v + q(t), \quad v \in R^+$$

with $p(t)$ and $q(t)$ real positive integrable functions on $[a,b)$, or if

$$g(t,v) = p(t)\frac{1}{v} + q(t)$$

with $p(t)$ and $q(t)$ real positive integrable functions on $[a,b)$.

§ 2.

Consider the linear boundary value problem

$$(5)\qquad \begin{cases} \dot{x}-A(t)x = f(t,x) & \\ Tx = r & r \in R^m, \quad m \le n \end{cases}$$

where

$A(t)$ is an $n \times n$ continuous matrix such that the linear system

$$(6)\qquad \dot{y} - A(t)y = 0$$

is stable, that is, the space D of all solutions of (6) that belongs to BC has dimension n,

$f : [a,b) \times A \to R^n$

is continuous and

$T : \operatorname{dom} T \subseteq BC \to R^m$

is a continuous linear operator such that

$D \subset \operatorname{dom} T$, $T(D)$ has dimension m.

This last condition asserts that the linear problem associated to (5) for $f(t,x) = 0$ has a solution for any $r \in R^m$.

Consider the abstract equation associated to (5):

(7) $Lx = Nx$

where:

$L : x \to (\dot{x}-A(t)x, Tx)$

$N : x \to (f(t,x), r)$

Note that the range of the linear operator L may have infinite codimension and may not be closed. Therefore we can not use the usual methods.

However is always possible to reduce equation (7) to a fixed point problem using the equivalence theorem in [9]. Then equation (7) reduces to

$$(8) \quad \begin{cases} x = Mx = Px + K_p Nx \\ x \in \Omega \end{cases}$$

where P is a projection onto the kernel of L, K_p is the right inverse of L associated to P and

$\Omega = \{x \in BC$ such that $Nx \in \operatorname{Im} L\}$.

The questions that arise now are the following:

1) How is Ω ?

2) Is the operator M completely continuous ?

We have the following corollary that we state for $A = R^n$:

Corollary 1. If dom T = BC, the operator M in (8) is defined on BC and is completely continuous if one of the following conditions are satisfied, where $X(t)$ is a fundamental matrix of (6):

(a) System (6) is stable

$$||X^{-1}f(t,u)|| \le p(t)||u|| + q(t)$$

with $p(t)$, $q(t)$ positive real functions for which

$$(9) \qquad \int_a^b p(t)dt < +\infty, \qquad \int_a^b q(t)dt < +\infty.$$

(b) System (6) is uniformly asymptotically stable and

$$||f(t,u)|| \le p(t)||u|| + q(t)$$

with $p(t)$, $q(t)$ positive real function that satisfies (9).

(c) System (6) is exponentially stable and

$$||f(t,u)|| \le \alpha||u|| + \lambda$$

$$\text{with} \quad \sup_{t\in[a,b)} \int_a^t ||X(t)X^{-1}(s)||ds < \alpha^{-1}.$$

(d) System (6) is stable in the strict sense and

$$||f(t,u)|| \le p(t)||u|| + q(t)$$

with $p(t)$, $q(t)$ positive real functions that satisfies (9).

Remark. In (a) and (d) we have $\operatorname{Im} M \subset BC$, while in (b) and (c) we have $\operatorname{Im} M \subset BC_o$, where $BC_o = \{x \in BC_\ell \text{ such that } \lim_{t\to b} x(t) = 0\}$.

When

$$S = \{u \in R^n \text{ such that } \rho_1 \le ||u|| \le \rho_2\}$$

with ρ_1, $\rho_2 \in R^+$, the following is true

Corollary 2. If dom T = BC the operator M is completely continuous in BC(S) if

$$||X^{-1}(t)f(t,u)|| \le p(t)\frac{1}{||u||} + q(t)$$

with $p(t)$, $q(t)$ positive real functions satisfying (9).

REFERENCES

[1] - AVRAMESCU, C., Sur l'éxistence des solutions convergentes des systèmes d'équations différentielles non linéaires, Ann. Mat. Pura Appl. 4(1969).

[2] - CECCHI, M., MARINI, M. and ZEZZA, P.L., Linear boundary value problems for systems of ordinary differential equations on non-compact intervals, Ann. Mat. Pura Appl., (to appear).

[3] - CECCHI, M., MARINI, M. and ZEZZA, P.L., Un metodo astratto per problemi ai limiti non lineari su intervalli non comapatti, Comunicazione a Equadiff 78, Firenze, 24-30 Maggio 1978, Conti, Sestini, Villari eds.

[4] - CECCHI, M., MARINI, M. and ZEZZA, P.L., Linear boundary value problems for systems of ordinary differential equations on non-compact intervals. Part II: Stability and bounded perturbations, Ann. Mat. Pura Appl., (to appear).

[5] - CESARI, L., Functional analysis, non linear differential equations and the alternative method, in Nonlinear Funct. Analysis and Differential Eqns., L. Cesari, R. Kannan, J. D. Schuur eds., Dekker, New York, (1977), 1-197.

[6] - CORDUNEANU, C., Integral Equations and Stability of Feedback Systems, Math. Sc. Eng. 104, Ac. Press, New York, (1973).

[7] - MAWHIN, J., Topological degree methods in nonlinear boundary value problems, Ref. conf. series in math. (40), Amer. Math. Soc., Providence, R.I., (1979).

[8] - KARTSATOS, A.G., The Leray-Schauder Theorem and the existence of solutions to boundary value problems on infinite intervals, Ind. Un. Math. J., 23, 11(1974).

[9] - ZEZZA, P.L., An equivalence theorem for non linear operator equations and an extension of Leray-Sachauder's continuation theorem, Boll. U. M. I. (5) 15-A (1978).

PERIODIC SOLUTIONS OF NONLINEAR AUTONOMOUS HYPERBOLIC EQUATIONS

by Shui-Nee Chow*

§1. Introduction.

The purpose of this paper is to show the existence of time periodic soltuions of a class of nonlinear hyperbolic equations. To be more precise, consider the nonlinear one-dimensional wave equation

(1.1) $u_{tt} - u_{xx} - \lambda u = f(\lambda,u), \quad x \in (0,\pi), \quad t \in R$

with the Dirichlet boundary conditions

(1.2) $u(t,0) = u(t,\pi) = 0.$

We are interested in the number of branches of time periodic solutions bifurcating from the equilibrium solution $u \equiv 0$ for certain values of λ in (1.1). For example, if

$$f(\lambda,u) = \lambda \sin u - \lambda u$$

then (1.1) is the well-known sine-Gordon equation. By one of the results in this paper (Theorem 6.2), one obtains that there exists a dense set $\Lambda \subset (0,1)$ such that for every $\lambda_o \in \Lambda$ there exists at least d_o analytic branches of time periodic solutions of period $T_o > 0$ (not necessarily the least period) bifurcating from $u \equiv 0$ at $\lambda = \lambda_o$, where both $d_o \geq 1$ and T_o depend only on λ_o. In general, $d_o > 1$.

Existence of time periodic solutions of nonlinear wave equations with boundary conditions (1.2) has been studied by many authors. A certain amount of the studies has been related to the nonautonomous case, i.e., the nonlinearity f in (1.1) is time dependent. See, for example [2], [5], [9] and references therein. Apparently, there has not been much work done for the autonomous case, especially for

*Partially supported by NSF Grant MCS 76-06739.

equations such as (1.1).

Recently, Rabinowitz [10] proved that there are periodic solutions for (1.1) and (1.2) under mild assumptions on f. His results are global. In [6], Kielhöfer considered the problem locally near $u \equiv 0$ as a bifurcation problem and proved the existence of a branch of periodic solutions bifurcating from $u \equiv 0$ at certain values of λ. Such problems have also been considered earlier by Melrose and Pemberton [7]. These results are in some sense close to the Hopf bifurcation theorem.

In this paper, we also consider the problem as a bifurcation problem. However, we only assume that the linearized equation of (1.1) has finitely many distinct periodic orbits with a common period $T_o > 0$ instead of a single orbit as in [6], [7]. The proof is drastically different from those in [6], [7]. In fact, analyticity is very essential in our proof. Similar situations also occur in finite dimensional Hamiltonian systems. Liapunov center theorem was extended to include the resonance case in [8], [12]. If the system is real analytic, then it was shown in [3] that there are distinct analytic branches of periodic solutions emanating from the equilibrium solution. If the system is C^∞, then we can only conclude [1] that there is a nonempty connected component K of the set of nontrivial periodic orbits such that the equilibrium is in the closure of K.

§2. Homogeneous equation.

Consider the linear wave equation with a real parameter λ

(2.1) $u_{tt} - u_{xx} - \lambda u = 0.$

We impose the following boundary conditions

(2.2) $u(t,0) = u(t,\pi) = 0.$

By separation of variable, we assume that

$$u(t,x) = \phi(t)\psi(x)$$

is a solution of (2.1), (2.2). It follows that

$$\psi(x) = \sin kx, \quad k = 1, 2, \ldots,$$

$$\Psi''(t) + (k^2-\lambda)\psi(t) = 0.$$

Thus, there are nontrivial time periodic solutions time periodic solutions of (2.1), (2.2) with period

$$(2.3) \quad T_k = \frac{2\pi}{\sqrt{k^2-\lambda}}, \quad k = 1, 2, \ldots,$$

provided $k^2-\lambda > 0$.

We now restrict our considerations to rational λ's, $0 < \lambda < 1$. Assume

$$(2.4) \quad \lambda = \frac{p}{q}, \quad 1 \le p < q, \quad p,q \text{ relative primes.}$$

Lemma 2.1. There exist a most finitely many integers m, $r > 0$ such that

$$(2.5) \quad q^2m^2 - pq^2 = r^2.$$

Proof. By (2.5),

$$(qm+r)(qm-r) = pq.$$

Since p,q are fixed, there are at most finitely many different ways to rewrite pq as a product of two integers.

Lemma 2.2. Suppose $\lambda = p/q$ satisfies (2.5). Let m,r in (2.5) be fixed but arbitrary. Then, there exist at most finitely many integers $k,j > 0$ such that

$$(2.6) \quad T_m = jT_k.$$

Proof. By (2.3) and (2.6),

$$j^2(m^2-\lambda) = k^2-\lambda.$$

By (2.5),

$$q^2k^2-q^2j^2r^2 = pq.$$

We may now argue exactly as in Lemma 2.1.

It is clear that the set of λ's satisfying (2.5) is nonempty and is in fact dense in $(0,1)$. We now fix m,r in (2.5), Lemma 2.1 says that there are exactly d $(1 \le d < \infty)$ time periodic orbits with period T_m which is not necessarily the least period. If $d=1$, then it was shown by Kielhöfer [6] that there exists a branch of periodic solutions of certain nonlinear perturbations of (2.1) with period T_m. We will consider similar questions for $d \ge 1$ but finite.

§3. Nonhomogeneous equation.

Consider the nonhomogeneous equation

$$(3.1) \quad u_{tt} - u_{xx} - \lambda u = g(t,x)$$

with boundary conditions:

$$(3.2) \quad u(t,0) = u(t,\pi) = 0.$$

Let $\lambda_o = p_o/q_o$ satisfy (2.4) and (2.5). Let m_o, r_o satisfy (2.5) and be fixed.

Let N denote the linear span of functions

$$\cos\frac{2\pi}{T_k} t \sin kx, \qquad \sin\frac{2\pi}{T_k} t \sin kx$$

where $T_{m_o} = jT_k$ for some integer $j > 0$. Let

$$\dim N = 2d < \infty.$$

We seek formally a periodic solution of (3.1), (3.2) with period T_{m_o} for T_m-periodic $g(t,x)$. Obviously, $g(t,x)$ could not have terms which belong to N in its Fourier expansion. Thus, let

$$g(t,x) = \sum_{k,j} g_{kj} e^{ij\sqrt{m_o^2-\lambda_o}\,t} \sin kx$$

where

$$g_{kj} = \bar{g}_{jk}; \quad \text{and}$$

$$g_{kj} = 0, \quad \text{if} \quad j^2(m_o^2-\lambda_o) = k^2-\lambda_o.$$

Hence, we have a formal solution of (3.1) and (3.2)

$$u(t,x) = \sum_{k,j} u_{kj} e^{ij\sqrt{m_o^2-\lambda_o}\,t} \sin kx$$

where

$$u_{kj} = \frac{g_{kj}}{k^2-j^2(m_o^2-\lambda_o)-\lambda_o} = \frac{q_o^2}{q_o^2k^2-j^2r_o^2-p_oq_o} g_{kj},$$

$$\text{if } j^2(m_o^2-\lambda_o) \neq k^2-\lambda_o$$

$$u_{kj} = 0, \text{ if } j^2(m_o^2-\lambda_o) \neq k^2-\lambda_o.$$

The above discussions yield the following result.

First, we need some notations.

Let C^∞ be the set of all continuously infinitely differentiable functions in t and x, and T_m periodic in t, $0 \leq x \leq \pi$ and C_o^∞ be the subset of C^∞ consisting of functions with compact support in $(0,\pi)$ with respect to x. Let $s \geq 0$ be an integer and for $\varphi, \psi \in C^\infty$, define

$$<\varphi,\psi>_s = \int_o^{2\pi}\int_o^{\pi} \sum_{|\alpha|=D}^{s} [(D^\alpha\varphi)(D^\alpha\psi)]dxdt$$

$$|\varphi|_s^2 = <\varphi,\varphi>_s$$

where

$$\alpha = (\alpha_1,\alpha_2), \quad |\alpha| = \alpha_1+\alpha_2, \quad D^\alpha = \frac{\partial^{\alpha_1+\alpha_2}}{\partial t^{\alpha_1}\partial x^{\alpha_2}}.$$

Let H^s (resp. H_o^s) be the completion of C^∞ (resp. C_o^∞) with respect to $|\cdot|_s$.

Let $N^\perp$ denote the orthogonal complement of N in $H^o = H_o^o$.

<u>Theorem</u> 3.1. If $g \in H^s \cap N^\perp$, $s \geq 1$, then there exists a unique generalized solution $w \in H_o^1 \cap N^\perp \cap H^s$ of (3.1), (3.2). Moreover, there exists a constant $c = c(s) > 0$ such that

$$|w|_s \leq c|g|_s.$$

<u>Proof</u>. This essentially follows from the work of Rabinowitz [11]. See also Kielhöfer [6].

§4. Bifurcation equation.

Consider now the nonlinear wave equation:

(4.1) $u_{tt} - u_{xx} - \lambda_o u - \mu u = f(\mu,u)$

with boundary conditions

(4.2) $u(t,0) = u(t,\pi) = 0$

where λ_o is as in §3, μ is real, $f(\mu,u)$ is sufficiently smooth and $f(\mu,u) = 0(|\mu|^2)$ uniformly for $\mu \in [-1,1]$. We are interested in the existence of T_{m_o}-periodic solutions of (4.1) and (4.2) near $\mu = 0$ and $u = 0$. Thus, we consider the problem (4.1) and (4.2) as an operator equation in $H^s \cap H^1_o$, $s \geq 1$, and define

$$Au = u_{tt} - u_{xx} - \lambda_o u$$

$$B(\mu,u) = f(\mu,u).$$

Consider now the operator equation

(4.3) $Au - \mu u = B(u).$

Since f is smooth

$$B : R \times (H^s \cap H^1_o) \to R \times H^s$$

is also smooth.

We may now apply Liapunov-Schmidt procedure to (4.3) to obtain an equivalent finite dimensional problem. Let P be the orthogonal projection of H^o onto N. Then (4.3) is equivalent to the following equations:

(4.4) $-\mu v = PB(\mu,v+w)$

(4.5) $Aw - \mu w = (I-P)B(\mu,v+w)$

where I is the identity, $Pu = v$ and $w = (I-P)u = u-v$.

By Theorem 3.1 and the implicit function theorem, there exists a unique solution $w = w^*(\mu,v) \in H^s \cap H^1_o \cap N^{\perp}$ of (4.5) for

$|\mu|$, $|v|_s << 1$ such that $w^*(0,0) = 0$. Moreover

$$w^*(\mu,v) = 0(|\mu v|_s+|v|_s^2).$$

Hence, (4.3) is equivalent to:

$$(4.6) \quad -\mu v = PB(\mu,v+w^*(\mu,v)).$$

Since (4.6) is a system of nonlinear equation in N which is isomorphic to R^{2d}, it is difficult to obtain nontrivial solutions without further considerations.

Note that (4.1) is actually the Euler-Lagrange equation for the functional

$$(4.7) \quad \Phi(\mu,u) = \int_o^{2\pi}\int_o^{\pi}\{\tfrac{1}{2}[u_t^2-u_x^2+\lambda_o u^2] + \tfrac{1}{2}\mu u^2+F(\mu,u)\}dxdt$$

defined on $R \times H_o^1$ with

$$F(\mu,u) = \int^u f(\mu,t)dt.$$

It is therefore natural to expect tnat (4.6) is related to the following scalar-valued function

$$(4.8) \quad \psi(\mu,v) = \Phi(\mu,v+w^*(\mu,v)).$$

We suppose for some fixed μ, $\bar{v}$ is a critical point of ψ. We have

$$\frac{\partial}{\partial v}\psi(\mu,\bar{v}) = 0.$$

In terms of the functional Φ, we obtain the following by using "D" as the differentiation operator,

$$(4.9) \quad D_u\Phi(\mu,\bar{v}+w^*(\mu,\bar{v})) \;.\; (\xi+D_v w^*(\mu,\bar{v})\xi) = 0$$

where $\xi \in N$ is arbitrary. Since f is smooth, $w^*(\mu,\bar{v})$ is also a smooth function by Sobolev's embedding theorem. Thus (4.9) says that

$$\langle A(\bar{v}+w^*(\mu,\bar{u}))-\mu(\bar{v}+w^*(\mu,\bar{v}))-B(\mu,\bar{v}+w^*(\mu,\bar{v})),$$

$$\xi + D_v w^*(\mu,\bar{v})\xi\rangle_o = 0.$$

Since $D_v w^*(\mu,\bar{v})\xi \in N^{\perp}$, by using (4.5) we have

$$<-\mu\bar{v}-PB(\mu,\bar{v}+w^*(\mu,\bar{v})),\xi> = 0.$$

We now have the following

<u>Theorem</u> 4.1. For any fixed μ, $\bar{v}$ is a solution of

$$-\mu v = PB(\mu,\bar{v}+w^*(\mu,\bar{v}))$$

if and only if $\bar{v}$ is a critical point of

$$\psi(\mu,v) = \Phi(\mu,v+w^*(\mu,v)).$$

Next, we note that equation (4.1) is autonomous. This implies that a phase shift will yield the same periodic orbit in the phase space. In order to obtain distinct periodic orbits, we have to identify these phase shifts. More precisely, let $\theta \in [0,T_{m_o}]$ and $u \in H^s \cap H^1_o$, define

$$L_\theta u(t,x) = u(t+\theta,x).$$

This defines an S^1 action on $H^s \cap H^1_o$. It is not difficult to see that $Au-\mu u-Bu$ commutes with L_θ and both N and $N^{\perp}$ are invariant under L_θ. Moreover, by the implicit function theorem $L_\theta w^*(\mu,u) = w^*(\mu,L_\theta u)$. Hence, we have the following

<u>Theorem</u> 4.2. For any fixed μ, ψ defined by (4.8) satisfies

$$\frac{\partial}{\partial v}\psi(\mu,L_\theta v) = L_\theta\frac{\partial}{\partial v}\psi(\mu,v).$$

The above lemma says that critical points must occur on circles. Moreover critical points on the same circle will yield the same periodic orbit. Hence, we are really searching for distinct critical circles of $\psi(\mu,v)$.

Finally, we note that if $d = 1$, then (4.6) is two-dimensional. If we take into account of the circle action, then it is actually a one-dimensional problem. By using implicit function theorem, one may solve (4.6) directly. We refer the reader to [6], [7] for details. It is also interesting to note that the same arguments may be applied to Hopf bifurcation theorem.

§5. Smooth nonlinearity.

By the definitions of $w^*(\mu,v)$ and Φ, we have the following estimate:

$$\psi(\mu,v) = \frac{\mu}{2}\int_0^{2\pi}\int_0^{\pi} v^2 dxdt + O(|v|_1^4).$$

Since $v \in N$ and $\dim N = 2d < \infty$, $|\cdot|_0$-norm is equivalent to $|\cdot|_1$-norm in N. Therefore, there exists a linear isomorphism from N to R^{2d}, say $v(a)$ with $a \in R^{2d}$, such that

$$\begin{aligned} (5.1)\quad & \psi(\mu,v(a)) \\ & = \psi(\mu,a), \quad \text{(definition)} \\ & = \tfrac{1}{2}\mu|a|^2 + O(|a|^4) \end{aligned}$$

where $|\cdot|$ denotes the euclidean norm in R^{2d}. Moreover, $\psi(\mu,a)$ is equivariant with respect to an S^1 action.

Note that the period T_{m_0} is not the least period for the functions in N (i.e., the space of T_{m_0}-periodic solutions of the linearized problem (3.1) and (3.2)) unless $d = 1$. This implies that the circle action L_θ is not free in the bifurcation equation (4.6). Hence, $\psi(\mu,a)$ is equivariant with respect to a non-free circle action.

If the circle action were free, then it is possible to apply Lusternik and Schnirelman theory to the quotient space (obtained by identifying each orbit under the circle action to a point) to get lower bounds of the number of distinct critical circles of $\psi(\mu,a)$. In [3], [8], such problems were considered. Fortunately, there is a generalized cohomological index theory developed recently by Fadell and Rabinowitz [4] which can be used together with the minimax principle [3], [4] to obtain the desired lower bounds.

We will use precisely these techniques to get the lower bounds of the number of critical circles of $\psi(\mu,a)$. The result is the following

Theorem 5.1. If $a = 0$ is an isolated critical point of $\psi(0,a)$, then there exist integers $\ell_+ \geq 0$, $\ell_- \geq 0$ such that for any $\varepsilon > 0$ sufficiently small there are ℓ_+ distinct critical circles of $\psi(\varepsilon,a)$ and ℓ_- distinct critical circles of $\psi(-\varepsilon,a)$ near $a = 0$. Moreover,

$$\ell_+ + \ell_- \geq d.$$

The proof is similar in spirit to the proof of Theorem A in [3] where a free circle action was considered. In fact, we could use the arguments in [4] where a non-free circle action was considered. For this reason, we will not give a proof here.

In terms of the differential equation, we have the following

Theorem 5.2. Consider the problem

$$(5.2)\quad u_{tt} - u_{xx} - \lambda_o u - \mu u = f(\mu,u)$$

$$(5.3)\quad u(t,0) = u(t,\pi) = 0$$

where $\mu \in R$, $f(\mu,u) = 0(|u|^2)$ uniformly for $\mu \in [-1,1]$ as $u \to 0$ and f is smooth. Suppose that $0 < \lambda_o = p_o/q_o < 1$, p_o, q_o are relative primes, and there exist positive integers $m_o, r_o > 0$ such that $q_o^2 m_o^2 - p_o q_o^2 = r_o^2$. Let $T_{m_o} = 2\pi/\sqrt{m_o^2 - \lambda_o}$. If there are no nontrivial T_{m_o}-periodic solutions of (5.1) and (5.2) with $\mu = 0$ near $u = 0$, then there exist integers $\ell_+, \ell_- \geq 0$ such that for any $\varepsilon > 0$ sufficiently small there are at least ℓ_+ (resp. ℓ_-) T_{m_o}-periodic solutions of (5.1) and (5.2) with $\mu = \varepsilon$ (resp. $\mu = -\varepsilon$). Moreover $\ell_+ + \ell_- \geq d$, where d is the number of pair of integers $j,k > 0$ such that $j^2(m_o^2 - \lambda_o) = k^2 - \lambda_o$.

§6. Analytic nonlinearity.

There is one hypothesis in Theorem 5.1 which is difficult to verify, namely, the nonexistence of T_{m_o}-periodic solutions near $u = 0$ at $\mu = 0$. In this section, we will show that we could drop this hypothesis if f is real analytic and has some special dependence on μ. In fact, we have the following.

Theorem 6.1. Consider the problem (5.2) and (5.3). Let λ_o, m_o and d be as in Theorem 5.1. Suppose that $f(\mu,u) = f(u)$, i.e., independent of μ, $f(u) = 0(|u|^2)$ as $u \to 0$ and f is a real analytic in u. Then, there exist at least d distinct analytic branches of periodic solutions of (5.1) and (5.2) from the equilibrium $u = 0$ with period T_{m_o}.

Proof. We observe from the construction of $\psi(\mu,v)$ (see (4.6)) or $\psi(\mu,a)$ (see (5.1)) that the analyticity of f implies that ψ is a real analytic function of the "polar" coordinates (μ,ρ,θ) where $\rho = |a|$ and $\theta = a/|a|$. Either there are finitely many analytic branches of critical circles of ψ each of the form

$$\rho = \rho_1\varepsilon + \rho_2\varepsilon^2 + \dots > 0$$

$$(6.1) \qquad \theta = \theta_o\varepsilon + \theta_1\varepsilon^2 + \dots$$

$$\mu = \mu_1\varepsilon + \mu_2\varepsilon^2 + \dots$$

for $0 < \varepsilon << 1$, or there are infinitely many in which case we are done. Assume then there are finitely many such branches. Moreover, we may assume that (6.1) accounts for all the non-zero critical points of ψ exactly once after an appropriate rotation (which depends on the least period of the periodic solution).

If none of the functions $\mu(\varepsilon)$ in (6.1) is identically zero, then $a = 0$ is an isolated critical point of $\psi(0,a)$. Thus, the result follows from Theorem 5.1.

Now, suppose some of the functions $\mu(\varepsilon)$ in (6.1) are identically zero. By the definition of ψ.

$$\psi(\mu,v) = \int_0^{2\pi}\int_0^{\pi}\frac{1}{2}[u_t^2-u_x^2 + \lambda_o u^2 + F(u)] - \frac{1}{2}\int_0^{2\pi}\int_0^{\pi}\mu u^2 dxdt$$

where $u = v+w^*(\mu,v)$. In terms of ψ, we have

$$\psi(\mu,a) = \theta(a,\mu) - \frac{\mu|a|^2}{2}, \quad \theta(a,\mu) = 0(|a|^3).$$

By the implicit function theorem, there exists a function $\mu(a)$ such that for $|a| << 1$

$$D_a\theta(a,\mu) \cdot a - \mu|a|^2 = 0.$$

Moreover, μ is real analytic in the polar coordinates $(\rho,\theta) = (|a|, a/|a|)$. It is also clear that μ is equivariant with respect to the circle action. Consider now the function

$$\xi(\tau,a) = \theta(a,\mu(a)) - \tau\frac{|a|^2}{2}.$$

Let τ be fixed. If $\tilde{a}$ is a critical point of $\xi(\tau,a)$ with $|\tilde{a}| << 1$, then $\tau = \mu(\tilde{a})$. Since $F(u)$ is independent of μ, a direct computation shows that $\tilde{a}$ is also a critical point of $\psi(\mu,a)$ at $\mu = \tau$. Hence the analytic branches of critical points of $\xi(\tau,a)$ have the following form

$$\begin{aligned} \rho &= \rho_1\varepsilon + \rho_2\varepsilon^2 + \ldots > 0 \\ \theta &= \theta_1\varepsilon + \theta_2\varepsilon^2 + \ldots \\ \tau &= \tau_1\varepsilon + \tau_2\varepsilon^2 + \ldots \end{aligned} \tag{6.2}$$

for $0 < \varepsilon << 1$.

We chose an integer $\ell > 0$ so large that along each of the branch in (6.2)

$$\tau(\varepsilon) - \ell|a(\varepsilon)|^{2(\ell-1)} \neq 0 \quad \text{for} \quad 0 < \varepsilon << 1. \tag{6.3}$$

Consider the function

$$\eta(\tau,a) = \theta(a,\mu(a)) - \tau\frac{|a|^2}{2} + |a|^{2\ell}.$$

Since

$$D_a\eta(\tau,a) = D_a\theta(a,\mu(a)) - [\tau - \ell|a|^{2(\ell-1)}]a$$

there is a one-to-one correspondence

$$(\tau,a) \to (\tau - \ell|a|^{2(\ell-1)},a)$$

between critical points of ξ and η. The condition (6.3) says that $\eta(0,a)$ has an isolated critical point at the origin. By applying Theorem 5.1 to $\eta(\tau,a)$ and using the correspondence (6.4) we obtain the desired result.

We now give another theorem which will include the sine-Gordon equation as a special case.

<u>Theorem</u> 6.2. If we replace the condition that $f(\mu,u) = f(u)$ in Theorem 6.1 by the following

$$f(\mu,u) = (\lambda_o+\mu)f(u)$$

where $f(u)$ is as in Theorem 6.1, then Theorem 6.1 still holds.

<u>Proof</u>. By the definition of ψ,

$$\psi(\mu,v) = \int_o^{2\pi}\int_o^{\pi}\frac{1}{2}[u_t^2-u_x^2 + \lambda_o u + \lambda_o F(u)]dxdt$$

$$(6.4)\qquad - \frac{1}{2}\int_o^{2\pi}\int_o^{2\pi}\mu[u^2 + F(u)]dxdt$$

where $u = v+w^*(\mu,v)$. In terms of ψ, we have

$$\psi(\mu,a) = \theta(a,\mu) - \frac{1}{2}\mu h(a)$$

where θ is derived from the first term on the right hand side of (6.4) and

$$h(a,\mu) = \frac{|a|^2}{2} + 0(|a|^3).$$

We may define the following functions as in the proof of Theorem 6.1.

$$\xi(\tau,a) = \theta(a,\mu(a)) - \tau h(a,\mu(a))$$

$$\eta(\tau,a) = \theta(a,\mu(a)) - \tau h(a,\mu(a)) + [h(a,\mu(a)]^{\ell}$$

where $\ell > 0$ is a large integer and $\mu(a)$ satisfies

$$D_a\theta(a,\mu(a)).a - \mu(a)D_a h(a,\mu(a)) = 0.$$

Since $F(u)$ is independent of μ, it is easily shown that for any fixed τ if $\tilde{a}$ is a critical point of $\xi(\tau,a)$ then $\tau = \mu(a)$ and $\tilde{a}$ is also a critical point of $\psi(\mu,a)$ at $\mu = \tau$. The rest of the proof is similar to that of Theorem 6.1.

REFERENCES

[1] - CHOW, S.N., MALLET-PARET, J. and YORKE, J.A., Global Hopf bifurcation from a multiple eigenvalue, Nonlinear Anal. Theory, Methods and Appl., 2 (1978), 753-763.

[2] - CHOW, S.N., MALLET-PARET, J. and HALE, J.K., Applications of generic bifurcations I, Arch. Rat. Mech. Anal., 59 (1975), 159-188.

[3] - CHOW, S.N. and MALLET-PARET, J., Periodic solutions of near an equilibrium of a non-positive definite Hamiltonian system, Preprint.

[4] - FADELL, E.R. and RABINOWITZ, P.H., Generalized cohomological index theories for Lie group actions with an application to bifurcation questions for Hamiltonian systems, MRC TR 1769, Univ. Wisc., (1977).

[5] - HALE, J.K., Periodic solutions of a class of hyperbolic equations containing a small parameter, Arch. Rat. Mech. Anal. 23 (1967), 380-398.

[6] - KIELHÖFER, H., Bifurcation of periodic solutions for a semilinear wave equation, MRC TR 1817, Univ. Wisc., (1978).

[7] - MELROSE, R.B. and PEMBERTON, M., Periodic solutions of certain nonlinear autonomous wave equations, Math. Proc. Camb. Phil. Soc., 78 (1975), 137-143.

[8] - MOSER, J., Periodic orbits near an equilibrium and a theorem by Alan Weinstein, Comm. Pure Appl. Math., 29 (1976), 727-746.

[9] - RABINOWITZ, P.H., Time periodic solutions of a nonlinear wave equation, Manus. Math. 5 (1971), 165-194.

[10] - RABINOWITZ, P.H., Free vibrations for a semilinear wave equation, MRC TR 1742, Univ. Wisc., (1977).

[11] - RABINOWITZ, P.H., Periodic solutions of nonlinear hyperbolic partial differential equations, Comm. Pure Appl. Math., 20 (1967), 145-205.

[12] - WEINSTEIN, A., Lagrangian submanifolds and Hamiltonian systems, Ann. Math., 98 (1973), 377-410.

CONTACT EQUIVALENCE AND BIFURCATION THEORY

by Luiz Carlos Guimarães*

1. Introduction.

The study of the changes on the set of zeros of a parametrized equation $F_\lambda(x) = 0$ is a problem common to the theory of singularities of maps and to bifurcation theory. Traditionally both theories used quite distinct methods to pursue this study but recently there has been a spate of works that recognise the similarities and use methods and ideas of singularity theory to treat bifurcation problems. For references, see the paper by Magnus in this collection. See also Golubitsky and Schaeffer [3] and Marsden [8] as well as Guimarães [4].

Let us review some of these concepts of singularity theory that can prove useful for bifurcation theory. Suppose we look at the equation

(1.1) $f(\lambda,a) = 0 \ , \ f(0,0) = 0$

where Λ, A, B are Banach spaces, $f: \Lambda \times A \to B$ is a C^∞ germ at the origin and $D_a f(0,0)$ is a Fredholm operator. We think of f as a 'deformation' of the germ $a \to f_o(a) = f(0,a)$. From now on we shall denote by $\mathcal{E}(A,B)$, where A,B are arbitrary Banach spaces, the real vector space of the germs at the origin of C^∞ maps $A \to B$, and by $m(A,B)$ the subspace of those germs in $\mathcal{E}(A,B)$ which vanish at the origin.

To detect changes on the set of zeros one needs a way of comparing these for two diferent germs. There are three equivalence relations used in singularity theory which are adequate for this purpose:

(i) $f_o, g_o \in m(A,B)$ are *contact equivalent* if, in a neighbourhood of the origin, $g_o(a) = K(a).f_o(\phi(a))$, where $\phi \in m(A,A)$ is a germ

* Research partially supported by CNPq

of a C^∞ diffeomorphism keeping the origin fixed, and $K \in E(A,L(B))$ is such that $K(0)$ is an isomorphism ($L(B)$ denotes the space of continuous linear operators in B).

(ii) $f_o, g_o \in m(A,B)$ are *C^μ-contact equivalent* if the equality above holds, but with ϕ and K required only to be germs of C^μ maps/diffeomorphism, $\mu \geq 0$.

(iii) $f_o, g_o \in m(A,B)$ are v-equivalent if $f_o^{-1}(0)$, $g_o^{-1}(0)$ are germs of homeomorphic sets.

The choice of an equivalence relation allows the idea of reducing a given non-linear equation to a simpler equation. For instance, standard in bifurcation theory is the idea of using the hypotesis of $Df_o(0)$ being Fredholm to reduce f_o to a germ defined on finite-dimensional spaces. The contact-equivalence version is:

(1.2) Theorem. *If* $f_o \in m(A,B)$ *is such that* $Df_o(0)$ *is a Fredholm operator, then* f_o *is contact equivalent to a germ of the for* $(u,x) \to (u,\overline{f}_o(x))$, *where* $\overline{f}_o$ *is a germ taking* $\ker Df_o(0)$ *into coker* $Df_o(0)$.

The proof of this result is straightforward (see Guimarães [4]). We recall the two steps involved: the Implicit Function Theorem (i.e., the standard Liapunov-Schmidt reduction) gives, through changes of coordinates, the equivalence of f_o to a germ of the form

$$U \oplus \ker Df_o(0) \to U \oplus \operatorname{coker} Df_o(0)$$

$$(u,x) \to (u,f(u,x))$$

where $A = U \otimes \ker Df_o(0)$ and $B = U \oplus \operatorname{coker} Df_o(0)$ are the splitings induced by $Df_o(0)$. Some linear algebra (similar to the proof of theorem 3.3, pg. 171 in Golubitsky-Guilemin [2]) is then used to show that $(u,x) \to (u,f(u,x))$ is always contact equivalent to $(u,x) \to (u,f(0,x))$. The following example emphasises that right-equivalence or right-left equivalence i.e., changes of coordinates (even C^0!) in the domain and counter domain, are too strong for a result like the theorem above to hold.

Equation (3.15) is equivalent to

$$\frac{\partial \overline{K}}{\partial t}(t,\lambda,x).F(t_o+t,\lambda,\phi) + \overline{K}(t,\lambda,x).[\frac{\partial F}{\partial t}(t_o+t,\lambda,\phi) +$$

$$+ D_xF'(t_o+t,\lambda,\phi)\frac{\partial \phi}{\partial t}(t,\lambda,x)] = 0$$

subject to the same initial conditions and restrictions, and since $K(t,\lambda,x)$ and $\phi(t,\lambda,.)$ are invertible for small t,λ,x, we can replace the last equation by

$$(3.16) \quad K(t,\lambda,x).F(t_o+t,\lambda,x) + \frac{\partial F}{\partial t}(t_o+t,\lambda,x) + D_xF(t_o+t,\lambda,x).H(t,\lambda,x) = 0$$

subject to the restriction that $K \in E(\mathbb{R}\times\Lambda\times\mathbb{R}^n, L(\mathbb{R}^p))$ and $H \in E(\mathbb{R}\times\Lambda\times\mathbb{R}^n,\mathbb{R}^n)$ satisfy

$$(3.17) \quad K(t,0,x) = 0, \quad H(t,0,x) = 0.$$

(3.15) follows from (3.16), (3.17) if we apply the theorem on the existence and uniqueness of a local flow to the system of differential equations

$$(3.18) \quad \begin{aligned} \frac{\partial \phi}{\partial t}(t,\lambda,x) &= H(t,\lambda,\phi(t,\lambda,x)) \\ \frac{\partial \overline{K}}{\partial t}(t,\lambda,x) &= \overline{K}(t,\lambda,x).K(t,\lambda,\phi(t,\lambda,x)) \end{aligned}$$

subject to the initial conditions $\phi(0,\lambda,x) = x$, $K(0,\lambda,x) = Id_{\mathbb{R}^p}$.

To show the existence of H and K satisfying (3.16) and (3.17) we use Nakayama's Lemma. Define $\tau_x(F_{t_o})$ to be

$$\tau_x(F_{t_o}) = \{D_xF(t_o+t,\lambda,x).H(t,\lambda,x) + K(t,\lambda,x).F(t_o+t,\lambda,x):$$

$$H \in E(\mathbb{R}\times\Lambda\times\mathbb{R}^n,\mathbb{R}^n), K \in E(\mathbb{R}\times\Lambda\times\mathbb{R}^n, L(\mathbb{R}^p))\}.$$

From our assumption on k and from the special form of $r(\lambda,x)$ (see (3.14)) it follows

$$m^{k-1}(\mathbb{R}^n)E(\mathbb{R}\times\Lambda\times\mathbb{R}^n,\mathbb{R}^p) \subset \tau_x(F_{t_o}) + m(\mathbb{R}\times\Lambda\times\mathbb{R}^m)m^{k-1}(\mathbb{R}^n)E(\mathbb{R}\times\Lambda\times\mathbb{R}^n,\mathbb{R}^p).$$

Nakayama's Lemma then implies

Here $x^{(i)}$ is the i-vector $(x,\dots,x)$. Observe that we can write

$$\sum_{i=1}^{k} \frac{1}{i!} D_x^i \bar{f}(\lambda,0).x^{(i)} = \sum_{i=1}^{p} \sum_{0\le|\sigma|\le k} \beta_\sigma^i(\lambda) x^\sigma e_i$$

where the e_i, $1 \le i \le p$, are the standard basis vectors in $\mathbb{R}^p$, $\beta_\sigma^i \in m(\Lambda)$ and σ is an n-vector $(i_1,\dots,i_n)$, $i_j \in N$, $|\sigma|=i_1+\dots+i_n$ and $x^\sigma = x_1^{i_1} x_2^{i_2}\dots x_n^{i_n}$. We can also write

$$(3.14) \qquad r(\lambda,x) = [\int_0^1 \frac{(1-t)^k}{k!} D_k^{k+1}\bar{f}(\lambda,tx)dt].x^{(k+1)}$$

$$= \sum_{i=1}^{p} \sum_{|\sigma|=k+1} \gamma_\sigma^i(\lambda,x)x^\sigma e_i$$

where each $\gamma_\sigma^i \in E(\Lambda\times\mathbb{R}^n)$ satisfies $\gamma_\sigma^i(0,x) \equiv 0$. Hence we have $r \in m^{k+1}(\mathbb{R}^n)E(\Lambda\times\mathbb{R}^n,\mathbb{R}^p)$ and $r(0,x) \equiv 0$. Therefore in order to show that there is an equivalence between $f(\lambda,x)$ and some germ of the form $f_o + \sum_{\text{finite}} \delta_i h_i$ we only have to show that f_o+r is equivalent to the trivial suspension $(\lambda,x) \to f_o(x)$ of f_o, if k is chosen large enough.

From Lemma (3.12) we know that we can choose k such that $m^{k-1}(\mathbb{R}^n)E(\mathbb{R}^n,\mathbb{R}^p) \subset \tau(f_o)$, and from now on we assume this is the case. By analogy with what we did in the proof of Theorem (2.3) consider the one-parameter family of germs given by $F_t(\lambda,x) = F(t,\lambda,x) = f_o(x) + tr(\lambda,x)$. As before, it is enough to show that F is locally trivial in t up to equivalence. We will show that given $t_o \in [0,1]$ we can find germs $\bar{K} \in E(\mathbb{R}\times\Lambda\times\mathbb{R}^n,L(\mathbb{R}^p))$, $\phi \in E(\mathbb{R}\times\Lambda\times\mathbb{R}^n,\mathbb{R}^p)$ such that

$$\bar{K}(t,\lambda,x).F(t_o+t,\lambda,\phi(t,\lambda,x)) = F(t_o,\lambda,x)$$

$$(3.15) \qquad \bar{K}(0,\lambda,x) = Id_{\mathbb{R}^p},\ \phi(0,\lambda,x) = x, \quad \text{and}$$

$$\bar{K}(t,0,x) = Id_{\mathbb{R}^p},\ \phi(t,0,x) = x.$$

$$D_\lambda F(t_o,0,x)\lambda_i = t_o D_\lambda f(0,x)\lambda_i + (1-t_o)D_u f_c(0,x)e_i$$

$$= t_o w_i(x) + (1-t_o)w_i(x)$$

$$= w_i(x).$$

Therefore, by Lemma (3.3), any element of $E(\mathbb{R}\times\Lambda\times\mathbb{R}^n,\mathbb{R}^p)$ can be written as

(3.11) $$(t,\lambda,x) \to D_x F(t_o+t,\lambda,x)H_1(t,\lambda,x) + K_1(t,\lambda,x).F(t_o+t,\lambda,x)$$
$$+ D_\lambda F(t_o+t,\lambda,x).U_1(t,\lambda)$$

(where $U_1(t,\lambda) = \sum_{i=1}^{c} \alpha_i(t,\lambda)\lambda_i$). So, as $\frac{\partial F}{\partial t}(t_o+t,\lambda,x) \in m(\Lambda)E(\mathbb{R}\times\Lambda\times\mathbb{R}^n,\mathbb{R}^p)$ is a finite sum of elements of the form $\delta(\lambda)h(t,\lambda,x)$ where $\delta \in m(\Lambda)$, and $h \in E(\mathbb{R}\times\Lambda\times\mathbb{R}^n,\mathbb{R}^p)$ can be written as in (3.12), it follows that there exist U,H,K which satisfy (3.8) and (3.9) and the proof of Theorem (2.3) is complete. QED

To prove Theorem (3.1) we need the follwing well result from singularity theory (see Mather [9]):

(3.12) Lemma. *If the codimension of* $f_o \in m(\mathbb{R}^n,\mathbb{R}^p)$ *is finite then for some* $k \in N$ *we have*

$$m^k(\mathbb{R}^n)E(R^n,R^p) \subset \tau(f_o)$$

where $m^k(\mathbb{R}^n)$ *denotes the subset of the germs in* $E(\mathbb{R}^n,\mathbb{R})$ *which vanish at the origin to order* k.

(3.13) Proof of Theorem (3.1). Put $\overline{f}(\lambda,x) = f(\lambda,x) - f_o(x)$. Then by Taylor's Theorem we have for any $k \geq 0$

$$f(\lambda,x) = f_o(x) + \overline{f}(\lambda,x)$$

$$= f_o(x) + \sum_{i=1}^{k} \frac{1}{i!}D_x^i\overline{f}(\lambda,0).x^{(i)} +$$

$$+ \left[\int_0^1 \frac{(1-t)^k}{k!} D_x^{k+1}\overline{f}(\lambda,tx)dt\right].x^{(k+1)}$$

The existence of $U \in \mathcal{E}(\mathbb{R}\times\Lambda,\Lambda)$ $H \in \mathcal{E}(\mathbb{R}\times\Lambda\times\mathbb{R}^n,\mathbb{R}^n)$, $K \in \mathcal{E}(\mathbb{R}\times\Lambda\times\mathbb{R}^n,L(\mathbb{R}^p))$ satifying (3.8) and the additional restrictions

$$(3.9) \quad U(t,0) = 0, \quad H(t,0,x) = 0, \quad K(t,0,x) = 0$$

then implies the existence of $\Psi,\phi,\overline{K}$ satisfying (3.5), (3.6), (3.7). For, if we take $\Psi,\phi,\overline{K}$ to be the solutions of the system of differential equations

$$\frac{\partial\Psi}{\partial t}(t,\lambda) = U(t,\Psi(t,\lambda))$$

$$(3.10) \quad \frac{\partial\phi}{\partial t}(t,\lambda,x) = H(t,\Psi(t,\lambda),\phi(t,\lambda,x))$$

$$\frac{\partial\overline{K}}{\partial t}(t,\lambda,x) = \overline{K}(t,\lambda,x).K(t,\Psi(t,\lambda),\phi(t,\lambda,x))$$

subject to the initial conditions (3.5), the existence and uniqueness theorem for ordinary differential equations provides us with the desired local equivalences ((3.9) takes care of the possibility of making the solution for any representative into a well defined germ at (0,0,0)).

To prove the existence of U,H,K we use Lemma (3.3). Here the assumption on the form of f made possible by Theorem (3.1) is crucial. We know that

$$\frac{\partial F}{\partial t}(t_o+t,\lambda,x) = f(\lambda,x) - f_c(u,x) \quad (\text{here } \lambda = (\lambda',u))$$
$$= \sum_{i=1}^{k} \delta_i(\lambda)h_i(\lambda,x) - \sum_{i=1}^{k} u_i w_i(x).$$

Hence $\frac{\partial F}{\partial t}(t_o+t,\lambda,x) \in m(\Lambda)\mathcal{E}(\mathbb{R}\times\Lambda\times\mathbb{R}^p)$. Also, F satisfies $F(t_o,0,x) = f_o(x)$, so that $(t,\lambda,x) \to F(t_o+t,\lambda,x)$ is a deformation of f_o with parameter space $\mathbb{R}\times\Lambda$. If we put $\lambda_i = (0,e_i)$ (where $\{e_i\}_{i=1}^{c}$ is the basis of $\mathbb{R}^c$ we chose after (2.2)) then the germs $(t,\lambda,x) \to D_\lambda F(t_o+t,\lambda,x).\lambda_i$ are deformations of the $w_i(x)$, $1 \le i \le c$, since

To prove (2.3) we assume that f has the form (3.2). We also assume that $\Lambda = \ker T_f \oplus \mathbb{R}^c$ and write $\lambda = (\lambda',u)$.

Let $F:[0,1]\times E(\Lambda\times\mathbb{R}^n,\mathbb{R}^p) \to E(\Lambda\times\mathbb{R}^n,\mathbb{R}^p)$ be the homotopy $F(t,\lambda,x) = tf(\lambda,x) + (1-t)f_c(u,x)$. Because $[0,1]$ is connected, to show that f is equivalent to $(\lambda,x) \to f_c(u,x)$ it is enough to show that F is locally constant in t up to equivalence. That is, for each $t_o \in [0,1]$ we want to find germs $\overline{K} \in E(\mathbb{R}\times\Lambda\times\mathbb{R}^n, L(\mathbb{R}^p))$, $\phi \in E(\mathbb{R}\times\Lambda\times\mathbb{R}^n,\mathbb{R}^p)$, $\Psi \in E(\mathbb{R}\times\Lambda,\Lambda)$ such that

(3.4) $$\overline{K}(t,\lambda,x).F(t_o+t,\Psi(t,\lambda),\phi(t,\lambda,x)) = F(t_o,\lambda,x)$$

as germs at $(t,\lambda,x) = 0$, and such that

(3.5) $$\overline{K}(0,\lambda,x) = \mathrm{Id}_{\mathbb{R}^p},\ \phi(0,\lambda,x) = x, \quad \Psi(0,\lambda) = \lambda,$$

(3.6) $$\overline{K}(t,0,x) = \mathrm{Id}_{\mathbb{R}^p},\ \phi(t,0,x) = x, \quad \Psi(t,0) = 0.$$

When we differentiate equation (3.4) with respect to t we find that it is equivalent to the equation

(3.7) $$\frac{\partial\overline{K}}{\partial t}(t,\lambda,x).F(t_o+t,\Psi,\phi) + \overline{K}(t,\lambda,x).[\frac{\partial F}{\partial t}(t_o+t,\Psi,\phi) + D_\lambda F(t_o+t,\Psi,\phi).\frac{\partial\Psi}{\partial t}(t,\lambda) + D_x F(t_o+t,\Psi,\phi).\frac{\partial\phi}{\partial t}(t,\lambda,x)] = 0$$

subject to the initial conditions (3.5) and the extra conditions (3.6) (Ψ,ϕ abbreviate $\Psi(t,\lambda)$, $\phi(t,\lambda,x)$ respectively). Now, (3.6) implies that $\overline{K}$ is a germ with image in $GL(\mathbb{R}^p)$ and that, for small enough t, $(\lambda,x) \to (\Psi(t,x),\phi(t,\lambda,x))$ is invertible. Hence we can change (3.7) to the equation

(3.8) $$K(t,\lambda,x).F(t_o+t,\lambda,x) + \frac{\partial F}{\partial t}(t_o+t,\lambda,x) + D_\lambda F(t_o+t,\lambda,x).U(t,\lambda) + D_x F(t_o+t,\lambda,x).H(t,\lambda,x) = 0.$$

induces g. If $\mu = (\mu_1,\dots,\mu_k) \in \mathbb{R}^k$ we can use the hypothesis on f and Malgrande's Division Theorem to show that $\overline{f}$ is equivalent to the deformation $(\lambda,\mu,a) \to f(\lambda,a) + (g(\mu_1,\dots,\mu_{k-1},0,a) - f_o(a))$. So, by induction, we see that $\overline{f}$ is equivalent to f, and therefore f induces g.

Michor [10] has shown that the Division Theorem is true also in a Banach space, but the proof above does not work. To prove our theorem (2.3) we have to devise a different way of applying the Division Theorem, following an idea of Magnus [7].

3. Proof of Theorem (2.3).

As a consequence of the Liapunov-Schmidt reduction (1.2) we can assume A,B to be finite dimensional spaces.

The first step on the proof is a result interesting enough in itself to be stated here as a theorem:

(3.1) Theorem. *Suposse* $f_o \in m(\mathbb{R}^n,\mathbb{R}^p)$ *has finite codimension. Then any deformation* $f \in E(\Lambda\times\mathbb{R}^n,\mathbb{R}^p)$ *of* f_o *is equivalent to a deformation of the form*

$$(3.2) \qquad (\lambda,x) \to f_o(x) + \sum_{i=1}^{k} \gamma_i(\lambda) h_i(\lambda,x)$$

for suitable $h_i \in E(\Lambda\times\mathbb{R}^n,\mathbb{R}^p)$, $\gamma_i \in m(\Lambda,\mathbb{R})$, $1 \le i \le k$.

We postpone the proof of (3.1) for the moment. The proof of (2.3) will also depend on the following consequence of the Division Theorem:

(3.3) Lemma: *Let* $f_o \in m(A,B)$ *have finite codimension and suppose* $w_1,\dots,w_c \in E(\mathbb{R}^n,\mathbb{R}^p)$ *are such that their projections form a basis of* $E(\mathbb{R}^n,\mathbb{R}^p)/\tau(f_o)$. *If* $g, v_i \in E(\Lambda\times\mathbb{R}^n,\mathbb{R}^p)$ *are deformations of* f_o, w_i *respectively* $(1 \le i \le c)$ *then any element of* $E(\Lambda\times\mathbb{R}^n,\mathbb{R}^p)$ *can be written in the form*

$$(\lambda,x) \to D_x g(\lambda,x).H(\lambda,x) + K(\lambda,x).g(\lambda,x) + \sum_{i=1}^{c} \alpha_i(\lambda) v_i(\lambda,x),$$

for suitable germs $H \in E(\Lambda\times\mathbb{R}^n,\mathbb{R}^n)$, $K \in E(\Lambda\times\mathbb{R}^n,L(\mathbb{R}^p))$, $\alpha_i \in E(\Lambda,\mathbb{R})$, $1 \le i \le c$.

(2.2) Lemma: *If the codimension of* f_o *is finite the operator* T_f *is continuous.*

This is a consequence of Lemma (3.3) below.

Let us assume for the moment that the map T_f is onto $E(A,B)/\tau(f_o)$, i.e. there exist $e_1,\dots,e_c \in \Lambda$ such that the images under Π of the germs $w_i: a \to D_\lambda F(0,a).e_i$, $1 \le i \le c$, form basis of $E(A,B)/\tau(f_o)$. Hence we can consider $\Lambda = \ker T_f \oplus \mathbb{R}^c$.

We now state the main result of this paper.

(2.3) Theorem. *Let* f_1, f_o *be as above and assume* T_f *is surjective Then* f *is equivalent to the deformation of* f_o *given by*
$(\lambda,a) \to f_c(u,a) = f_o(a) + \sum_{i=1}^{c} u_i D_\lambda f(0,a).e_i$ *where* $\lambda = (x,u)$ *is the spliting of* Λ *referred to above.*

A consequence of this is the following strong version of the Liapunov-Schmidt reduction:

(2.4) Theorem. *Suppose that* $f_o \in m(A,B)$ *is such that* $Df_o(0)$ *is a Fredholm operator and* f_o *has finite codimension. Then any deformation of* f_o *is induced by a germ that depends only on a finite number of variables.*

Another consequence is a characterization of versal deformations:

(2.5) Theorem. *Suppose* $f_o \in m(A,B)$ *is a germ with finite contact codimension and such that* $Df_o(0)$ *is a Fredholm operator. Then a deformation* $f \in E(\Lambda\times A,B)$ *of* f_o *is versal if and only if the operator* $T_f:\Lambda \to E(\Lambda,B)/\tau(f_o)$ *is onto.*

The proof of this theorem, in the finite-dimensional case, was first given by Mather. We sketch here the idea of it, so as to point out why it does not work in the case of an infinite-dimensional parameter space. Suppose f, f_o are as in the statement of the theorem (2.5), and that $(\mu,a) \to g(\mu,a)$ is a deformation of f_o. Clearly $\bar{f}:(\lambda,\mu,a) \to f(\lambda,a) + (g(\mu,a) - f_o(a))$ is a deformation of f_o that (i.e., two C^∞ germs $g \in m(\Lambda_2 \times A,B)$ $f \in m(\Lambda_1 \times A,B)$), which preserves the special role of the parameter space Λ: f and g are two *equivalent deformations* if in a neighbourhood of the origin.

$$(2.1) \quad g(\lambda,a) = K(\lambda,a).f(\Psi(\lambda),\phi(\lambda,a))$$

where $\Psi \in m(\Lambda_2,\Lambda_1)$ is a germ of a C^∞ diffeomorphism, $\phi \in m(\Lambda_2 \times A,A)$ is such that $a \to \phi(0,a)$ is a local C^∞ diffeomorphism of A, and and $K \in E(\Lambda_2 \times A,L(B))$ with $K(0,0)$ invertible.

If equality (2.1) holds but Ψ is not necessarily a germ of a dipheomorphism, we say that g is *induced* by f. Clearly knowledge of Ψ allows us to deduce the bifurcation set of g from that of f and, in this sense, $f_\lambda = 0$ is a more general equation than $g_\lambda = 0$.

We say that F is a *versal* (i.e., the most general) deformation of $f_o \in m(A,B)$ if any deformation of f_o is induced by f.

Theorem (2.1) can be extended to reduce a deformation of a Fredholm germ f_o (i.e., one with $Df_o(0)$ a Fredholm operator) to a deformation of a germ of an application between finite-dimensional spaces, parametrized by a Banach space. A reduction of the parameter space is also possible if we restrict the class of germs we are considering. We will need the following definition:

If $f_o \in m(A,B)$ the (contact) *tangent space* to f_o is the subspace $\tau(f_o)$ of $m(A,B)$ of the germs that can be put in the form $a \to Df_o(a).h(a) + K(a).f_o(a)$, for suitable $h \in E(A,A)$ and $K \in E(A,L(B))$. We say that f_o has finite codimension if the dimension of the quotient $E(A,B)/\tau(f_o)$, considered as a real vector space, is finite.

Given a deformation f of f_o, with parameter space Λ, we can define a linear operator.

$$T_f : \Lambda \to E(A,B)/\tau(f_o)$$

given by $T_f.\lambda = \Pi(D_\lambda f(0,.).\lambda)$, where Π is the canonical projection $E(A,B) \to E(A,B)/\tau(f_o)$.

Take $f_o:(x,y,z) \to (x^2+y^3+yz,z)$, $g_o:(x,y,z) \to (x^2+y^3,z)$. That they are contact equivalent is easily checked. But they can't be right or right-left equivalent, for this would imply that the level surfaces of both maps be locally homeomorphic, which isn't true.

It is obvious that for C^μ-contact equivalence and v-equivalence, they being weaker than contact equivalence, a 'Liapunov-Schmidt' reduction also holds.

The concept of determinacy in singularity theory takes this idea of reducing to a simpler problem further. A germ f_o in $m(A,B)$ is said k-determined with respect to a given equivalence relation if is equivalent to f_o any germ whose Taylor expansion at the origin to order k coincides with that of f_o.

If so, one needs only to consider the $k^{\underline{th}}$-oder Taylor polynomial of f_o which, hopefully, can be put into a standard form for which the bifurcation set is known.

Mather [9] has given sufficient (algebraic) conditions for a given germ to be k-determined with respect to contact equivalence. The case of C^μ-contact equivalence was studied by Bochnak-Kuo [1] who give necessary and sufficient conditions for a germ to be finitely determined with respect to it; and in [5] Kuo gives sufficient conditions for k-determinacy with respect to v-equivalence.

In the context of bifurcation problems, equivalence (i) was considered by Golubitsky-Schaeffer [3], and determinacy with respect to a modified version of (i), (ii) was the subject of the talk by Magnus referred to earlier. The concept of v-equivalence is implicit for instance in Magnus [6], Marsden [8] and Shearer [11].

For the remaining part of this talk we restrict our attention to contact equivalence.

2. <u>Versality</u>.

We return now to the parametrized equation (1.1). We consider here a contact-equivalence relation between two such deformations

(3.19) $m^{k-1}(R^n)\mathcal{E}(R\times\Lambda\times R^n,R^p) \subset \tau_x(F_{t_o})$.

Looking again at (3.19) we see that

$$\frac{\partial F}{\partial t}(t_o+t,\lambda,x) = r(\lambda,x) = \sum_{i=1}^{p} \sum_{|\sigma|=k+1} \gamma_\sigma^i(\lambda,x)x^\sigma e_i.$$

Hence (3.10) and the fact that $\gamma_\sigma^i(0,x) = 0 \ \forall\ i,\sigma$ imply the existence of H and K which satisfy (3.16) and (3.17). QED.

REFERENCES

[1] - BOCHNAK, J. and KUO, T.C., Rigid and finitely v-determined germs of C^∞ maps, Canadian J. Math. XXV(1973), 727-732.

[2] - GOLUBITSKY, M. and GUILLEMIN, V., Stable Mappings and their Singularities, Graduate Texts in Math., Springer-Verlag, (1973).

[3] - GOLUBITSKY, M. and SCHAEFFER, D., A theory for imperfect bifucartions via singularity theory, Comm. Pure Appl. Math. 32(1979), 21-97.

[4] - GUIMARÃES, L., Contact equivalence and bifurcation theory, University of Southampton, Faculty of Mathematics Preprint Series, Southompton, 1979.

[5] - KUO, T.C., Characterizations of v-sufficiency of jets, Topology 11(1972), 115-131.

[6] - MAGNUS, R. J., On the local structure of the set of zeros of a Banach space valued mapping, J. Func. Anal. 22(1976), 58-72.

[7] - MAGNUS, R.J., Universal unfoldings in Banach spaces: reduction and stability, Battelle-Geneva Nº 107(1977). (To appear in Math. Proc. Camb. Phil. Soc.).

[8] - MARSDEN, J., Qualitative methods in bifurcation theory, Bull. of the A.M.S. 84(1978), 1125-1148.

[9] - MATHER, J. N., Stability of C^∞ mappings III, Publ. I.H.E.S. 35(1969), 127-156.

[10] - MICHOR, P., The preparation theorem on Banach spaces, Mathematisches Institut der Universitat, Wein.

[11] - SHEARER, M., Small solutions of a nonlinear equation in Banach space for a degenerate case, Proc.Roy.Soc. Edinburgh, 79A(1977), 58-73.

SOME RECENT RESULTS ON DISSIPATIVE PROCESSES*

by Jack K. Hale

1. Introduction.

Suppose X is a complete metric space and $T : X \to X$ is a continuous mapping. If S is a family of subsets of X and $B \subseteq X$ is a fiven set, we say B *dissipates sets of* S *relative to* T if, for any $A \in S$, there is an integer N such that $T^N A \subseteq B$ for $n \geq N$. We say a set C *attracts sets of* S *relative to* T if, for any $\varepsilon > 0$, $B = \mathcal{B}(C,\varepsilon)$ dissipates sets of S relative to T, where $\mathcal{B}(C,\varepsilon)$ is the open ε-neighborhood of C. A set $J \subseteq X$ is invariant under T if $TJ = J$ and it is *maximal* if, for any set $H \subseteq X$ invariant under T, we have $H \subseteq J$.

The map T is *point dissipative* if there is a bounded set B that dissipates points of X relative to T; *compact dissipative* if a bounded set B dissipates compact sets of X relative to T; *bounded dissipative* or *uniformly ultimately bounded* if a bounded set B dissipates bounded sets of X relative to T; *local dissipative* if there is a bounded set B such that, for any $x \in X$, there is a neighborhood V_x of x such that B dissipates V_x relative to T; *local compact dissipative* if, there is a bounded set B such that, for any compact set $K \subseteq X$, there is a neighborhood V_K of K such that B dissipates V_K relative to T.

The literature on dissipative processes (that is, maps T satisfying one of the properties above) is mainly concerned with the following questions:

(1) What conditions imply the existence of maximal compact invariant set?

(2) What are the stability properties of this set ?

(3) What is the relationship between different types of attractivity of this maximal compact invariant set ? For example, when is

* This research was supported in part by the Air Force Office of Scientific Research under AF-AFOSR 76-3092C, in part by the National Science Foundation under NSF-MCS 78-18858, and in part by the United States Army under ARO-D-31-124-73-G130.

asymptotic stability (stable and point dissipative) equivalent to uniform asymptotic stability (stable and local compact dissipative) ? When is the property of this maximal compact invariant set being asymptotically stable equivalent to the property that it attracts bounded sets of X.

(4) When does T have a fixed point ?

The purpose of this paper is to present some of the recent results on the above questions; especially, those of Cooperman [7] and Massatt [15], [16]. At the same time, we take the opportunity to correct a few mistakes in Hale [10] and give a few new implications of the results. For a brief bibliographical discussion of the history of the subject, see [10]. The hypotheses for the results will involve certain attractivity and dissipative properties as well as some smoothness properties of T.

If $U(t) : X \to X$, $t \geq 0$, is a strongly continuous semigroup on X and $T = U(r)$ for some fixed $r > 0$, then the above questions (1)-(3) have obvious implications to the asymptotic behavior of the orbits of $U(t)$. If it happens that $U(r)$ has a fixed point for every r and certain compactness conditions are satisfied, then there will be an x_o such that $U(t)x_o = x_o$ for all $t \geq 0$; that is, x_o is an equilibrium point of $U(t)$.

Strongly continuous semigroups $U(t)$ arise in a natural way as the solution operator of an autonomous evolutionary equation. Our smoothness properties on $T = U(r)$ imposed below will be satisfied for retarded functional differential equations with finite delay, certain types of retarded equations with infinite delay (for example, the ones for which the initial space satisfies the axioms in Hale and Kato [12]), neutral functional differential equations with a stable D operator (see Hale [10]) and quasilinear parabolic equations. For the latter problem, the existence of an equilibrium point of $U(t)$ implies the existence of a solution of an elliptic boundary value problem (for some reference on this latter problem, see Amann [1]).

For evolutionary equations which are periodic in the independent variable, one can choose T as the period map and obtain results on asymptotic behavior. Fixed points of T now correspond to periodic solutions of the equation of the same period as the vector field.

Semigroups of transformations can also be associated with nonautonomous evolutionary equations via the skew product flow (see, e.g., Sell [18], Dafermos [8], Artstein [2]). In Section 3, we show how the general results on dissipative systems yield information about stability in nonautonomous equations.

2. Maximal Compact Invariant Sets.

In this section, we give conditions on the map T to insure the existence and uniform asymptotic stability of a maximal compact invariant set. We need a few definitions.

Definition 2.1. The map T is asymptotically smooth if for any bounded set $B \subseteq X$, there is a compact set $J \subseteq X$ such that, for any $\varepsilon > 0$, there is an integer $n_o(\varepsilon,B) > 0$ such that, if $T^n x \in B$ for $n \geq 0$, then $T^n x \in \mathcal{B}(J,\varepsilon)$ for $n \geq n_o(\varepsilon,B)$.

An equivalent definition of asymptotically smooth due to Cooperman [6] is contained in the following result.

Lemma 2.1. $T : X \to X$ is asymptotically smooth if and only if for any closed bounded set $B \subseteq X$ such that $TB \subseteq B$, there is a compact set $J \subseteq B$ such that J attracts B.

Proof. Suppose T is asymptotically smooth, $B \subseteq X$ is bounded and $TB \subseteq B$. Then $J \subseteq B$ attracts B. Conversely, if B is any closed bounded set, $L = \{x : T^n x \in B \text{ for } n \geq 0\}$, then $TL \subseteq L$ $T\bar{L} \subseteq \bar{L}$, $\bar{L}$ is closed. The hypothesis implies there is a compact set J in $\bar{L}$ that attracts $\bar{L}$. This shows T is asymptotically smooth.

<u>Definition</u> 2.2. An invariant set J is stable if, for any neighborhood V of J, there is a neighborhood U of J such that $T^n x \in V$, $n \geq 0$, if $x \in V$. The set J is asymptotically stable if it is stable and there is a neighborhood W of J such that J attracts W.

<u>Remark</u> 2.1. An equivalent definition of stable is the following. An invariant set J is stable, if for any neighborhood V of J, there is a neighborhood $V' \subset V$ of J such that $TV' \subseteq V'$. In fact, it is clear that this definition implies the previous one. Conversely, let U be such that $T^n x \in V$ if $x \in U$ and define $V' = \bigcup_{n\geq 0} T^n U$. Then $V' \subseteq V$ is open $J \subseteq V'$ and $TV' \subseteq V'$.

With this remark, relative to the set V', asymptotic stability corresponds to point dissipative and uniform asymptotic stability to local dissipative.

In the theory of dissipative processes, the existence of a maximal compact invariant set is fundamental since it contains the limit sets of all orbits of T. The first result in this direction is contained in the following theorem.

<u>Theorem</u> 2.1. If $T : X \to X$ is continuous and there is a compact set K which attracts compact sets of X and $J = \bigcap_n T^n K$, then

(i) J is independent of K;

(ii) J is maximal, compact, invariant;

(iii) J is stable and attracts compact sets of X.

If, in addition, T is asymptotically smooth, then

(iv) for any compact set $H \subseteq X$, there is a neighborhood H_1 of H such that $\bigcup_{n\geq 0} T^n H_1$ is bounded and J attracts H_1. In particular, J is uniformly asymptotically stable.

<u>Proof</u>. We give an outline of the proof of this theorem. The first remark is that the orbit $\gamma^+(H) = \bigcup_{n\geq 0} T^n H$ through H is precompact

for any compact set H. Since this is true, the ω-limit set $\omega(K)$ of K exists, is nonempty, compact and invariant. Furthermore, $\omega(K) = \bigcup_{n\geq 0} T^n K \overset{\text{def}}{=} J(K)$ since $\omega(K) \subseteq K$. If K_1 is any other compact set that attracts compact sets of X, then one shows that $J(K) \subseteq K_1$, $J(K_1) \subseteq K$ and $J(K) \subseteq T^n K_1$, $J(K_1) \subseteq T^n K$ for all n. Thus $J(K)$ is independent of K. In a similar way, one shows J is maximal. This proves (i), (ii).

To prove J is stable, suppose the contrary. Then the compactness of J implies there is an $\varepsilon > 0$ (as small as desired), a sequence of integers $n_j \to \infty$, $y_j \to y \in J$ as $j \to \infty$ such that $T^n y_j \in \mathcal{B}(J,\varepsilon)$, $0 \leq n \leq n_j$, $T^{n_j+1} y_j \notin \mathcal{B}(J,\varepsilon)$, where $\mathcal{B}(J,\varepsilon)$ is the ε-neighborhood of J. The set $\{y, y^j, j \geq 1\} = H$ is compact, $\gamma^+(H)$ is precompact, J attracts H and $\omega(H) \subseteq J$. Thus, we may assume $T^{n_j} y_j \to z$ as $j \to \infty$. But $z \in \omega(H) \subseteq J$ which is a contradiction since $Tz \in \omega(H) \subseteq J$ and $Tz \notin \mathcal{B}(J,\varepsilon)$.

To prove J attracts compact sets, suppose H is an arbitrary compact set. Then $\gamma^+(H)$ is precompact, $\omega(H)$ exists and $\omega(H) \subseteq J$. Since $\omega(H)$ attracts H, the set J attracts H. This proves J attracts compact sets.

Since J is stable, for any $\varepsilon > 0$, $\exists\ \delta = \delta(\varepsilon)$, such that $x \in \mathcal{B}(J,\delta)$ implies $T^n x \in \mathcal{B}(J,\varepsilon)$ for $n \geq 0$. Since J attracts points of X, for any compact set H of X, there is an integer $N(H,\varepsilon)$ and a neighborhood H_1 of H such that $T^{N(H,\delta)} H_1 \subset \mathcal{B}(J,\delta)$. Consequently, $T^n H_1 \subseteq \mathcal{B}(J,\varepsilon)$ for $n \geq H(H,\delta)$. Thus, $\mathcal{B}(J,\varepsilon)$ dissipates neighborhoods of compact sets. Consequently, for any compact set $H \subset X$, there is a neighborhood H_1 of H such that $\bigcup_{n\geq 0} T^n H_1$ is bounded and $T^n H_1 \subseteq \mathcal{B}(J,\varepsilon)$ for $n \geq N(H,\varepsilon)$. Let $B = \{x \in \mathcal{B}(J,\varepsilon), T^n x \in \mathcal{B}(J,\varepsilon), n \geq 0\}$. Then $T\overline{B} \subseteq B$ and asymptotic smoothness implies there is a compact set $K \subseteq \overline{B}$ such that K attracts $\overline{B}$. But then K attracts H_1. Obviously, $K \subseteq J$ and the theorem is proved.

Corollary 2.1. For asymptotically smooth maps T, the following statements are equivalent:

(i) There is a compact set K which attracts compac sets of X.

(ii) There is a compact set K which attracts neighborhoods of compact sets of X.

In [10, Theorem 3.1, p. 84], it was stated that conclusion (iv) in Theorem 2.1 was valid without the hypothesis that T is asymptotically smooth. Cooperman [7] pointed out that the proof in [9] yielded only properties (i)-(iii). He also gave an example of a linear operator in a Hilbert space for which conclusion (iv) is not valid without some additional smoothness hypothesis on T. Another counterexample is also available for a specific linear autonomous neutral differential difference equation. In fact, Brumley [6] gave such a linear equation for which the zero solution is asymptotically stable but not uniformly asymptotically stable.

Conclusion (iv) in Theorem 2.1 is due to Cooperman [7].

To apply Theorem 2.1 in the applications, one must have some practical procedure for the verification of the hypotheses. A step in this direction is the following result of Cooperman [7].

Theorem 2.2. If T is asymptotically smooth and T is compact dissipative, then there exists a compact invariant set which attracts compact sets and the conclusions of Theorem 1 hold. In addition, if S is any family of sets of X such that $\bigcup_{n\geq 0} T^n A$ is bounded for each $A \in S$, then the maximal compact set J in Theorem 2.1 attracts elements of S.

Proof. Let C be a bounded set which dissipates compact sets. Let $B = \{x \in C : T^n x \in C, \ n \geq 0\}$. Then $T\overline{B} \subseteq \overline{B}$ and there is a compact set K which attracts $\overline{B}$. This set K attracts compact sets. Now use Theorem 2.1. For the proof of the last part, follow the argument used in the proof of (iv) of Theorem 2.1.

<u>Remark</u> 2.2. It is not difficult to show that compact dissipative is equivalent to local compact dissipative which is equivalent to local dissipative. However, compact dissipative alone has nothing to do with the fact that there exists a compact set which attracts compact sets.

<u>Corollary</u> 2.2. Suppose T is asymptotically smooth, T is compact dissipative and $\bigcup_{n\geq 0} T^n B$ is bounded for every bounded subset $B \subseteq X$. Then T is bounded dissipative or uniformly ultimately bounded. More specifically, there is a maximal compact invariant set J which attracts bounded sets of X.

To say more about stability, we need the following result (see [7]).

<u>Lemma</u> 2.2. A compact invariant set K is stable and attracts points if and only if K attracts compact sets. In this case, it is necessarily true that K is a maximal compact invariant.

<u>Proof</u>. Suppose K is stable and attracts points. Then K attracts compact sets by the same proof as in the beginning of the proof of part (iv) of Theorem 2.1. That K attracts compact sets implies K is stable and attracts points is the same as the proof of the first part of part (iii) of Theorem 2.1. The last part is almost trivial since, for any compact set L, $T^n L \subseteq B(K,\varepsilon)$ for $n \geq N(L,\varepsilon)$. Thus, L invariant implies $L \subseteq B(K,\varepsilon)$ for every $\varepsilon > 0$ and $L \subseteq K$ since L, K are compact. This proves the lemma.

Suppose K is a compact invariant set which is asymptotically stable. From Remark 2.1, we can find a neighborhood (which may be taken closed) of K such that $Y = \bigcup_{n\geq 0} T^n U$ satisfies $T : Y \to Y$ and K attracts points of Y. From Lemma 2.2, K attracts compact sets of Y. Thus, T is compact dissipative when restricted to Y. From Theorem 2.2, we obtain the following result

Corollary 2.3. If a compact invariant set K is asymptotically stable and T is asymptotically smooth, then K is uniformly asymptotically stable.

Similar results have been obtained by Izé and dos Reis [9].

Corollary 2.3 is very important in the applications because it relates attracting points to attracting neighborhoods of compact sets under the hypothesis of stability and T asymptotically smooth. Therefore, it becomes important to know simple criteria for determining when a map is asymptotically smooth. A general criterion will be given in Section 4. If T is completely continuous, then T is obviously asymptotically smooth. In this case, one can improve on Theorem 2.2 with the following result of Billotti and LaSalle [5].

Theorem 2.3. If $T : X \to X$ is completely continuous and point dissipative, then there is a maximal compact invariant set which attracts bounded sets of X.

Proof. One shows (see Hale [10]) that the hypotheses imply there is a compact set which attracts compact sets. Since T is asymptotically smooth, Theorem 2.1 may be applied. For any bounded set $B \subseteq X$, the set TB is precompact, so that the last conclusion of the theorem holds.

The next result from [7] relates the concept of compact dissipative to the existence of certain types of Liapunov functions

Theorem 2.4. Suppose X is a Banach space, $T : X \to X$ is continuous and asymptotically smooth. Let $V : X \to R$ be a continuous functional such that for all $c \in R$, $\{x \in X : V(x) < c\}$ is bounded. Further, suppose $V(T(x)) < V(x)$ unless x is in some bounded set B, in which case $V(T(x)) \leq V(x)$. Then T is local dissipative which is equivalent to local compact dissipative. If, in addition, $m(\gamma) = \sup_{|x|\leq\gamma} V(x) < \infty$ for each γ, then T is bounded dissipative or uniformly ultimately bounded.

The proof is complicated althought the reasoning is similar to the usual ones in Liapunov theory (see [7]).

3. Skew Product Flows.

In this section, we give some implications of the results of the previous section to the relationship between asymptotic stability and uniform asymptotic stability for nonautonous evolutionary equations through the skew product flow.

This flow is defined in the following manner. Let $\phi_f(t_o,x_o,t)$, $t \geq 0$, $\phi_f(t_o,x_o,0) = x_o$, be the solution of the equation

$$\frac{dx}{dt} = f(t+t_o,x)$$

with $x(0) = x_o$, $x \in X$. The function f is assumed to belong to some metric space of functions F and for simplicity in notation, assume $\phi_f(t_o,x_o,t)$ is defined for all $t \geq 0$. If, for each $t \geq 0$, f_t defined by $f_t(s,x) = f(t+s,x)$ belongs to F, then we can define the semigroup of transformations

$$(3.1)\qquad \begin{aligned} &U(t) : X \times F \to X \times F \\ &U(t)(x,f) = (\phi_f(0,x;t),f_t), \quad t \geq 0. \end{aligned}$$

If we assume that $U(t)$ is a strongly continuous semigroup, then this is called the *skew product flow* associated with the nonautonomous equation

$$(3.2)\qquad \frac{dx}{dt} = f(t,x).$$

One of the interesting aspects of skew product flows is to determine those properties of the space F which will imply that $U(t)$ is strongly continuous. To obtain precise information about (3.2) from the associated skew product flow, authors have always imposed the condition that $H(f) = \text{closure}\{f_t,\ t \geq 0\}$ is a compact set of F. The best conditions which ensure this latter property are nontrivial. The paper of Artstein [2] gives a very complete discussion of the spaces F which satisfy the above properties when the underlying space X is R^n. Palmer [17] treats the analogous problemas when (3.2) is a functional differential equation on R^n. We do not discuss these problems here, but only point out a few general implications of the previous theory to skew product flows. However, it should be remarked that $f(t,x)$ periodic in t, almost periodic in t uniformly in x and $f(t,x)$ asymptotic in an appropriate sense to a function $g(x)$ are special cases where $H(f)$ is compact.

The results below are based on Cooperman [7].

Lemma 3.1. Let X be a Banach space, $f : R \times X \to X$ be a function such that $f \in F$, $U(t)$ in (3.1) be asymptotically smooth and $H(f)$ be compact. If $f(t,0) = 0$, then the solution $x = 0$ of (3.2) is uniformly stable (uniformly asymptotically stable) if and only if $\{0\} \times H(f)$ is stable (uniformly asymptotically stable) for the skew product flow.

Proof. Suppose the solution $x = 0$ of (3.2) is uniformly stable and let $F = \{f_t, t \geq 0\}$. Then, for any neighborhood V of $x = 0$, there is a neighborhood U of $x = 0$ such that $\gamma^+(U \times F) \subseteq V \times F$. Since F is dense in $H(f)$, continuity implies $\gamma^+(U \times H(f)) \subseteq V \times H(f)$. This proves $\{0\} \times H(f)$ is stable. The converse is trivial.

Now, suppose the solution $x = 0$ of (3.2) is uniformly asymptotically stable. Then $x = 0$ is uniformly stable and $\{0\} \times H(f)$ is uniformly stable for the skew product flow. Also, there is a neighborhood W of $x = 0$ such that for any neighborhood V of $x = 0$ there is a $T > 0$ such that, for any $t_o \geq 0$, $x_o \in W$ we have $\phi_f(t_o,x_o,t) \in V$ for $t \geq T$; that is, $\cup_{t \geq T} U(t)(W \times F) \subseteq V \times F$. Consequently, $\cup_{t \geq T} U(t)(W \times H(f)) \subseteq V \times H(f)$ and we have $\{0\} \times H(f)$ is uniformly asymptotically stable for the skew product flow.

Theorem 3.1. Let X be a Banach space, $f : R \times X \to X$, $f \in F$,

$f(t,0) = 0$, $U(t)$ in (3.1) be asymptotically smooth and $H(f)$ be compact. The solution $x = 0$ of (3.2) is uniformly asymptotically stable if and only if $x = 0$ is uniformly stable and there is a neighborhood U of $x = 0$ such that, for all $x_o \in U$, and all $g \in H(f)$, $\phi_g(0,x_o,t) \to 0$ as $t \to \infty$.

Proof. Suppose the latter statement of the theorem is satisfied. Since $x = 0$ is uniformly stable, Lemma 3.1 implies $\{0\} \times H(f)$ is stable. We may choose U so that $Z = \text{closure}\{\gamma^+(U \times H(f))\}$ is bounded and $\{0\} \times H(f)$ attracts points of Z. A simple compactness argument shows that $\{0\} \times H(f)$ attracts compact sets of Z. Theorem 2.2 implies $\{0\} \times H(f)$ attracts bounded sets of Z since $\{0\} \times H(f)$ is maximal compact invariant with respect to Z. This proves $\{0\} \times H(f)$ is uniformly asymptotically stable. Lemma 3.1 implies the solution $x = 0$ is uniformly asymptotically stable. Conversely, if $x = 0$ is uniformly asymptotically stable, the $\{0\} \times H(f)$ is uniformly asymptotically stable from Lemma 3.1 and this clearly implies the last assertion of the theorem. The theorem is proved.

This generalizes a result of Artstein [3] corresponding to $x \in R^n$. The conclusion of Theorem 3.1 is also true if the skew product flow arises from functional differential equations as in Palmer [17]. Applications may be found in Artstein [3], Palmer [17].

4. Criteria for Asymptotically Smooth Maps.

To apply the previous results to differential equations, we need some easy criteria to determine when a map T is asymptotically smooth. As remarked earlier, any completely continuous map is asymptotically smooth. Completely continuous maps are sufficiently general to treat most problems of asymptotic behavior for ordinary differential equations, quasilinear parabolic equations and retarded equations with finite retardations. Retarded equations with infinite delay, neutral equations, and certain hyperbolic problems require the more sophisticated concept of β-contraction which will now be discussed.

Definition 4.1. A measure of noncompactness β on a metric space X is a function β from the bounded sets of X to the nonnegative real numbers satisfying

(i) $\beta(A) = 0$ for $A \subseteq X$ if and only if A is precompct.

(ii) $\beta(A \cup B) = \max[\beta(A), \beta(B)]$.

The Kuratowskii measure of noncompactness α is defined by

$$\alpha(A) = \inf\{d : A \text{ has a finite cover of diameter } < d\}.$$

Definition 4.2. A continuous map $T : X \to X$ is a β-contraction of order $k < 1$ with respect to the measure of noncompactness β if $\beta(TA) \leq k\beta(A)$ for all bounded sets $A \subseteq X$. The map T is β-condensing if $\beta(TA) < \beta(A)$ for all bounded sets $A \subseteq X$ with $\beta(A) > 0$.

The following result is due to Hale and Lopes [13].

Theorem 4.1. β-contractions are asymptotically smooth.

Proof. Suppose $B \subseteq X$ is bounded, $TB \subseteq B$. Then T an α-contraction implies $\gamma^+(x)$ is precompact for any $x \in B$. Let $B^* = \bigcup_{x \in B} \omega(x)$. Then $T\overline{B}^* = \overline{B}^*$ and $\alpha(T\overline{B}^*) \leq k\,\alpha(B^*)$, $k \in [0,1]$ implies $\overline{B}^*$ is compact. But $\overline{B}^*$ attracts B and the theorem is proved.

Theorem 4.2. β-condensing maps are asymptotically smooth.

Proof. We begin by showing that, for any bounded $B \subset X$ with $TB \subseteq B$ and $\{k_j\}$ integers with $k_j \to \infty$, the set $\{T^{k_j}x_j\}$ is precompact. Let $C = \{\{T^{k_j}x_j\} : \{x_j\} \subseteq B, \{k_j\} \text{ integers}, k_j \to \infty\}$. Let $\eta = \sup\{\beta(h) : h \in C\}$. Since $h \subseteq B$ if $h \in C$, we have $\eta < \beta(B)$. We claim there is an $h^* \in C$ such that $\beta(h^*) = \eta$. Let

$\{h_\ell\}$ be a sequence of elements of C such that $\beta(h_\ell) \to \eta$. Define $\tilde{h}_\ell = \{T^{k_j}x_j : T^{k_j}x_j \in h_\ell,\ k_j > \ell\}$. Then $\beta(\tilde{h}_\ell) = \beta(h_\ell)$. If $\tilde{h}^* = \cup_\ell \tilde{h}_\ell$ ordered in any way, then $h^* \in C$ and $\beta(h^*) \geq \beta(h_\ell)$ for each ℓ. So $\beta(h^*) = \eta$. Let $\tilde{h}^* = \{T^{k_j-1}x_j : T^{k_j}x_j \in h^*\}$. Then $\tilde{h}^* \in C$ and $\beta(\tilde{h}^*) \leq \eta$. But $T(\tilde{h}^*) = h^*$ so $\beta(T\tilde{h}^*) \geq \beta(\tilde{h}^*)$ which implies $\tilde{\beta}(\tilde{h}^*) = 0$ since T is β-condensing. Thus, $\tilde{h}^*$ and h^* are precompact and $\eta = 0$. Thus, every sequence in C is precompact.

This shows $\omega(B) = \cap_{n\geq 0} T^n B$ is nonempty. Since it is invariant, $T\omega(B) = \omega(B)$ and it follows that $\omega(B)$ is compact. The set $\omega(B)$ always attracts B and the theorem is proved.

Theorem 4.2 is due to Massatt [15]. Exploiting the ideas in the above proof, Massatt [15] has been able to prove certain results on continuous dependence of fixed points on parameters generalizing results of Hale [11], Artstein [4]. Similar results were first obtained by Cooperman for β-condensing mappings using the following interesting concept.

Definition 4.3. Let β be a general measure of noncompactness on a metric space X. The derived measure of noncompactness β' on X is defined by

$$\beta'(A) = \sup\{\beta(B) : B \text{ is a countable subset of } A\}.$$

Using this concept, Cooperman [7] proved the following

Lemma 4.1. Suppose X, Y are metric spaces and $T : X \to Y$ is α-condensing. If $A_1 \supseteq A_2 \supseteq \ldots$, $\alpha(A_j) \to \delta$, $\alpha(T(A_j)) \to \delta$ as $j \to \infty$, then $\delta = 0$.

This lemma was systematically used by Cooperman [7] to prove the above results for α-condensing maps. Massatt [15] has shown that Lemma 4.1 does not hold for general measures of noncompactness and, thus, the method of Cooperman could not be used for general measures of noncompactness. However, the Lemma 4.1 is very useful for other purposes. In fact, Cooperman used this lemma to prove the following interesting fixed point theorem.

Theorem 4.3. If E is a Banach space $\dim E = \infty$, $S = \{x \in E : |x| = 1\}$ and $T : S \to S$ is α-condensing, then T has a fixed point in S.

The proof was presented at the Symposium on Functional Differential Equations and Approximation of Fixed Points, Bonn, Germany, July, 1978.

5. Dependence on Parameters.

Suppose $T : \Lambda \times X \to X$ is continuous, Λ is a metric space and X is a complete metric space. Also suppose $T(\lambda,\cdot) : X \to X$ has a maximal compact invariant set $J(\lambda)$ for each $\lambda \in \Lambda$. Our objective in this section is to give sufficient conditions on T which will ensure that $J(\lambda)$ is upper semi-continuous in λ. We say $T : \Lambda \times X \to X$ is *collectively* β-*condensing* if for all bounded sets B, $\beta(B) > 0$, one has $\beta(\bigcup_{\lambda\in\Lambda} T(\lambda,B)) < \alpha(B)$. The following result is contained in [7].

Theorem 5.1. Let X be a complete metric space, Λ a metric space, $T : \Lambda \times X \to X$ continuous and suppose there is a bounded set B independent of $\lambda \in \Lambda$ such that B is compact dissipative under $T(\lambda,\cdot)$ for every $\lambda \in \Lambda$. If T is collectively β-condensing, then the maximal compact invariant set $J(\lambda)$ of $T(\lambda,\cdot)$ is upper semi-continuous in λ.

Proof. Without loss in generality, suppose B is closed. Theorems 4.2 and 2.2 imply each $J(\lambda)$ exists. The hypothesis implies $J(\lambda) \subseteq B$ and $V \overset{\text{def}}{=} \bigcup_{\lambda\in\Lambda} J(\lambda)$ is bounded. Also, $J(\lambda)$ invariant implies

$$\beta(V) = \beta(\bigcup_{\lambda\in\Lambda} T(\lambda,J(\lambda))) \le \beta(\bigcup_{\lambda\in\Lambda} T(\lambda,V)).$$

Since T is collectively β-condensing, this implies V is precompact.

We must show that, if $\{\lambda_j\} \subseteq \Lambda$, $\{x_j\} \subseteq X$ are sequences such that $x_j \in J(\lambda_j)$, $\lambda_j \to \tilde{\lambda}$, $x_j \to \tilde{x}$, then $\tilde{x} \in J(\tilde{\lambda})$. Since V is precompact and B is closed, we may assume $x_j' \in J(\lambda_j)$, $x_j = T(\lambda_j, x_j')$, $x_j' \to \tilde{x}' \in B$ as $j \to \infty$. Since T is continuous $T(\tilde{\lambda}, \tilde{x}') = \tilde{x}$. Now $\gamma^+(\tilde{x})$ is invariant under $T(\tilde{\lambda}, \cdot)$ and precompact which implies $\gamma^+(\tilde{x}) \subset J(\tilde{\lambda})$ and, in particular, $\tilde{x} \in J(\tilde{\lambda})$. This proves the theorem.

An interesting application of Theorem 5.1 gives a proof of a result of Walther [19]. Consider the scalar equation

$$(5.1) \quad \dot{x}(t) = -x(t-1)[1+x(t)], \quad x(0) > -1,$$

where $0 < \lambda < \pi/2$. Let $X = \{\phi \in C([-1,0], R) : \phi(0) > -1\}$. For any $\phi \in X$, let $x(\phi)$ designate the solution through ϕ and define $T_\lambda \phi = x_1(\phi)$. The map T_λ is completely continuous and collectively α-condensing. For any $\lambda > 0$, the solution $x(\phi)(t)$ satisfies $-1 < x(\phi)(t) < e^\lambda - 1$ for $t \geq t_o(\phi)$. Thus, T is point dissipative. It is compact dissipative from Theorem 2.3. For any given $\lambda_o \in (0, \pi/2)$, there is a neighborhood V of λ_o and a bounded set B such that B is compact dissipative for each $\lambda \in V$. Theorem 5.1 implies the maximal compact invariant set $J(\lambda)$ is upper semicontinuous. Let $S = \{\lambda \in (0, \pi/2) :$ all solutions of the above equation approach zero as $t \to \infty\}$. Since the solutions z of $z + \lambda \exp(-z) = 0$ satisfy $\operatorname{Re} z < 0$ for each $\lambda \in (0, \pi/2)$, the solution $x = 0$ of Equation (1) is uniformly asymptotically stable. If $\lambda_o \in S$, it follows that the set $J(\lambda_o) = \{0\}$. Since $J(\lambda)$ is upper semicontinuous, $J(\lambda) = \{0\}$ for λ in a neighborhood of λ_o and S is an open set. The set S is also nonempty. This may be proved by the method of integral manifolds near $\lambda = 0$ (see Kurzweil [14]) or by referring to Wright [20] where be proved the interval $(0, 37/24) \subseteq S$. It is not known if S is closed so that $S = (0, \pi/2)$. It is known that S cannot contain any $\lambda > \pi/2$

since there is a nonconstant periodic solution of Equation (5.1).

6. Implications of Point Dissipative.

In this section, we assume T is point dissipative and impose further conditions on T which will imply a stronger form of dissipation.

In Theorem 2.3, we gave the following results of Billotti and LaSalle [5].

Theorem 6.1. If $T : X \to X$ is completely continuous and point dissipative, then there is a maximal compact invariant set which attracts bounded sets of X.

It is not known if the above result is true for T a β-contraction. If it were known that T a point dissipative β-contraction implies the existence of a maximal compact invariant set J, then J would be stable and attract points of X (this is not trivial, see Cooperman [7]). Thus, K would attract compact sets. Since T is asymptotically smooth, this would imply J attracts neighborhoods of compact sets. Thus, Theorem 2.2 would imply J attracts bounded sets of X if the orbits of bounded sets are bounded.

From the above discussion, we see that the validity of Theorem 6.1 for β-contractions depends essentially upon showing that point dissipative implies the existence of a maximal compact invariant set and we pose this as an interesting unsolved problem.

Problem. If $T : X \to X$ is a point dissipative β-contraction, does there exist a maximal compact invariant set ?

We now give a result of Massatt [16] which asserts that point dissipative in one space may imply bounded dissipative in another space provided the map T satisfies some additional hypotheses.

Theorem 6.2. Suppose X_1, X_2 are Banach spaces with norms $|\cdot|_1$, $|\cdot|_2$ and X_1 compactly imbedded in X_2. Suppose $T, C, U : X_j \to X_j$,

$j = 1,2$, are continuous operators, $T = C + U$, $C(0) = 0$, C a contraction on X_1, $U : (X_1,\tau_2) \to (X_1,\tau_1)$ takes bounded sets to bounded sets, where τ_j is the topology on X_j, $j = 1,2$. If $B_R^j = \{x \in X_j : |x|_j < R\}$, then the following conclusions hold:

(i) For any $L > 0$, $R > 0$, there is a $K = K(L,R)$ such that, for any $A \subseteq B_L^1$ and any n, $0 \le n \le \infty$, the relation $\bigcup_{0\le m\le n} T^m(A) \subseteq B_R^2$ implies $\bigcup_{0\le m\le n} T^m(A) \subset B_K^1$.

(ii) For any $L > 0$, $R > 0$, there is an $n_1(L,R)$ and $Q(R)$ such that for any n, $0 \le n \le \infty$, if $\bigcup_{0\le m\le n} T^m(A) \subseteq B_L^1 \cap B_R^2$ then $\bigcup_{n_1(L,R)\le m\le n} T^m(A) \subseteq B_{Q(R)}^1$.

(iii) If T is point dissipative in X_2, then T is bounded dissipative in X_1.

<u>Proof</u>. (i) There is a continuous function $h : [0,\infty) \to [0,\infty)$ such that $U(X_1 \cap B_R^2) \subseteq B_{h(R)}^1$. Let $\lambda \in [0,1)$ be the contraction constant for C, $K = K(L,R) = (1-\lambda)^{-1}(h(R) + L)$. Suppose $A \subseteq B_L^1$ and $\bigcup_{0\le m\le n} T^m(A) \subseteq B_R^2$. Since $K(L,R) \ge L$, we have $T^0(A) = A \subseteq B_K^1$. Suppose $0 \le j < n$ is given and we know that $T^j(A) \subseteq B_K^1$. We want to show that $T^{j+1}(A) \subseteq B_K^1$. For $x \in A$, we have

$$
\begin{aligned}
|T^{j+1}x|_1 &\le \lambda|T^j x|_1 + h(R) \\
&\le \lambda K + h(R) \\
&\le K + h(R) + L = K.
\end{aligned}
$$

This proves $\bigcup_{0\le m\le n} T^m(A) \subset B_K^1$.

(ii) For any $\delta_0 > 0$, let $Q(R) = (1-\lambda)^{-1}h(R) + \delta_0$. If $A \subseteq X_j$ is bounded, let $|A|_j = \sup\{|x|_j : x \in A\}$. Let $K(0,R) = (1-\lambda)^{-1}h(R)$. If $L \le K(0,R)$, then the same argument as in (i) proves the assertion with $n_1(L,R) = 0$. If $L > K(0,R)$, then observing that

$$|T^j B_L^1|_1 \le \lambda L + h(R) = \lambda(L-K(0,R)) + K(0,R)$$

one obtains by induction

$$|T^j B_L^1|_1 \le \lambda^j (L-K(0,R)) + K(0,R).$$

If n_1 is chosen so that $\lambda^{n_1}(L-K(0,R)) < \delta_o$, then the assertion in (ii) follows.

(iii) Let B_R^2 dissipate points in X_2. By part (i), for any $x \in X_1$, the orbit through x, $\gamma^+(x)$, is bounded in X_1. Furthermore, part (ii) implies $B_{Q(R)}^1$ dissipates points in X_1.

We next show that $\gamma^+(B_{Q(R)}^1) = \bigcup_{n\ge 0} T^n B_{Q(R)}^1$ is bounded in X_1. For any $x \in B_{Q(R)}^1$, the boundedness of $\gamma^+(x)$ in X_2 implies there is a $c(x)$ such that $\gamma^+(x) \in B_{c(x)}^2$ and so $\gamma^+(x) \in B_{K(Q(R),c(x))}^1$ by part (i). Let $n_o(x)$ be chosen so that $n > n_o(x)$ implies $T^n x \in B_R^2$, $n_1(x) = n_1(K(Q(R),c(x)),R)$, $n^*(x) = n_o(x) + n_1(x)$. For any $\delta > 0$, let $V(\delta,x) = B_{Q(R)}^1 \cap \{y : |y-x|_2 < \delta\}$. By continuity, there is a $\delta(x) > 0$ such that

$$\bigcup_{0\le m\le n^*(x)} T^m V(\delta(x),x)) \subset B_{c(x)}^2.$$

$$\bigcup_{n_o(x)\le m\le n^*(x)} T^m V(\delta(x),x)) \subset B_R^2.$$

Part (i) implies $\bigcup_{0\le m\le n^*(x)} T^m V(\delta(x),x)) \subset B_{K(Q(R),c(x))}^1$ and part (ii) implies $T^{n^*(x)} V(\delta(x),x)) \subset B_{Q(R)}^1$.

Since $B_{Q(R)}^1$ is a compact set in X_2, the sets $V(\delta(x),x)$ form an open cover of $B_{Q(R)}^1$ for which there is a finite subcover $V(\delta(x_i),x_i)$, $i = 1, 2, \ldots, p$. If $N = \max\{n^*(x_i),\ i = 1,2,\ldots,p\}$, then

$$\gamma^+(B_{Q(R)}^1) \subset \bigcup_{0\le m\le N} T^m (B_{Q(R)}^1) \subseteq X_1$$

since any point $B_{Q(R)}^1$ returns to $B_{Q(R)}^1$ by the Nth iteration. This proves $\gamma^+(B_{Q(R)}^1)$ is bounded in X_1.

One uses a similar argument to show that $\gamma^+(B_{Q(R)}^1)$ dissipates bounded sets in X_1 to complete the proof of the theorem.

Corollary 6.1. If the assumptions of Theorem 6.2 are satisfied, T is point dissipative in X_2 and β-condensing in X_1, then there is a maximal compact invariant set in X_1 which is stable and attracts bounded sets.

Proof. This is a consequence of Theorem 6.2, Theorem 2.1 and Theorem 2.2.

Corollary 6.2. If the hypotheses of Corollary 6.1 are satisfied and the measure of noncompactness satisfies the additional property $\beta(\overline{co}\ A) = \beta(A)$ where $\overline{co}$ is the closed convex hull then T has a fixed point.

This is a consequence of Theorem 6.2 and the following fixed point theorem of Hale and Lopes [13] and Massatt [15] which is stated without proof.

Theorem 6.3. If X is a Banach space, $T : X \to X$ is β-condensing and compact dissipative, then T has a fixed point. Here the measure of noncompactness satisfies $\beta(A) = 0$ if A is precompact, $\beta(A \cup B) = \max[\beta(A), \beta(B)]$, $\beta(\overline{co}\ A) = \beta(A)$.

For neutral functional differential equations with a stable D operator, it is known (see Hale [10]) that the solution operator on the initial space C of continuous functions is a contraction plus a completely continuous operator. It can also be shown the same is true on W_1^∞. The previous results are now immediately applicable with $X_1 = W_1^\infty$, $X_2 = C$. One obtains point dissipative in C implies bounded dissipative in W_1^∞. Also, for a periodic system of period p, point dissipative implies the existence of a periodic solution of the same period p.

For retarded equations with infinite delay satisfying the axioms in Hale and Kato [12], one also knows the solution operator is a contraction plus a completely continuous operator. To apply the previous results one needs to have the equation well defined on two different Banach spaces of initial data corresponding to X_1, X_2. See Massatt [16] for some illustrations of appropriate initial spaces.

REFERENCES

[1] - AMANN, H., Periodic solutions of semilinear parabolic equations. pp. 1-29 in Nonlinear Analysis, Academic Press, 1978.

[2] - ARTSTEIN, Z., The limiting equations of nonautonomous differential equations, J. Differential Eqs., 25(1977), 184-201.

[3] - ARTSTEIN, Z., Uniform asymptotic stability via the limiting equations, J. Differential Eqs., 27(1978), 172-189.

[4] - ARTSTEIN, Z., On continuous dependence of fixed points condensing maps. Dynamical Systems, An International Symposium, Vol. II, 73-75.

[5] - BILLOTTI, J. and LASALLE, J.P., Periodic dissipative processes. Bull. Am. Math. Soc., 6(1971), 1082-1089.

[6] - BRUMLEY, W.E., On the asymptotic behavior of solutions of differential-difference equations of neutral type, J. Differential Eqs., 7(1970), 175-188.

[7] - COOPERMAN, G.D., α-condensing maps and dissipative systems, Ph.D. Thesis, Brown University, June, 1978.

[8] - DAFERMOS, C., Semiflows associated with compact and uniform processes, Math. Systems Theory, 8(1974), 142-149.

[9] - IZÉ, A.F. and DOS REIS, J.G., Contributions to stability of neutral functional differential equations, J. Differential Eqs., Vol. 29, 1(1978).

[10] - HALE, J.K., Theory of Functional Differential Equations, Appl. Math. Series, Vol. 3, 2nd edition, Springer-Verlag, 1977.

[11] - HALE, J.K., Continuous dependence of fixed points of condensing maps, J. Math. An. Appl., 46(1974), 388-394.

[12] - HALE, J.K. and KATO, J., Phase space for retarded equations with infinite delays, Funkc. Ekvacioj 21(1978), 11-41.

[13] - HALE, J.K. and LOPES, O.F., Fixed point theorems and dissipative processes, J. Differential Eqs., 13(1973), 391-402.

[14] - KURZWEIL, J., Global solutions of functional differential equations, In Lecture Notes in Math., Vol. 144, Springer--Verlag, 1970.

[15] - MASSATT, P., Some properties of condensing maps, Annali di Mat. Pura Appl.. To appear.

[16] - MASSATT, P., Stability and fixed points of point dissipative systems, J. Differential Eqs.. Submitted.

[17] - PALMER, J.W., Liapunov stability theory for nonautonomous functional differential equations, Ph.D. Thesis, Brown University, June, 1978.

[18] - SELL, G., Lecture on Topological Dynamics and Differential Equations, van Nostrand, 1971.

[19] - WALTHER, H., Stability of attractivity regions for autonomous functional differential equations, Manuscripta Math., 15(1975), 349-363.

[20] - WRIGHT, E.M., A nonlinear differential-difference equation. J. Reine Anafw. Math., 194(1955), 66-87.

VOLTERRA STIELTJES-INTEGRAL EQUATIONS

by Chaim Samuel Hönig

INTRODUCTION

In this article we study linear Volterra Stieltjes-integral equations

$$\text{(K)} \qquad y(t) - x + \int_a^t d_s K(t,s)\cdot y(s) = f(t) - f(a), \qquad a \leq t \leq b$$

in a Banach space context. X is a Banach space and $y,f\colon [a,b] \rightarrow X$ are regulated functions (i.e., functions that have only discontinuities of the first kind). $K(t,s) \in L(X)$ and K satisfies a necessary and sufficient condition that assures that the integral in (K) is a regulated function whenever y is regulated (see Theorem 2.6). As particular instances of equation (K) we have linear delay differential equations (Cf.§4.1), linear Volterra integral equations (Cf.§4.3)

$$y(t) - x + \int_a^t H(t,s)\cdot y(s)ds = f(t) - f(a), \qquad a \leq t \leq b$$

and linear differential equations, or more generally, linear Stieltjes integro-differential equations (Cf.§4.2)

$$\text{(L)} \qquad y(t) - x + \int_{t_0}^t dA(s)\cdot y(s) = f(t) - f(t_0), \qquad a \leq t \leq b.$$

We prove several equivalent necessary and sufficient conditions for the existence of a resolvent of equation (K) (see Theorem 3.4); in particular we prove that the following condition is necessary:

(I) For every $t \in [a,b[$ the operator $I_X + K(t+,t+) - K(t+,t)$ is invertible.

This condition is also sufficient if the operator defined by the kernel

K is compact (see III of Theorem 3.4).

Notice that since K and y in (K) may have common points of discontinuity we have to replace the usual Riemann-Stieltjes operator integral by the interior or Dushnik integral (see §1).

The first proof of the existence of the resolvent for an equation of type (K) was given by Hinton, [13]. He supposes that K is regulated as a function of the first variable and that there exists an increasing function $g: [a,b] \longrightarrow \mathbb{R}$ such that for all $t \in [a,b]$ and $a \leq s_1 < s_2 \leq b$ we have $\|K(t,s_2) - K(t,s_1)\| \leq g(s_2) - g(s_1)$. Hinton works with the left and right Riemann-Stieltjes integrals and obtains the resolvent as a Neumann series (see [13],Theorem 3.1). His results were extended by Bitzer, [2].

In the case where $\dim X < \infty$ and the kernel K is of finite two-dimensional Vitali variation, Schwabik ([17],Theorem 3.1 and [18],Theorem 4) gave a necessary and sufficient condition (equivalent to (*I*)) for (K) to have a resolvent (y and f are functions of bounded variation and the integral he works with is equivalent to the Young integral - Cf.§4.4).

In [4], Theorems III.1.5, III.1.22 etc., we gave sufficient conditions for the existence of a resolvent of (K) in the case where K as a function of the second variable is continuous and of bounded semivariation (see §1). Our results were extended, in a noticible way, by Arbex in [1]. In [3] Gomes extends the results of Hinton and Schwabik.

In [10] we apply the results of this article to equation (L) and in particular we characterize completely its resolvent. For results on equation (K) with linear constraints, i.e., generalized boundary conditions, see [4]; for results on the adjoint system see [11]. Elsewhere we will apply the results of this article to the study of the stability of solutions, regularity of solutions, convolution equations, non-linear equations etc. Finally let us mention that the closed interval [a,b] we work with may be infinite and that along this article we mention several open problems.

Let us give a birdseye view of the content of this article. In §§ 1 and 2 we bring the auxiliary results we need. Our main results are in §3. In §4 we sketch some applications. The reader that is only interested in the case of bounded variation, or more particularly, in the case where $\dim X < \infty$ should first read the final remark of §3.

§1 - NOTATIONS AND AUXILIARY RESULTS

We follow the notations of [4], [5] and [8] but for some small changes.

We consider always complex vector spaces but all our results are also true for real ones. W, X, Y, Z denote Banach spaces; $L(X,Y)$ denotes the Banach space of all linear continuous mappings $u: X \to Y$. We write $X' = L(X,\mathbb{C})$, and $L(X) = L(X,X)$. By I_X we denote the identical automorphism of X. By χ_E we denote the characteristic function of the set E: $\chi_E(t) = 1$ if $t \in E$ and $\chi_E(t) = 0$ if $t \notin E$.

A *division* of an interval $[a,b]$ is a finite sequence

$$d: t_0 = a < t_1 < t_2 < \ldots < t_n = b.$$

We write $|d| = n$ and $\Delta d = \sup_{1 \le i \le |d|} |t_i - t_{i-1}|$. We denote by $D_{[a,b]}$ or simply by D the set of all divisions of $[a,b]$. We say that a division $\bar{d}$ is *finer* than a division d, and we write $\bar{d} \ge d$, if every point t_i of d is a point of $\bar{d}$. Given points x, $(x_d)_{d \in D}$ of a topological space E, we write $x = \lim_{d \in D} x_d$ if for every neighborhood V of x there exists $d_V \in D$ such that for every $d \in D$ with $d \ge d_V$ we have $x_d \in V$. The meaning of $x = \lim_{\Delta d \to 0} x_d$ is obvious.

If a function $f:[a,b] \to X$ is *regulated* (i.e. if for every $t \in [a,b[$ $[t \in]a,b]]$ there exists $f(t+) = \lim_{\tau \downarrow t} f(\tau)$ $[f(t-) = \lim_{\tau \uparrow t} f(\tau)]$) we write $f \in G([a,b],X)$.

1.1 - $G([a,b],X)$ is Banach space when endowed with the sup norm (i.e., the norm $\|f\| = \sup_{a \le t \le b} \|f(t)\|$) - See [4, Theorem 1.3.6].

1.2 - $G^{-}([a,b],X) = \{f \in G([a,b],X) \mid f(a+) = f(a), f(t-) = f(t) \text{ for } a<t\leq b\}$ is a closed vector subspace of $G([a,b],X)$ - See [4,I.3.11].

We consider as *equivalent* regulated functions f,g such that for every $t\in]a,b]$ we have $f(t-) = g(t-)$ (hence also $f(t+) = g(t+)$).

1.3 - Every equivalence class contains one and only one element of $G^{-}([a,b],X)$. Given $f\in G([a,b],X)$ the element of the equivalence class of f that is in $G^{-}([a,b],X)$ is f^{-}, where $f^{-}(a) = f(a+)$ and $f^{-}(t) = f(t-)$ for $a<t\leq b$ - See [4,p.20].

We say that a function $f:[a,b] \longrightarrow L(X,Y)$ is *simply regulated*, and we write $f\in G^{\sigma}([a,b],L(W,X))$, if for every $w\in W$ we have $f\cdot w\in G([a,b],X)$, where $(f\cdot w)(t) = f(t)w$. From the Banach-Steinhaus theorem it follows then that for every $t\in [a,b[$ $[t\in]a,b]]$ there exists an element of $L(W,X)$, that we denote by $f(t\dot{+})$ $[f(t\dot{-})]$, such that $\lim_{\tau\downarrow t} f(t)w = f(t\dot{+})w$ $[\lim_{\tau\uparrow t} f(t)w = f(t\dot{-})w]$ for every $w\in W$.

From the principle of uniform boundedness it follows that

1.4 - Every simply regulated function is bounded and $G^{\sigma}([a,b],L(W,X))$ is a Banach space when endowed with the sup norm.

1.5 - We have $G([a,b],L(W,X))\subset G^{\sigma}([a,b],L(W,X))$; $G^{\sigma}([a,b],L(W,X)) = G([a,b],L(W,X))$ iff $\dim W < \infty$.

Given a function α: $[a,b] \longmapsto L(X,Y)$ we define its *semivariation* (on $[a,b]$) by

$$SV[\alpha] = SV_{[a,b]}[\alpha] = \sup_{d\in D} SV_d[\alpha]$$

where

$$SV_d[\alpha] = \sup\left\{\Big\| \sum_{i=1}^{|d|} [\alpha(t_i) - \alpha(t_{i-1})]\cdot x_i\Big\| \,\Big|\, x_i\in X, \|x_i\|\leq 1\right\}.$$

If $SV[\alpha] < \infty$ we say that α is a *function of bounded semivariation* and we write $\alpha\in SV([a,b],L(X,Y))$. If we have furthermore $\alpha(c) = 0$ for some $c\in [a,b]$ we write $\alpha\in SV_c([a,b],L(X,Y))$.

1.6 - Every function of bounded semivariation is bounded

PROOF: we have $\|\alpha(t)\| \leq \|\alpha(a)\| + \|\alpha(t)-\alpha(a)\| \leq \|\alpha(a)\| + SV[\alpha]$.

THEOREM 1.7 - $SV_a([a,b],L(X,Y))$ is a Banach space when endowed with the norm $\alpha \mapsto SV[\alpha]$.

PROOF: see Theorem 1.12 that follows or [8, I.3.3].

If $Y = \mathbb{C}$ we have

$$SV_d[\alpha] = V_d[\alpha] = \sum_{i=1}^{|d|} \|\alpha(t_i)-\alpha(t_{i-1})\|$$

hence $V[\alpha] = \sup_{d\in D} V_d[\alpha]$ is the usual *variation* of α. If $V[\alpha] < \infty$ we say that α is a *function of bounded variation* and we write $\alpha \in BV([a,b],X')$. In an analogous way we define $BV([a,b],X)$.

1.8 - We have $BV([a,b],L(X,Y)) \subset SV([a,b],L(X,Y))$; $SV([a,b],L(X,Y)) = BV([a,b],L(X,Y))$ iff $\dim Y < \infty$.

THEOREM 1.9 - Given $\alpha \in SV([a,b],L(X,Y))$ and $f \in G([a,b],X)$ we have

a) There exists the *interior integral*

$$\int_a^b \cdot d\alpha(t)\cdot f(t) = \lim_{d\in D} \sum_{i=1}^{|d|} [\alpha(t_i)-\alpha(t_{i-1})]\cdot f(\xi_i^\bullet) \in Y,$$

where $\xi_i^\bullet \in]t_{i-1},t_i[$.

b) The integral depends only on the equivalence class of f.

c) $\left\| \int_a^b \cdot d\alpha(t)\cdot f(t)\right\| \leq SV[\alpha]\|f^-\| \leq SV[\alpha]\|f\|$.

PROOF: see [4, Theorem I.4.12].

THEOREM 1.10 - For $\alpha \in SV([a,b],L(X,Y))$ and $g \in G^\sigma([a,b],L(W,X))$ there exists the *simple interior integral*

$$I = {}^\sigma\!\!\int_a^b \cdot d\alpha(t)\circ g(t) \in L(W,Y) \quad \text{defined by} \quad Iw = \int_a^b \cdot d\alpha(t)\cdot g(t)w$$

where $w \in W$. We have $\|I\| \leq SV[\alpha]\|g\|$. Sometimes we suppress the sign σ.

THEOREM 1.11 - If $\alpha:[a,b] \to L(X,Y)$ is such that for every $f \in G([a,b],X)$

$[g\in G^{\sigma}([a,b],L(X))]$ there exists

$$\int_a^b \cdot d\alpha(t)\cdot f(t) \qquad \left[{}^{\sigma}\!\int_a^b \cdot d\alpha(t)\circ g(t)\right]$$

then we have $\alpha\in SV([a,b],L(X,Y))$.

PROOF: see [4, Theorem I.4.20].

If α and f have no common points of discontinuity (for instance, if one of them is continuous) the interior integral reduces to the usual Riemann-Stieltjes operator integral

$$\int_a^b d\alpha(t)\cdot f(t) = \lim_{\Delta d\to 0}\sum_{i=1}^{|d|}[\alpha(t_i)-\alpha(t_{i-1})]\cdot f(\xi_i).$$

where $\xi_i\in[t_{i-1},t_i]$ - See [4, Theorem I.2.3]. We recall that in this case we have the integration by parts formula

$$\int_a^b d\alpha(t)\cdot f(t) + \int_a^b \alpha(t)\cdot df(t) = \alpha(b)\cdot f(b) - \alpha(a)\cdot f(a)$$

which, in general, is not true for the interior integral (Cf. Theorem 4.7).

The main justification of the notions of function of bounded semi-variation and interior integral lies in the following generalization of the Riesz representation theorem for $F\in C([a,b])'$:

THEOREM 1.12 - The mapping

$$\alpha\in SV_a([a,b],L(X,Y) \longmapsto F_\alpha\in L[G^-([a,b],X),Y]$$

is an isometry (i.e, $\|F_\alpha\| = SV[\alpha]$) of the first Banach space onto the second, where for every $f\in G([a,b],X)$ we define

$$F_\alpha[f] = \int_a^b \cdot d\alpha(t)\cdot f(t).$$

For $t\in\,]a,b]$ and $x\in X$ we have $\alpha(t)\cdot x = F_\alpha[\chi_{[a,t]}x]$ - See [4, Theorem I.5.1].

THEOREM 1.13 (Helly) - If the sequence $\alpha_n\in SV([a,b],L(X,Y))$ is such that

$SV[\alpha_n] \leq M$ for all n and if there exists $\alpha: [a,b] \longrightarrow L(X,Y)$ such that $\alpha_n(t)x \longrightarrow \alpha(t)x$ for all $t\in[a,b]$ and $x\in X$ then

a) $\alpha\in SV([a,b],L(X,Y))$ and $SV[\alpha] \leq \liminf SV[\alpha_n]$

b) For every $f\in G([a,b],X)$ we have

$$\int_a^b \cdot d\alpha_n(t)\cdot f(t) \longrightarrow \int_a^b \cdot d\alpha(t)\cdot f(t)$$

c) For every $g\in G^\sigma([a,b],L(W,X))$ we have

$$\left[\int_a^{\sigma b} \cdot d\alpha_n(t)\circ g(t)\right]w \longrightarrow \left[\int_a^{\sigma b} \cdot d\alpha(t)\circ g(t)\right]w \quad \text{for all } w\in W.$$

PROOF: See [4, Theorem I.5.8].

§2 - INTEGRAL REPRESENTATION FOR OPERATORS

Given a function $K: T\times S \longrightarrow Z$, for every $(t,s)\in T\times S$ we define

$$K^t(s) = K_s(t) = K(t,s).$$

For $K: [c,d]\times[a,b] \longrightarrow L(X,Y)$ we consider the following properties:

(G^σ) - For every $s\in[a,b]$ we have $K_s\in G^\sigma([c,d],L(X,Y))$

(G) - For every $s\in[a,b]$ we have $K_s\in G([c,d],L(X,Y))$.

(SV^u) - K is uniformly of bounded semivariation as a function of the second variable, i.e., we have $SV^u[K] = \sup_{c\leq t\leq d} SV[K^t] < \infty$.

(SV^u_a) - K satisfies (SV^u) and $K(t,a) = 0$ for every $t\in[c,d]$.

We write $K\in G^\sigma\cdot SV^u_a([c,d]\times[a,b],L(X,Y))$ if K satisfies (G^σ) and (SV^u_a) In an analogous way we define

$$G^\sigma\cdot BV^u_a([c,d]\times[a,b],L(X,Y)),\ G\cdot BV^u_a([c,d]\times[a,b],L(X,Y)),\ \text{etc.}$$

By Theorem 2.2 that follows we have

THEOREM 2.1 - $G^\sigma\cdot SV^u_a([c,d]\times[a,b],L(X,Y))$ is a Banach space when endowed with the norm $K \longmapsto SV^u[K]$.

We have the following representation theorem:

THEOREM 2.2 - The mapping

$$K \in G^{\sigma} \cdot SV_a^u([c,d]\times[a,b],L(X,Y)) \longmapsto F_K \in L[G^-([a,b],X),G([c,d],Y)]$$

is an isometry (i.e., $\|F_K\| = SV^u[K]$) of the first Banach space onto the the second, where for every $f \in G([a,b],X)$ we define

$$F_K[f](t) = \int_a^b \cdot d_s K(t,s)\cdot f(x), \quad c \le t \le d;$$

we have $K(t,s)x = F_K[\chi_{[a,s]}x](t)$ where $s \in]a,b]$, $t \in [c,d]$ and $x \in X$.

PROOF: See [4], Theorem I.5.10 and Remark 8 that follows it.Reciprocally:

THEOREM 2.3 - Let $K:[c,d]\times[a,b] \longrightarrow L(X,Y)$ be such that $K_{s_0} \in G^{\sigma}([c,d],L(X,Y))$ for some $s_0 \in [a,b]$ (for instance, $K_a = 0$) and such that for every $f \in G([a,b],X)$ [$g \in G^{\sigma}([a,b],L(X))$] there exists

$$(F_K f)(t) = \int_a^b \cdot d_\tau K(t,\tau)\cdot f(\tau) \quad [(F_K g)(t) = \sigma\!\!\int_a^b \cdot d_\tau K(t,\tau)\circ g(\tau)]$$

with $F_K f \in G([c,d],Y)$ [$F_K g \in G^{\sigma}([c,d],L(X,Y))$] then we have

$$K \in G^{\sigma}\cdot SV^u([c,d]\times[a,b],L(X,Y)).$$

PROOF: Let us take $s \in [a,b]$, $s > s_0$ and $x \in X$. We define $f = \chi_{[s_0,s]}x$, hence

$$(F_K f)(t) = \int_{s_0}^{s} \cdot\, d_\tau K(t,\tau)x = K(t,s)x - K(t,s_0)x$$

and the function $K_s x$ is regulated. Analogously for $s < s_0$, hence K satisfies (G^{σ})

On the other hand for $t \in [c,d]$ there exists

$$\int_a^b \cdot d_\tau K(t,\tau)\cdot f(\tau),$$

for all $f \in G([a,b],X)$; from Theorem 1.11 it follows that $K^t \in SV([a,b],L(X,Y))$. From the closed graph theorem it follows that the operator F_K is continuous: indeed, if the sequence $f_n \in G([a,b],X)$ is such that $\|f_n\| \longrightarrow 0$ and

$\|F_K f_n - g\| \longrightarrow 0$ we have to prove that $g = 0$ and this follows from the fact that for every $t \in [c,d]$ we have $K^t \in SV([a,b], L(X,Y))$, hence, by c of Theorem 1.9 we have

$$\int_a^b \cdot d_\tau K(t,\tau) \cdot f_n(\tau) \longrightarrow 0,$$

i.e., $g(t) = 0$. Hence $\|F_K\| < \infty$ and therefore $K \in G^\sigma \cdot SV^u([c,d]\times[a,b], L(X,Y))$ since

$$SV^u[K] = \sup_{c\le t\le d} SV[K^t] = \sup_{c\le t\le d} \sup\left\{ \left\| \int_a^b \cdot d_\tau K(t,\tau)\cdot f(\tau) \right\| \;\middle|\; f \in G([a,b],X), \|f\| \le 1 \right\}$$

$$= \sup\left\{ \sup_{c\le t\le d} \|(F_K f)(t)\| \;\middle|\; f\in G([a,b],X), \|f\|\le 1\right\}$$

$$= \sup\left\{ \|F_K f\| \;\middle|\; f\in G([a,b],X), \|f\|\le 1\right\} = \|F_K\|.$$

In the second case the proof is analogous (take $g = \chi_{[s_0,s]} I_X$, etc.)

For $K \in G^\sigma \cdot SV^u([a,b]\times[a,b], L(X,Y))$ we write $K \in G^\sigma_\Delta \cdot SV^u([a,b]\times[a,b], L(X,Y))$ if K satisfies the property

(G^σ_Δ) - K satisfies (G^σ) and K is simply regulated along the diagonal, i.e., for every $x \in X$ the function $K_\Delta x : t \in [a,b] \longmapsto K(t,t)x \in Y$ is reguted.

Given $K \in G^\sigma \cdot SV^u([a,b]\times[a,b], L(X,Y))$ and $t_0 \in [a,b]$ for every $f \in G([a,b],X)$ $[g \in G^\sigma([a,b], L(W,X))]$ we define

$$(k_{t_0} f)(t) = \int_{t_0}^t \cdot\, d_\tau K(t,\tau)\cdot f(\tau) \quad \left[(k_{t_0} g)(t) = {}^\sigma\!\!\int_{t_0}^t \cdot\, d_\tau K(t,\tau) \circ g(\tau) \right]$$

$a \le t \le b$; in general the function $k_{t_0} f$ $[k_{t_0} g]$ is not regulated [simply regulated].

More generally for $t_0 \in [a,b]$ we define

$$\Gamma_{t_0} = \left\{ (t,s) \in [a,b]\times[a,b] \;\middle|\; t\le s\le t_0 \text{ or } t_0 \le s \le t \right\}$$

For $K:\Gamma_{t_0} \longrightarrow L(X,Y)$ we consider the following properties

(G^σ) - $K_s \in G^\sigma([a,s],L(X,Y))$ if $s \le t_0$ and $K_s \in G^\sigma([s,b],L(X,Y))$ if $s \ge t_0$.

(SV^u) - $\sup[\sup_{a\le t\le t_0} SV_{[t,t_0]}[K^t], \sup_{t_0\le t\le b} SV_{[t_0,t]}[K^t]] < \infty$

We write $K\in G^{(\sigma,\cdot)}(\Gamma_{t_0},L(X,Y))$ if K satisfies (G^σ). If K satisfies (G^σ) and (SV^u) we write $K\in G^\sigma\cdot SV^u(\Gamma_{t_0},L(X,Y))$; if furthermore K is simply regulated along the diagonal we write $K\in G^\sigma_\Delta\cdot SV^u(\Gamma_{t_0},L(X,Y))$. It is obvious that given $K\in G^\sigma_\Delta\cdot SV^u([a,b]\times[a,b],L(X,Y))$ $[K\in G^\sigma\cdot SV^u([a,b]\times[a,b],L(X,Y))]$ then for every $t_0\in[a,b]$ we have $K\in G^\sigma_\Delta\cdot SV^u(\Gamma_{t_0},L(X,Y))$ $[K\in G^\sigma\cdot SV^u(\Gamma_{t_0},L(X,Y))]$.

For $K: \Gamma_{t_0} \longrightarrow L(X,Y)$ we define $\tilde{K}: [a,b]\times[a,b] \longrightarrow L(X,Y)$ by

$$\tilde{K}(t,s) = \begin{cases} K(t,s) & \text{if } (t,s)\in\Gamma \\ K(t,t) & \text{if } s\le t\le t_0 \text{ or } t_0\le t\le s \\ K(t,t_0) & \text{if } t\le t_0\le s \text{ or } s\le t_0\le t \end{cases}$$

It is immediate that

2.4 - $\tilde{K}\in G^\sigma_\Delta\cdot SV^u([a,b]\times[a,b],L(X,Y))$ if $K\in G^\sigma_\Delta\cdot SV^u(\Gamma_{t_0},L(X,Y))$ and

$$SV^u[\tilde{K}] = \sup[\sup_{a\le t\le t_0} SV_{[t,t_0]}[K^t], \sup_{t_0\le t\le b} SV_{[t_0,t]}[K^t]].$$

2.5 - If $K:[a,b]\times[a,b] \to L(X,Y)$ is such that $K\in G^\sigma_\Delta\cdot SV^u(\Gamma_{t_0},L(X,Y))$ for $t_0 = a$ and $t_0 = b$ then we have $K\in G^\sigma_\Delta\cdot SV^u([a,b]\times[a,b],L(X,Y))$.

THEOREM 2.6 - For $K\in G^\sigma\cdot SV^u(\Gamma_{t_0},L(X,Y))$ the following properties are equivalent:

a) K is simply regulated along the diagonal (hence
$$K\in G^\sigma_\Delta\cdot SV^u(\Gamma_{t_0},L(X,Y))).$$
b) For every $f\in G([a,b],X)$ we have $k_{t_0} f\in G([a,b],Y)$.

c) For any Banach space W and every $g\in G^\sigma([a,b],L(W,X))$ we have
$$k_{t_0} g\in G^\sigma([a,b],L(W,Y)).$$

PROOF - c) $\Longrightarrow$ b): we take $W=\mathbb{C}$ and by 1.5 we have,

$$G^{\sigma}([a,b],L(\mathbb{C},Z)) = G([a,b],Z).$$

b) $\Longrightarrow$ c): $g \in G^{\sigma}([a,t_0],L(W,X)) \Longleftrightarrow g\cdot w \in G([a,t_0],X)$ for all $w \in W \Longrightarrow$

$\Longrightarrow (k_{t_0}g)w = k_{t_0}(g\cdot w) \in G([a,t_0],Y)$ for all $w \in W \Longleftrightarrow k_{t_0}g \in G^{\sigma}([a,t_0],L(W,Y))$

and analogously in $[t_0,b]$.

a) $\Longrightarrow$ b): We have

$$(k_{t_0}f)(t) = \int_{t_0}^{t} \cdot\, d_{\tau}K(t,\tau)\cdot f(\tau) = \int_a^b \cdot d_{\tau}\tilde{K}(t,\tau)\cdot f(\tau).$$

By 2.4 we have $\tilde{K} \in G^{\sigma}_{\Delta}\cdot SV^{u}([a,b]\times[a,b],L(X,Y))$ hence by Theorem 2.2 it follows that $k_{t_0}f \in G([a,b],Y)$ for every $f \in G([a,b],X)$.

b) $\Longrightarrow$ a): We take $f(\tau) \equiv x$; then we have

$$(k_{t_0}f)(t) = \int_{t_0}^{t} \cdot\, d_{\tau}K(t,\tau)x = K(t,t)x - K(t,t_0)x$$

hence the result since by the hypothesis (G^{σ}) we have

$$K_{t_0} \in G^{\sigma}([a,b],L(X,Y)).$$

The preceding result is essentially a particular case of a theorem of Arbex, [1,p.30].

COROLLARY 2.7 - Let $K: \Gamma_{t_0} \longrightarrow L(X,Y)$ satisfy (G^{σ}) and be simply regulated along the diagonal; if for every $f \in G([a,b],X)$ $[f \in G^{\sigma}([a,b],L(X))]$ there exists $k_{t_0}f \in G([a,b],X)$ $[k_{t_0}f \in G^{\sigma}([a,b],L(X,Y))]$ then we have

$$K \in G^{\sigma}_{\Delta}\cdot SV^{u}(\Gamma_{t_0},L(X,Y)).$$

PROOF: We have $\tilde{K}_s \in G^{\sigma}([a,b],L(X,Y)$ for every $s \in [a,b]$ as follows from (G^{σ}) and from the hypothesis that K is simply regulated along the diagonal. Since we have

$$(k_{t_0}f)(t) = \int_a^b \cdot d_{\tau}\tilde{K}(t,\tau)\cdot f(\tau)$$

the result follows from Theorem 2.3 and from 2.4.

For $K: T\times S \longrightarrow Z$ we define $K^{\square}: T \longrightarrow Z^{S}$ by $(K^{\square})(t) = K^{t}$. We have

THEOREM 2.8 - a) If $K \in G^{\sigma}\cdot SV^{u}_{a}([c,d]\times[a,b],L(X,Y))$ is such that

$F_K \in K[G([a,b],X),G([c,d],Y)]$ then $K^{\square} \in G([c,d],SV_a([a,b],K(X,Y)))$.

b) If $K \in G^{\sigma} \cdot SV_a^u([a,b] \times [\dot{a},b],K(X))$ is such that

$K^{\square} \in G([a,b],SV_a([a,b],K(X)))$ then $F_K^2 \in K[G([a,b],X)]$.

PROOF: See [7], Theorem 10 and corollary of Theorem 13.

THEOREM 2.9 - For $K \in G_{\Delta}^{\sigma} \cdot SV_a^u(\Gamma_{t_0},K(X))$ we have

a) If $k_{t_0} \in K[G([a,b],X)]$ then $\tilde{K}^{\square} \in G([a,b],SV_a([a,b],K(X)))$

b) If $\tilde{K}^{\square} \in G([a,b],SV_a([a,b],K(X)))$ then $(k_{t_0})^2 \in K[G([a,b],X)]$

PROOF: Since for every $y \in G([a,b],X)$ we have

$$(k_{t_0} y)(t) = \int_{t_0}^{t} \cdot\, d_{\sigma} K(t,\sigma) \cdot y(\sigma) = \int_a^b \cdot d_{\sigma} \tilde{K}(t,\sigma) \cdot y(\sigma) = (F_{\tilde{K}} y)(t), \quad a \le t \le b$$

the result follows from Theorem 2.8 and 2.4.

We say that an operator $F \in L[G([a,b],X),G([a,b],Y)]$ is *causal* if for each $t \in [a,b]$ the value $(Ff)(t)$ depends only on the values of f in $[a,t]$, for every $f \in G([a,b],X)$. We have the following representation theorem:

THEOREM 2.10 - a) If $K \in G_{\Delta}^{\sigma} \cdot SV^u(\Gamma_a,L(X,Y))$, the operator is causal.

Reciprocally, if $F \in L[G^-([a,b],X),G([a,b],Y)]$ is a causal operator then

b) There exists $K \in G_{\Delta}^{\sigma} \cdot SV_a^u(\Gamma_a,L(X,Y))$ such that $F = k_a$.

c) There exists $\hat{K} \in G_{\Delta}^{\sigma} \cdot SV^u(\Gamma_a,L(X,Y))$ with $\hat{K}(t,t) \equiv 0$ such that $F = \hat{k}_a$.

PROOF - a) Follows from Theorem 2.6.

b): By Theorem 2.2 we have

$$(Ff)(t) = \int_a^b \cdot d_{\tau} K(t,\tau) \cdot f(\tau)$$

where $K(t,s)x = F[\chi_{[a,s]}x](t)$ for $s \in]a,b]$, $t \in [a,b]$ and $x \in X$. Since F is causal, for $s \ge t$ we have $F[\chi_{[a,s]}x](t) = F[\chi_{[a,t]}x](t)$, i.e., $K(t,s)x = K(t,t)x$, hence

$$(Ff)(t) = \int_a^t \cdot d_s K(t,s) \cdot f(s) = (k_a f)(t).$$

and by Theorem 2.6 we have $K\in G_\Delta^\sigma\cdot SV_a^u(\Gamma_a,L(X,Y))$.

c) :With the notations of b) we take $\hat{K}(t,s) = K(t,s) - K(t,t)$.

REMARK - From c) it follows that we may always suppose that

$$K\in G_\Delta^\sigma\cdot SV^u(\Gamma_a,L(X,Y))$$

is *normalized*, i.e., satisfies $K(t,t) = 0$ (we write $K\in G_0^\sigma\cdot SV^u(\Gamma_a,L(X,Y))$).

THEOREM 2.11 - Take $K\in G_\Delta^\sigma\cdot SV^u(\Gamma_a,L(X,Y))$, $f\in G([a,b],X)$ and $g\in G^\sigma([a,b],L(W,X))$.

a) For $t\in[t_0,b[$ we have

$$(k_{t_0}f)(t+) - (k_{t_0}f)(t) = \int_{t_0}^{t}\cdot\, d_\tau[K(t\dot{+},\tau) - K(t,\tau)]\cdot f(\tau) +$$

$$+ [K(t\dot{+},t\dot{+}) - K(t\dot{+},t)]\cdot f(t+)$$

$$(k_{t_0}g)(t\dot{+}) - (k_{t_0}g)(t) = {}^{\sigma}\!\!\int_{t_0}^{t}\cdot\, d_\tau[K(t\dot{+},\tau)-K(t,\tau)]\circ g(\tau) +$$

$$+ [K(t\dot{+},t\dot{+}) - K(t\dot{+},t)]\circ g(t\dot{+})$$

b) For $t\in]a,t_0]$ we have

$$(k_{t_0}f)(t) - (k_{t_0}f)(t-) = \int_{t_0}^{t}\cdot\, d_\tau[K(t,\tau)\cdot K(t\dot{-},\tau)]\cdot f(\tau) -$$

$$- [K(t\dot{-},t) - K(t\dot{-},t\dot{-})]\cdot f(t-)$$

$$(k_{t_0}g)(t) - (k_{t_0}g)(t\dot{-}) = {}^{\sigma}\!\!\int_{t_0}^{t}\cdot\, d_\tau[K(t,\tau) - K(t\dot{-},\tau)]\circ g(\tau) -$$

$$- [K(t\dot{-},t) - K(t\dot{-},t\dot{-})]\circ g(t\dot{-}).$$

PROOF: a) We have

$$(k_{t_0}f)(t+\varepsilon) = \int_{t_0}^{t+\varepsilon}\cdot\, d_\tau K(t+\varepsilon,\tau)\cdot f(\tau) = \int_{t_0}^{t}\cdot\, d_\tau K(t+\varepsilon,\tau)\cdot f(\tau) +$$

$$+ \int_{t}^{t+\varepsilon}\cdot\, d_\tau K(t+\varepsilon,\tau)\cdot f(\tau)$$

Since for every $\tau\in[t_0,t]$ and $x\in X$ we have by (G^σ) that $\lim_{\varepsilon\downarrow 0}K(t+\varepsilon,\tau)x = K(t\dot{+},\tau)x$ and $SV[K^{t+\varepsilon}] \le SV^u[\tilde{K}]$ it follows from Theorem 1.13 that

$$\lim_{\varepsilon\downarrow 0}\int_{t_0}^{t}\cdot\, d_\tau K(t+\varepsilon,\tau)\cdot f(\tau) = \int_{t_0}^{t}\cdot\, d_\tau K(t\dot{+},\tau)\cdot f(\tau).$$

On the other hand we have

$$\int_{t}^{t+\varepsilon}\cdot\, d_\tau K(t+\varepsilon,\tau)\cdot f(\tau) = \int_{t}^{t+\varepsilon}\cdot\, d_\tau K(t+\varepsilon,\tau)\cdot f(t+)+\int_{t}^{t+\varepsilon}\cdot\, d_\tau K(t+\varepsilon,\tau)\cdot[f(\tau)-f(t+)];$$

$$\lim_{\varepsilon\downarrow 0}\int_{t}^{t+\varepsilon}\cdot\, d_\tau K(t+\varepsilon,\tau)\cdot f(t+) = \lim_{\varepsilon\downarrow 0}\{[K(t+\varepsilon,t+\varepsilon)-K(t+\varepsilon,t)]\cdot f(t+)\} =$$

$$= [K(t\dot{+},t\dot{+}) - K(t\dot{+},t)]\cdot f(t+)$$

and by b) of Theorem 1.9 we have

$$\left\|\int_{t}^{t+\varepsilon}\cdot\, d_\tau K(t+\varepsilon,\tau)\cdot[f(\tau)-f(t+)]\right\| \le SV_{[t,t+\varepsilon]}[K^{t+\varepsilon}]\|f-f(t+)\|_{]t,t+\varepsilon]} \le$$

$$\le SV^u[\tilde{K}]\|f-f(t+)\|_{]t,t+\varepsilon]}$$

where $\|f-f(t+)\|_{]t,t+\varepsilon]} = \sup\{\|f(\tau)-f(t+)\| \mid t<\tau\le t+\varepsilon\} \longrightarrow 0$ when $\varepsilon\downarrow 0$ since $\lim_{\tau\downarrow t}f(\tau) = f(t+)$. Hence the result.

The second part follows from the first one if we take $f = g\cdot w$ where $w\in W$. In an analogous way we prove b).

THEOREM 2.12 (Bray) - For $\alpha\in SV([c,d],L(Y,Z))$, $K\in G^\sigma\cdot SV^u([c,d]\times[a,b],L(X,Y))$ and $f\in G([a,b],X)$ $[f\in G^\sigma([a,b],L(W,X))]$ we have

a) $F_\alpha K\in SV([a,b],L(X,Z))$ where

$$(F_\alpha K)(s) = {}^{\sigma}\!\!\int_{c}^{d}\cdot d\alpha(t)\circ K(t,s), \qquad a\le s\le b$$

b) $F_K f\in G([c,d],Y)$ $[F_K f\in G^\sigma([c,d],L(W,Y))]$ where

$$(F_K f)(t) = \int_a^b\cdot d_s K(t,s)\cdot f(s) \quad [(F_K f)(t) = {}^{\sigma}\!\!\int_a^b\cdot d_s K(t,s)\circ f(s)], \qquad c\le t\le d$$

c) $\int_a^b \cdot d_s\left(\int_a^{\sigma d} \cdot d\alpha(t)\circ K(t,s)\right)\cdot f(s) = \int_c^d \cdot d\alpha(t)\cdot\left(\int_a^b \cdot d_s K(t,s)\cdot f(s)\right)$

PROOF: See [4], Theorem II.1.1 and the remark 10 at the end of that §. See also [1, Theorem 2.14].

THEOREM 2.13 (The Dirichlet formula) - Under the hypothesis of Theorem 2.12 we suppose that $[c,d]=[a,b]$. We have

$$\int_a^b \cdot d_s\left(\int_a^s \cdot d\alpha(t)\circ K(t,s)\right)\cdot f(s) = \int_a^b \cdot d\alpha(t)\left(\int_t^b \cdot d_s K(t,s)\cdot f(s)+K(t,t)\cdot f(t)\right)$$

$$\int_a^b \cdot d_s\left(\int_s^b \cdot d\alpha(t)\circ K(t,s)\right)\cdot f(s) = \int_a^b \cdot d\alpha(t)\left(\int_a^t \cdot d_s K(t,s)\cdot f(s)-K(t,t)\cdot f(t)\right)$$

PROOF: See [4, Theorem II.1.6].

§3. LINEAR VOLTERRA STIELTJES-INTEGRAL EQUATIONS

In this § we prove our results on linear Volterra Stieltjes - integral equations

$$\text{(K)} \qquad y(t) - x + \int_{t_0}^t \cdot\, d_\tau K(t,\tau)\cdot y(\tau) = f(t) - f(t_0), \qquad a \le t \le b$$

We suppose that $y,f\in G([a,b],X)$, $x\in X$ and the point $t_0\in[a,b]$ is fixed; from Theorems 2.3, 2.6 and 2.7 it follows that in order to have $k_{t_0}y\in G([a,b],X)$ for every $y\in G([a,b],X)$ it is necessary and sufficient that $K\in G^\sigma_\Delta\cdot SV^u(\Gamma_{t_0},L(X))$ and we suppose from now on that this condition is satisfied. Obviously the study of (K) in $[a,b]$ reduces to the consideration of $[a,t_0]$ and $[t_0,b]$; since the study of $[a,t_0]$ is analogous to the one of $[t_0,b]$ we consider only the case $[t_0,b]$ with $t_0=a$.

For $a\le c<d\le b$, $x\in X$ and $f\in G([c,d],X)$ we consider

$$(K_c^d) \qquad y(t) - x + \int_c^t \cdot d_\tau K(t,\tau)\cdot y(\tau) = f(t) - f(c), \qquad c \le t \le d$$

$$(K_c^d)_0 \qquad y(t) - x + \int_c^t \cdot d_\tau K(t,\tau)\cdot y(\tau) = 0, \qquad c \le t \le d.$$

3.1 - Suppose that $K \in G^{\sigma}_{\Delta}\cdot SV^{u}(\Gamma_a, L(X))$ and $a \le c < d < e \le b$

a) If (K_c^e) has always (i.e., for every $x \in X$ and $f \in G([c,e],X)$) at most one solution then the same applies to (K_d^e).

b) If (K_c^e) admits always a solution then the same applies to (K_c^d).

c) If (K_c^d) and (K_d^e) have always at most one solution, then the same applies to (K_c^e).

d) If (K_c^d) and (K_d^e) admit always a solution then the same applies to (K_c^e).

PROOF - a): If $y_1, y_2 \in G([d,e],X)$ are solutions of (K_d^e) we define $z \in G([c,e],X)$ by $z(t) = 0$ for $c \le t < d$ and $z(t) = y_2(t) - y_1(t)$ for $d \le t \le e$; then $z_1 = 0$ and $z_2 = z$ are solutions of

$$y(t) + \int_c^t \cdot d_\tau K(t,\tau)\cdot y(\tau) = 0, \qquad c \le t \le e$$

hence $z_1 = z_2$, i.e., $z = 0$ and so we have $y_1 = y_2$.

b) is immediate: it is enough to extend $f \in G([c,d],X)$, as a regulated function, to $[c,e]$.

c): If $y_1, y_2 \in G([c,e],X)$ are solutions of (K_c^e) then $z = y_2 - y_1$ satisfies

$$z(t) + \int_c^t \cdot d_\tau K(t,\tau)\cdot z(\tau) = 0, \qquad c \le t \le e$$

hence by the unicity of the solution of (K_c^d) it follows that $z(t) = 0$ for $c \le t \le d$, hence

$$z(t) + \int_d^t \cdot d_\tau K(t,\tau)\cdot z(\tau) = 0 \quad \text{for} \quad d \le t \le e,$$

and by the unicity of the solution of (K_d^e) we have $z(t) = 0$ for $d \le t \le e$, hence $z(t) = 0$ for $c \le t \le e$, i.e., $y_1 = y_2$.

d): Given $x \in X$ and $f \in G([c,e],X)$ let $y_1 \in G([c,d],X)$ denote a solution of

$$y(t) - x + \int_c^t \cdot d_\tau K(t,\tau)\cdot y(\tau) = f(t) - f(c), \qquad c \le t \le d.$$

Then we have

$$y_1(d) = f(d) - f(c) + x - \int_c^d \cdot d_\tau K(d,\tau)\cdot y_1(\tau).$$

For $d \le t \le e$ the equation (K_c^e) is equivalent to

$$y(t) - [f(d) - f(c) + x - \int_c^d \cdot d_\tau K(d,\tau)\cdot y_1(\tau)] + \int_d^t \cdot d_\tau K(t,\tau)\cdot y(\tau) =$$

$$= [f(t) - \int_c^d \cdot d_\tau K(t,\tau)\cdot y_1(\tau)] - [f(d) - \int_c^d \cdot d_\tau K(d,\tau)\cdot y_1(\tau)]$$

that by the hypothesis on (K_d^e) admits a solution $y_2 \in G([d,e],X)$ and we have $y_2(d) = y_1(d)$. If we define $y(t) = y_1(t)$ for $c \le t \le d$ and $y(t) = y_2(t)$ for $d \le t \le e$ we have a regulated solution of (K_c^e).

The Example 1 that follows and the Example 2 that we bring later on show that for the equation (K) may arrive very peculiar situations.

EXAMPLE 1 - A kernel $K \in G_\Delta \cdot BV_a^u(\Gamma_a)$ [$= G_\Delta^\sigma \cdot SV_a^u(\Gamma_a, L(\mathbb{C}))$ - see 1.5 and 1.8] such that

a) For every $d \in]a,b]$ the equation $(K_a^d)_0$ has one and only one solution ($y(t) \equiv x$).

b) For $c \in]a,b[$ and $x \ne 0$ the equation $(K_c^b)_0$ has no solution.

c) There exists $d \in]a,b]$ such that (K_a^d) does not always have a solution.

Indeed, we take $[a,b] = [0,1]$ and for $0 \le s \le t \le 1$ the kernel is defined by

$$K(t,s) = \begin{cases} 0 \text{ for } s = 0 \text{ or } s = t \\ 1 - s \text{ for } 0 < s \le t/2 \\ 1 - t + s \text{ for } t/2 \le s < t \end{cases}$$

PROOF of a): The equation $(K_0^d)_0$ is equivalent to the system

$$y(0) = x,\ y(t) - x + y(0+) - y(t-) - \int_0^{t/2} y(s)ds + \int_{t/2}^{t} y(s)ds = 0 \text{ for } 0 < t \le d$$

Since the integrals are continuous functions of $t \in]0,d]$ it follows that $y(t-) = y(t)$ and hence

$$\int_0^{t/2} y(s)ds - \int_{t/2}^{t} y(s)ds = y(0+) - x \quad \text{for} \quad 0 < t \le d.$$

If we take the left derivative of this equation we obtain

$$\frac{1}{2}\, y(\tfrac{t}{2}-) - y(t-) + \frac{1}{2}\, y(\tfrac{t}{2}-) = 0$$

hence $y(\frac{t}{2}) = y(t) = y(0+) = x$.

PROOF of b): In $[c, \inf(2c,1)]$ the equation $(K_c^1)_0$ becomes

$$y(c) = x,\ y(t) - x - y(t-) + \int_c^t y(s)ds = 0 \quad \text{for} \quad c < t \le \inf(2c,1)$$

and, as in the proof of a), we have $y(t-) = y(t)$ hence

$$\int_c^t y(s)ds = x \quad \text{for} \quad c < t \le \inf(2c,1)$$

i.e. $x = 0$ and $y(t) = 0$ for $c < t \le \inf(2c,1)$.

If $2c < 1$ then for $t \in [2c,1]$ the equation $(K_c^1)_0$ becomes

$$y(t) - x - y(t-) - \int_c^{t/2} y(s) + \int_{t/2}^{t} y(s)ds = 0, \quad 2c \le t \le 1$$

and since we have $y(t) = 0$ for $c \le t \le 2c$ it follows that for $c \le \frac{t}{2} \le 2c$ we have

$$\int_c^{t/2} y(s)ds = 0 \text{ hence } \int_{t/2}^{t} y(s)ds = 0 \quad \text{for} \quad 2c \le t \le \inf(4c,1)$$

i.e. we have also $y(t) = 0$ for $2c \le t \le \inf(4c,1)$, in particular, if $4c \le 1$, we have $y(t) = 0$ for $2c \le t \le 4c$. If we proceed in this way we prove that $y(t) = 0$ for $t \in [c,1]$.

PROOF of c): Follows from b) by the equivalence of the properties 2) and 3) of Theorem 3.4 that follows.

REMARKS - 1) The Remark f that follows Theorem 3.4 implies that in the Example 1 the operator k_a defined by K is not compact and has no compact power.

2) By Theorem 3.2 it follows that the kernel K in Example 1 has no quasi-resolvent (hence no resolvent).

We write $G_a^{(\sigma,\cdot)} = G^{(\sigma,\cdot)}(\Gamma_a, L(X))$.

THEOREM 3.2 - For $K \in G_\Delta^\sigma \cdot SV^u(\Gamma_a, L(X))$ the following properties are equivalent:

1) For every $s \in [a,b]$ and $x \in X$ the homogeneous equation

$$(1) \qquad y(t) - x + \int_s^t \cdot d_\tau K(t,\tau) \cdot y(\tau) = 0, \qquad s \le t \le b$$

has only one regulated solution, $y_{s,x} \in G([s,b],X)$.

1^σ) For every $s \in [a,b]$ and $u \in L(X)$ the homogeneous equation

$$U(t) - u + \sigma\!\!\int_s^t \cdot d_\tau K(t,\tau) \circ U(\tau) = 0, \qquad s \le t \le b$$

has only one solution that is simply regulated, $U_{s,u} \in G^\sigma([s,b],L(X))$.

2) In $G_a^{(\sigma,\cdot)}$ the *resolvent equation*

$$(R^*) \qquad R(t,s) - I_X + \sigma\!\!\int_s^t \cdot d_\tau K(t,\tau) \circ R(\tau,s) = 0, \qquad a \le s \le t \le b$$

has only one solution.

PROOF - 1) $\Longrightarrow$ 2): If for every $s \in [a,b]$ and $x \in X$ the equation (1) in $G([s,b],X)$ has only one solution $y_{s,x}$, we define $R(t,s)$ by $R(t,s)x = y_{s,x}(t)$.

Let us prove that $R(t,s) \in L(X)$. It is obvious that $R(t,s)$ is linear;

we have to prove that it is continuous. For every $s\in[a,b]$

$$K_s = \{y_{s,x}\in G([s,b],X) \mid x\in X\}$$

is a closed vector subspace of $G([s,b],X)$ since if the sequence $y_n\in K_s$ is such that $\|y_n-y\| \longrightarrow 0$ then from

$$y_n(t) - y_n(s) + \int_s^t \cdot d_\tau K(t,\tau)\cdot y_n(\tau) = 0, \qquad s\le t\le b$$

it follows that

$$y(t) - y(s) + \int_s^t \cdot d_\tau K(t,\tau)\cdot y(\tau) = 0, \qquad s\le t\le b$$

hence $y\in K_s$. By the hypothesis 1), the mapping $y\in K_s \longmapsto y(s)\in X$ is bijective and since it is obviously continuous it follows from the open mapping theorem that it is bicontinuous, hence the inverse mapping

$$x\in X \longmapsto y_{s,x}\in G([s,b],X)$$

is continuous and so is the composed application

$$x\in X \longmapsto R(t,s)x = y_{s,x}(t)\in X$$

i.e, $R(t,s)\in L(X)$.

From the definition of R it follows that for every $s\in[a,b]$ we have $R_s\in G^\sigma([s,b],L(X))$ (i.e. $R\in G_a^{(\sigma,\cdot)}$) and that R satisfies (R*).

2) $\Longrightarrow$ 1): If we have 2) we define $y_{s,x}(t) = R(t,s)x$ and it is immediate that $y_{s,x}$ satisfies (1). Let us prove the unicity of the solution in $G([s,b],X)$. Let us suppose that for an $s_0\in[a,b]$ and $x_0\in X$ there exists $z\in G([s_0,b],X)$, $z\ne y_{s_0,x_0}$, that is a solution of

$$y(t) - x_0 + \int_{s_0}^t \cdot\ d_\tau K(t,\tau)\cdot y(\tau) = 0, \qquad s_0\le t\le b.$$

We will then define an $S\in G_a^{(\sigma,\cdot)}$, $S\ne R$, S solution of (R*) in contradiction to 2). We consider two cases:

i) $x_0 \neq 0$. Let us remark that λz is a solution of

$$y(t) - y(s_0) + \int_{t_0}^{t} d_\tau K(t,\tau)\, y(\tau) = 0, \qquad s_0 \leq t \leq b$$

with $\lambda z(s_0) = \lambda x_0$. By the Hahn-Banach theorem there exists a closed hyperplane H of X that is complementary to the "line" $\mathbb{C}\{x_0\}$. Hence for every $x \in X$ there is one and only one $\lambda_H \in \mathbb{C}$ and one and only one $x_H \in H$ such that $x = \lambda_H x_0 + x_H$. We define

$$z_{s,x}(t) = \begin{cases} y_{s,x}(t) \text{ if } s \neq s_0, \ s \leq t \leq b \\ \lambda_H z(t) + y_{s_0,x_H}(t) \text{ if } s = s_0 \text{ and } x = \lambda_H x_0 + x_H, \ s_0 \leq t \leq b \end{cases}$$

and $S(t,s)x = z_{s,x}(t)$. We have $S(t,s) \in L(X)$ since $S(t,s) = R(t,s)$ if $s \neq s_0$ and $S(t,s_0)x = \lambda_H z(t) + R(t,s_0)x_H$, and the mappings

$$x \in H \longmapsto (\lambda_H, x_H) \in \mathbb{C} \times X \text{ and } (\lambda_H, x_H) \in \mathbb{C} \times H \longmapsto \lambda_H z(t) + R(t,s_0)x_H \in X$$

are linear continuous.

From the definition of S it follows immediately that S satisfies (R*) and that $S \in G_a^{(\sigma,\cdot)}$. Finally we have $S \neq R$ since $S(t,s_0)x_0 \not\equiv R(t,s_0)x_0$ for some $t \in [s_0,b]$ because $S(t,s_0)x_0 = z(t)$ and $R(t,s_0)x_0 = y_{s_0,x_0}(t)$ and we made the hypothesis that $z \neq y_{s_0,x_0}$.

ii) $x_0 = 0$. We define

$$z_{s,x}(t) = \begin{cases} y_{s,x}(t) \text{ if } s \neq s_0 \text{ or } x \neq 0, \ s \leq t \leq b \\ z(t) \text{ if } s = s_0 \text{ and } x = 0, \ s_0 \leq t \leq b \end{cases}$$

and proceed analogously as in the case i).

$1^\sigma) \Longrightarrow 2)$: We just take $u = I_X$.

$1) + 2) \Longrightarrow 1^\sigma)$: By 2) we just take $U(t) = R(t,s) \circ u$. If there were another solution V, take $t_1 \in [s,b]$ such that $V(t_1) \neq U(t_1)$; hence there exists $x \in X$ such that $V(t_1)x \neq U(t_1)x$. Then y and z defined by $y(t) = U(t)x$

and $z(t) = V(t)x$ are different regulated solutions of (1) in contradiction to 1).

We say that a function $R: \Gamma_a \longrightarrow L(X)$ that satisfies (R^*) is a *quasi-resolvent* of (K) or of K.

REMARKS - 1) In the proof of Theorem 3.2 we saw that if the equivalent conditions of Theorem 3.2 are satisfied the mapping $x \in X \mapsto y_{s,x} \in G([s,x],X)$ is continuous.

2) When the hypothesis of Theorem 3.2 are satisfied we do not know if (R^*) allows solutions $R \notin G_a^{(\sigma,\cdot)}$, or equivalently, if (1) has non regulated solutions (but see the Remark at the end of §2 of [10]).

3) At the end of [10] we give an example of an equation (L) with more then one resolvent, hence quasi-resolvent.

THEOREM 3.3 - If $K \in G_\Delta^\sigma \cdot SV^u(\Gamma_a, L(X))$ has a quasi-resolvent $R \in G_a^{(\sigma,\cdot)}$ (not necessarely unique) the following properties are equivalent:

1_a) For every $x_0 \in X$ and $f_0 \in G([a,b],X)$ the function y defined by

$$(\rho_a) \qquad y(t) = f_0(t) + R(t,a)[x_0 - f_0(a)] - \int_a^t \cdot d_\tau R(t,\tau) \cdot f_0(\tau), \qquad a \le t \le b$$

is a regulated solution of (K_a^b).

1) For every $s \in [a,b[$, $x \in X$ and $f \in G([s,b],X)$ the function y defined by

$$(\rho_s) \qquad y(t) = f(t) + R(t,s)[x - f(s)] - \int_s^t \cdot d_\tau R(t,\tau) \cdot f(\tau), \qquad s \le t \le b$$

is a regulated solution of (K_s^b).

1_a^σ) For every $u_0 \in L(X)$ and $\Phi_0 \in G^\sigma([a,b],L(X))$ the function U_0 defined by

$$U_0(t) = \Phi_0(t) + R(t,a) \circ [u_0 - \Phi_0(a)] - {}^\sigma\!\!\int_a^t \cdot d_\tau R(t,\tau) \circ \Phi_0(\tau), \qquad a \le t \le b$$

is simply regulated and satisfies

$$U_0(t) - u_0 + \int_a^{\sigma,t} \cdot d_\tau K(t,\tau) \circ U_0(\tau) = \Phi_0(t) - \Phi_0(a), \quad a \le t \le b$$

1^σ) For every $s \in [a,b[$, $u \in L(X)$ and $\Phi \in G^\sigma([s,b],L(X))$ the function U defined by

$$U(t) = \Phi(t) + R(t,s) \circ [u - \Phi(s)] - \int_s^{\sigma,t} \cdot d_\tau R(t,\tau) \circ \Phi(\tau), \quad s \le t \le b$$

is simply regulated and satisfies

$$U(t) - u + \int_s^{\sigma,t} \cdot d_\tau K(t,\tau) \circ U(\tau) = \Phi(t) - \Phi(s), \quad s \le t \le b$$

2) $R \in G_\Delta^\sigma \cdot SV^u(\Gamma_a, L(X))$

PROOF - $1_a) \Longrightarrow 1)$: We take $x_0 = 0$ and $f_0(t) = f(s)$ if $a \le t < s$, $f_0(t) = f(t) + x$ if $s \le t \le b$. If y is a solution of (K_a^b) given by (ρ_a) then its restriction to $[s,b]$ (which is obviously a solution of (K_s^b)) is given by (ρ_s) since for $t \in [s,b]$ we have

$$y(t) = f_0(t) + R(t,a)[x - f_0(a)] - \int_a^t \cdot d_\tau R(t,\tau) \cdot f_0(\tau) = f(t) + x -$$

$$- R(t,a) \cdot f(s) - \int_a^s \cdot d_\tau R(t,\tau) \cdot f(\tau) - \int_s^t \cdot d_\tau R(t,\tau) \cdot [f(\tau) + x] =$$

$$= f(t) + R(t,s) \cdot [x - f(s)] - \int_s^t \cdot d_\tau R(t,\tau) \cdot f(\tau)$$

The proof of $1_a^\sigma) \Longrightarrow 1^\sigma)$ is similar to the preceding one.

The implications $1) \Longrightarrow 1_a)$ and $1^\sigma) \Longrightarrow 1_a^\sigma)$ are obvious.

$1_a) \Longrightarrow 2)$: From $1_a)$ it follows immediately that for every $f \in G([a,b],X)$ the function

$$t \in [a,b] \longmapsto \int_a^t \cdot d_\tau R(t,\tau) \cdot f(\tau) \in X$$

is regulated. The result follows from Theorem 2.7.

2) $\Longrightarrow$ 1): If $R\in G^{\sigma}_{\Delta}\cdot SV^{u}(\Gamma_a,L(X))$ then using (R*) and Theorem 2.13 it follows immediately that y defined by (ρ_a) is regulated.

The proofs of $1^{\sigma}_a) \Longrightarrow 2)$ and $2) \Longrightarrow 1^{\sigma})$ are similar, respectively, to the proofs of $1_a) \Longrightarrow 2)$ and $2) \Longrightarrow 1)$.

We say that a quasi-resolvent R of K is a *resolvent* of K or of (K) if it satisfies the equivalent properties of Theorem 3.3.

REMARKS - 1) In $1^{\sigma}_a)$ we may obviously replace $u_0\in L(X)$, $\Phi_0\in G^{\sigma}([a,b],L(X))$ by $u_0\in L(W,X)$, $\Phi_0\in G^{\sigma}([a,b],L(W,X))$, where W is a Banach space; analogously in $1^{\sigma})$.

2) If in Theorem 3.3 we have $K\in G^{\sigma}_{\Delta}\cdot BV^{u}(\Gamma_a,L(X))$ we do not have necessarely $R\in G^{\sigma}_{\Delta}\cdot BV^{u}(\Gamma_a,L(X))$ as shows the example at the end of [10]. We ignore what is the corresponding situation if the resolvent is unique (i.e., for Theorem 3.4).

3) We do not know if the hypothesis that K has a unique quasi-resolvent $R\in G^{(\sigma,\cdot)}_a$ implies that $R\in G^{\sigma}_{\Delta}\cdot SV^{u}(\Gamma_a,L(X))$ (i.e., that R is a resolvent of K). The answer is positive if the operator k_a defined by K is compact or has a compact power (see the Remark g that follows Theorem 3.4).

EXAMPLE 2 - A kernel $K\in G_{\Delta}\cdot BV^{u}_a(\Gamma_a)$ such that

a) For $x\in\mathbb{C}$ and $f\in G([a,b])$ the equation (K^b_a) has one and only one solution $y_{f,x}\in G([a,b])$.

b) The operator $f\in G([a,b]) \longmapsto y_f = y_{f,0}\in G([a,b])$ is not causal (see Theorem 2.10).

c) For every $d\in\,]a,b[$ the equation (K^d_a) has always more than one solution.

d) There exists $c\in\,]a,b[$ such that $(K^b_c)_0$ and (K^b_c) do not have always a solution.

Indeed, we take $[a,b]=[0,1]$ and $K(t,\tau)=\chi_{[t^2,t[}(\tau)$ for $0\le\tau\le t\le 1$.

PROOF of a): The equation

$$(2) \qquad y(t) - x + \int_0^t \cdot d_\tau \chi_{[t^2,t[}(\tau)\cdot y(\tau) = f(t) - f(0), \quad 0 \le t \le 1$$

has one and only one solution $y \in G([0,1])$, given by

$$(3) \qquad y(t) = f(t) - f^-(t) + f^-(\sqrt{t}) - f(0) + x, \quad 0 \le t \le 1.$$

Indeed, this follows from the fact that (2) is equivalent to

$$y(t) - x + y^-(t^2) - y^-(t) = f(t) - f(0), \quad 0 \le t \le 1$$

and this relation implies that $-x + y^-(t^2) = f^-(t) - f(0)$ for $0<t\le 1$, hence $y^-(t) = f^-(\sqrt{t}) - f(0) + x$ (for $0<t\le 1$), hence (3).

b): follows immediately from (3).

c): It is immediate that the solutions of (K_0^d) are given by

$$y(t) = \begin{cases} f(t) - f^-(t) + f^-(\sqrt{t}) - f(0) + x \text{ if } 0 \le t \le d^2 \\ f(t) - f^-(t) + g^-(t) \text{ if } d^2 < t \le d \text{ (for any } g \in G([d^2,d])) \end{cases}$$

d): For $c \in]0,1[$ the equation

$$y(t) - x + \int_c^t \cdot d_\tau \chi_{[t^2,t[}(\tau)\cdot y(\tau) = f(t) - f(c), \quad c \le t \le 1$$

is equivalent to $y(c) = x$, $y(t) - x - y^-(t) = f(t) - f(c)$ if $c < t \le \sqrt{c}$ and $y(t) - x + y^-(t^2) - y^-(t) = f(t) - f(c)$ if $\sqrt{c} < t \le 1$, hence for $c < t \le \sqrt{c}$ we must have $-x = f^-(t) - f(c)$ i.e. in $]c,\sqrt{c}]$ the function f^- must be constant and equal to $f(c) - x$. By b of 3.1 it follows that then (K_c^1) has at most one solution. Thus it is easy to verify that

The equation (K_c^1) *has a solution if and only if we have* $f^-(t) = f(c) - x$ *for* $c < t \le \sqrt{c}$ *and this solution is given by*

$$y(c) = x, \; y(t) = f(t) - f^-(t) + f^-(\sqrt{t}) - f(c) + x \text{ if } c < t \le 1.$$

REMARK - From d) and Theorem 3.2 it follows that the kernel K has no quasi-resolvent. However the equation (K_a^b) allows a resolvent *as a Fredholm Stieltjes-integral equation* (see [7], Theorem 16), hence there exists

$$R \in G\cdot BV_a^u([a,b]\times[a,b])$$

such that the solution of (K_a^b) is given by

$$y(t) = f(t) - f(a) + x - \int_a^b \cdot d_s R(t,s)\cdot[f(s)-f(a)+x].$$

THEOREM 3.4 - We give $K \in G_\Delta^\sigma \cdot SV^u(\Gamma_a, L(X))$.

I - The following properties are equivalent

1) For every $g \in G([a,b],X)$ the equation

$$(4) \qquad y(t) + \int_a^t \cdot d_\tau K(t,\tau)\cdot y(\tau) = g(t), \qquad a \le t \le b$$

has one and only one solution $y_g \in G([a,b],X)$ and the operator

$$g \in G([a,b],X) \longmapsto y_g \in G([a,b],X)$$

is causal (i.e., if $g_{|[a,t]} \equiv 0$ for some $t \in [a,b]$ then $y_g(\tau) = 0$ for $\tau \in [a,t]$).

2) For any $d \in]a,b]$, $x \in X$ and $f \in G([a,d],X)$ the equation (K_a^d) has one and only one regulated solution.

3) For any $c \in [a,b[$, $x \in X$ and $f \in G([c,b],X)$ the equation (K_c^b) has one and only one regulated solution.

4) For any $[c,d] \subset [a,b]$, $x \in X$ and $f \in G([c,d],X)$ the equation (K_c^d) has one and only one regulated solution.

5) For every $t \in [a,b]$ there exist $\varepsilon_t > 0$ such that

i) For any $c \in [a,b[$, $\varepsilon \in]0,\varepsilon_c]$, $x \in X$ and $f \in G([c,c+\varepsilon],X)$ the equation $(K_c^{c+\varepsilon})$ has one and only one regulated solution.

ii) For any $d \in]a,b]$, $\varepsilon \in]0,\varepsilon_d]$, $x \in X$ and $f \in G([d-\varepsilon,d],X)$ the equation $(K_{d-\varepsilon}^d)$ has one and only one regulated solution.

6) In $G_a^{(\sigma,\cdot)}$ the kernel has only one quasi-resolvent R and we have

$R \in G^{\sigma}_{\Delta} \cdot SV^{u}(\Gamma_a, L(X))$.

7) There exists $R \in G^{\sigma}_{\Delta} \cdot SV^{u}(\Gamma_a, L(X))$ that satisfies

$$(R^*) \qquad R(t,s) - I_X + \int_s^{\sigma,t} \cdot d_\tau K(t,\tau) \circ R(\tau,s) = 0, \quad a \le s \le t \le b$$

and (if K is normalized, i.e., $K(t,t) \equiv 0$ - see the Remark that follows Theorem 2.10)

$$(R_*) \qquad R(t,s) - K(t,s) - I_X + \int_s^{\sigma,t} \cdot d_\tau R(t,\tau) \circ K(\tau,s) = 0, \quad a \le s \le t \le b$$

II - If the equivalent properties of I are satisfied we have

8) For any $[c,d] \subset [a,b]$ the solution of (K_c^d) is given by

$$(\rho) \qquad y(t) = f(t) + R(t,c) \cdot [x - f(c)] - \int_c^t \cdot d_\tau R(t,\tau) \cdot f(\tau), \quad c \le t \le d$$

9) For every $t \in [a,b[$ the operator $I_X + K(t\dot{+},t\dot{+}) - K(t\dot{+},t)$ is invertible (in $L(X)$).

III - If the operator $k_a \in L[G([a,b],X)]$ defined by K is compact then the property 9) is equivalent to the properties of I.

PROOF - 1) $\Longrightarrow$ 4): Given $x \in X$ and $f \in G([c,d],X)$, we are looking for a solution of

$$(5) \qquad y(t) + \int_c^t \cdot d_\tau K(t,\tau) \cdot y(\tau) = f(t) - f(c) + x, \quad c \le t \le d$$

We define

$$g(t) = \begin{cases} 0 \text{ if } a \le t < c \text{ or } d < t \le b \\ \\ f(t) - f(c) + x \text{ if } c \le t \le d \end{cases};$$

then by 1) there exists one and only one regulated function $y_g \in G([a,b],X)$ solution of (4); since $g_{|[a,c[} = 0$ and $g \longmapsto y_g$ is causal, it follows that

$y_g(t) = 0$ for $t \in [a,c[$ hence $(y_g)_{|[c,d]}$ satisfies (5). In $[c,d]$ the solution of (5) is unique since otherwise the difference z of two solutions would satisfy

$$z(t) + \int_c^t \cdot d_\tau K(t,\tau) \cdot z(\tau) = 0, \qquad c \le t \le d,$$

and if we take z zero in $[a,b] \cap \complement [c,d]$ we have

$$z(t) + \int_a^t \cdot d_\tau K(t,\tau) \cdot z(\tau) = \chi_{]d,b]}(t) \cdot \int_c^d \cdot d_\tau K(t,\tau) \cdot z(\tau), \quad a \le t \le b$$

and 1) implies that $z_{|[c,d]} \equiv 0$.

4) $\Longrightarrow$ 5): is obvious.

5) $\Longrightarrow$ 4): Given $[c,d] \subset [a,b]$ we define

$$d_1 = \sup\left\{ d' \in]c,d] \;\middle|\; \begin{matrix} \text{for every } t \in [c,d'] \text{ the equation } (K_c^t) \\ \text{has always one and only one solution} \end{matrix} \right\}$$

Then $(K_c^{d_1})$ too has always one and only one solution since there exists $d_1' < d_1$ such that for $t \in]d_1',d]$ the equation $(K_t^{d_1})$ has always one and only one solution; hence if we take c_1 such that $c < \sup(c,d_1') < c_1 < d_1$ then $(K_c^{c_1})$ and $(K_{c_1}^{d_1})$ have always one and only one solution and by c) and d) of 3.1 then the same is true for $(K_c^{d_1})$. But by 5) there exists $d_2 > d_1$ such that for every $t \in]d_1,d_2]$ the equation $(K_{d_1}^t)$ has one and only one solution and the same applies to (K_c^t) with $c < t < d_1$, with $c = t_1$ and with $d_1 < t \le d_2$ (by c and d of 3.1) in contradiction to the definition of d_1.

It is immediate that 4) $\Longrightarrow$ 2) and 4) $\Longrightarrow$ 3).

2) $\Longrightarrow$ 1): By 2) the solution $y_g \in G([a,b],X)$ of (4) is unique. If for some $d \in]a,b[$ we had $g_{|[a,d]} = 0$ but $z = (y_g)_{|[a,d]} \ne 0$, then z would be a non-zero solution of

$$y(t) + \int_a^t \cdot d_\tau K(t,\tau) \cdot y(\tau) = 0, \qquad a \le t \le d,$$

in contradiction to 2).

3) $\Longrightarrow$ 6): By Theorem 3.2 in $G_a^{(\sigma,\cdot)}$ there exists a unique quasi-resolvent R; we want to prove that $R \in G_\Delta^\sigma \cdot SV^u(\Gamma_a, L(X))$. By hypothesis the equation

$$v(t) - f(c) + \int_c^t \cdot d_\tau K(t,\tau)\cdot v(\tau) = f(t) - f(c), \qquad c \le t \le b$$

i.e., the equation

$$(6) \qquad v(t) + \int_c^t \cdot d_\tau K(t,\tau)\cdot v(\tau) = f(t), \qquad c \le t \le b$$

has for every $f \in G([c,b],X)$ one and only one solution $v \in G([c,b],X)$. We take $c = a$; since the bijective mapping $v \in G([a,b],X) \mapsto f = v + k_a v \in G([a,b],X)$ is obviously continuous, it is bicontinuous by the open mapping theorem. Hence the inverse mapping is continuous. If we consider the restriction of this inverse mapping to $G^-([a,b],X)$ then by theorem 2.2 there exists one and only one

$$H \in G^\sigma \cdot SV_a^u([a,b]\times[a,b], L(X)) \text{ such that } v(t) = \int_a^b \cdot d_\sigma H(t,\sigma)\cdot f(\sigma)$$

for every $f \in G^-([a,b],X)$ and $a \le t \le b$. If we replace this expression in (6) and recall that $f(t) = \int_a^b \cdot d_\sigma Y(\sigma - t)\cdot f(\sigma)$ for $t \in]a,b]$ (where $Y = \chi_{[0,\infty[}$) and $f(a) = f(a+) = \int_a^b \cdot d\chi_{]a,\infty[}(\sigma)\cdot f(\sigma)$, we obtain for $t > a$

$$\int_a^b \cdot d_\sigma H(t,\sigma)\cdot f(\sigma) + \int_a^t \cdot d_\tau K(t,\tau)\left(\int_a^b \cdot d_\sigma H(\tau,\sigma)\cdot f(\sigma)\right) = \int_a^b \cdot d_\sigma Y(\sigma - t)\cdot f(\sigma)$$

If we apply c of Theorem 2.12 we have

$$\int_a^b \cdot d_\sigma [H(t,\sigma) + \int_a^{\sigma,t} \cdot d_\tau K(t,\tau)\circ H(\sigma,t) - Y(\sigma - t)] f(\sigma) = 0, \qquad a < t \le b$$

for every $f \in G^-([a,b],X)$; hence by Theorem 1.12 we have

$$H(t,\sigma) + \int_a^{\sigma,t} \cdot d_\tau K(t,\tau)\circ H(\tau,\sigma) - Y(\sigma - t) = 0, \quad a \le \sigma \le b,\ a < t \le b.$$

It is immediate that for $t = a$ we have $H(a,\sigma) = \chi_{]a,b]}(\sigma)$, $a \le \sigma \le b$, hence H satisfies

$$(7)\ H(t,\sigma) + \int_a^{\sigma,t} \cdot d_\tau K(t,\tau)\circ H(\sigma,t) = Y(\sigma - t) - Y(a-\sigma)\cdot Y(a-t), \ a \le \sigma \le t,\ a \le t \le b.$$

We assert that the solution of (7) is given by

$$(8) \qquad H(t,\sigma) = \begin{cases} R(t,a) - Y(a-\sigma)\cdot Y(a-t) & \text{if } \sigma \geq t \\ R(t,a) - R(t,\sigma) & \text{if } \sigma < t \end{cases}$$

Indeed, this is immediate for $\sigma = a$. For $\sigma > a$ we have to prove that H given by

$$H(t,\sigma) = \begin{cases} R(t,a) & \text{if } \sigma \geq t \\ R(t,a) - R(t,\sigma) & \text{if } \sigma < t \end{cases}$$

satisfies

$$(9) \qquad H(t,\sigma) + \int_a^t \cdot d_\tau K(t,\tau) \circ H(\tau,\sigma) = Y(\sigma - t), \quad a \leq t \leq b$$

By (R*) it is immediate that this is true for $t \leq \sigma$. For $t > \sigma$ if we replace $H(t,\sigma) = R(t,a) - R(t,\sigma)$ in (9) we obtain

$$H(t,\sigma) + \int_a^\sigma \cdot d_\tau K(t,\tau) \circ H(\tau,\sigma) + \int_\sigma^t \cdot d_\tau K(t,\tau) \circ H(\tau,\sigma) = R(t,a) - R(t,\sigma) +$$

$$+ \int_a^\sigma \cdot d_\tau K(t,\tau) \circ R(\tau,\sigma) + \int_\sigma^t \cdot d_\tau K(t,\tau) \circ [R(\tau,a) - R(\tau,\sigma)] = R(t,a) +$$

$$+ \int_a^t \cdot d_\tau K(t,\tau) \circ R(\tau,a) - R(t,\sigma) - \int_\sigma^t \cdot d_\tau K(t,\tau) \circ R(\tau,\sigma)$$

and by (R*) this expression is zero, i.e., we have (9). Since

$$H \in G^\sigma \cdot SV_a^u([a,b] \times [a,b], L(X))$$

it follows from (8) that $R \in G_\Delta^\sigma \cdot SV^u(\Gamma_a, L(X))$.

6 $\Longrightarrow$ 7): For every $s \in [a,b]$ we take $U_s = \chi_{[s,b]} I_X \in G^\sigma([s,b], L(X))$. We have

$$\chi_{[s,b]}(t) I_X - I_X + {}^\sigma\!\!\int_s^t \cdot d_\tau K(t,\tau) \circ \chi_{[s,b]}(\tau) I_X = \chi_{[s,b]}(t) I_X - I_X - K(t,s),$$

$s \leq t \leq b$, i.e., we define $\Phi(t) = \chi_{[s,b]}(t) I_X - K(t,s)$ then $\Phi \in G^\sigma([s,b], L(X))$ and the solution of

$$V(t) - I_X + {}^\sigma\!\!\int_s^t \cdot d_\tau K(t,\tau) \cdot V(\tau) = \Phi(t) - \Phi(s), \quad t \in [s,b]$$

is $V = U_s$. From 1^σ) of the Theorem 3.3 it follows that

$$U_s(t) = \phi(t) + R(t,s)\circ[I_X - \phi(s)] - \int_s^{\sigma,t} \cdot d_\tau R(t,\tau)\circ\Phi(\tau), \quad t\in[s,b]$$

which is equivalent to (R_*) by the definition of U_s and Φ.

7) $\Longrightarrow$ 6): Let us suppose that K has another quasi-resolvent $S\in G_a^{(\sigma,\cdot)}$. We extend R to $[a,b]\times[a,b]$ defining $R(t,s) = R(t,t)$ for $s \geq t$ and we do the same for S and K. Then the equation (R^*) for R and S and (R_*) for R are equivalent, respectively, to

$$R(t,s) - I_X + \int_a^{\sigma,b} \cdot d_\tau K(t,\tau)\circ[R(\tau,s) - I_X] + K(t,s) = 0$$

$$S(t,s) - I_X + \int_a^{\sigma,b} \cdot d_\tau K(t,\tau)\circ[S(\tau,s) - I_X] + K(t,s) = 0$$

$$R(t,s) - I_X + \int_a^{\sigma,b} \cdot d_\tau [R(t,\tau) - I_X]\circ K(\tau,s) + K(t,s) = 0$$

i.e., to $F_{R-I} + F_K\circ F_{R-I} + F_K = 0$, $F_{S-I} + F_K\circ F_{S-I} + F_K = 0$ and $F_{R-I} + F_{R-I}\circ F_K + F_K = 0$ respectively (see Theorem 2.2). Since $F_R = F_{R-I}$ and $F_S = F_{S-I}$ they are equivalent, respectively, to $(F_K+I)\circ F_R = I$, $(F_K+I)\circ F_S = I$ and $F_R\circ(F_K+I) = I$. So F_R and F_K+I are inverses of each other, hence $F_S = F_R$. Since $R(t,t) = S(t,t)$, by Theorem 2.2 we have $R = S$.

6) $\Longrightarrow$ 1): Given $g\in G([a,b],X)$, by Theorem 3.3 the equation (4) has a solution given by (ρ); by Theorem 3.2 this solution y_g is unique and (ρ) shows that the operator $g \longmapsto y_g$ is causal.

We thus completed the proof of part I. From 6) and from Theorem 3.3 follows 8).

PROOF of 9): If we apply a) of Theorem 2.11 to (R^*) and (R_*) (with $t_0 = s$, $t = t_0$ and $K(\tau,s) - K(\tau,\tau)$ instead of $K(\tau,s)$ in (R_*) since K, in (R_*), must be normalized) we obtain respectively

$$[I_X + K(\dot{s+},\dot{s+}) - K(\dot{s+},s)]\circ R(\dot{s+},s) = I_X \text{ and } R(\dot{s+},s)\circ[I_X + K(\dot{s+},\dot{s+}) - K(\dot{s+},s)] = I_X$$

hence the result.

PROOF of III: we will prove 4); we may suppose that $[c,d] = [a,b]$ and that $K(t,s) = 0$ for $s \geq t$ (by Theorems 2.10 and 2.4). To start with we prove that if $K^{\square} \in G([a,b], SV_b([a,b], L(X)))$ then property 9) implies that if $z \in G([a,b], X)$ satisfies

$$z(t) + \int_a^t \cdot d_\tau K(t,\tau) \cdot z(\tau) = 0, \quad a \leq t \leq b,$$

then $z \equiv 0$: it is obvious that $z(a) = 0$. Next we define

$$c = \sup\{\tau \in [a,b] \mid z_{|[a,\tau]} \equiv 0\};$$

it is immediate that $z|_{[a,c]} \equiv 0$; we will prove that $c=b$. For $t \geq c$ we have

$$z(t) = -\int_c^t \cdot d_\tau K(t,\tau) \cdot z(\tau)$$

and by a) of Theorem 2.11 we have

$$z(c+) = K(c\dot{+},c) \cdot z(c+) \quad \text{i.e.,} \quad [I_X - K(c\dot{+},c)] \cdot z(c+) = 0,$$

hence 9) implies that $z(c+) = 0$. On the other hand

$$z(c+\varepsilon) = -\int_c^{c+\varepsilon} \cdot \, d_\tau K(c+\varepsilon,\tau) \cdot z(\tau) = -\int_c^{c+\varepsilon} \cdot \, d_\tau K(c+\varepsilon,\tau) \cdot z(c+\varepsilon) +$$

$$+ \int_c^{c+\varepsilon} \cdot \, d_\tau K(c+\varepsilon,\tau) \cdot [z(c+\varepsilon) - z(\tau)] = K(c+\varepsilon,c) \cdot z(c+\varepsilon) +$$

$$+ \int_c^{c+\varepsilon} \cdot \, d_\tau K(c+\varepsilon,\tau) \cdot [z(c+\varepsilon) - z(\tau)]$$

i.e.,

$$[I_X - K(c+\varepsilon,c)] \cdot z(c+\varepsilon) = \int_c^{c+\varepsilon} \cdot \, d_\tau K(c+\varepsilon,\tau) \cdot [z(c+\varepsilon) - z(\tau)]$$

hence

$$[I_X - K(c+\varepsilon,c)] \cdot z(c+\varepsilon) = \int_c^{c+\varepsilon} \cdot \, d_\tau [K(c+\varepsilon,\tau) - K(c\dot{+},\tau)] \cdot [z(c+\varepsilon) - z(\tau)] +$$

$$+\int_c^{c+\varepsilon} \bullet\, d_\tau K(c\dot{+},\tau)\cdot[z(c+\varepsilon)-z(\tau)]$$

and since we have $K(c\dot{+},s)=0$ for $s>c$ and $z(c+)=0$ it follows that the last integral is equal to $-K(c\dot{+},c)\cdot z(c+\varepsilon)$. Hence we have

$$(11)\qquad [I_X+K(c\dot{+},c)-K(c+\varepsilon,c)]\cdot z(c+\varepsilon) = \int_c^{c+\varepsilon} \bullet\, d_s[K(c+\varepsilon,s)-K(c\dot{+},s)]\cdot[z(c+\varepsilon)-z(s)].$$

The hypothesis $K^{\square}\in G([a,b],SV_b([a,b],L(X)))$ implies that there exists $\delta>0$ such that for $0<\varepsilon\le\delta$ we have $SV[K^{c+\varepsilon}-K^{c\dot{+}}]\le\frac{1}{4}$ hence (by 1.6, since $K(c\dot{+},b)=$ $=K(c+\varepsilon,b)=0$) we have $\|K(c\dot{+},c)-K(c+\varepsilon,c)\|\le\frac{1}{4}$ and since

$$\left\|\int_c^{c+\varepsilon} \bullet\, d_s[K(c+\varepsilon,s)-K(c\dot{+},s)]\cdot[z(c+\varepsilon)-z(s)]\right\| \le SV[K^{c+\varepsilon}-K^{c\dot{+}}]\cdot 2\|z\|_{[c,c+\varepsilon]}$$

it follows that if we define $m=\sup_{0<\varepsilon\le\delta}\|z(c+\varepsilon)\|$ then by (11) we have $\frac{3}{4}m\le$ $\le\frac{1}{4}\cdot 2m$, hence $m=0$, i.e., $z_{|[c,c+\delta]}\equiv 0$ in contradiction to the definition of c, hence $z\equiv 0$.

Now, if the operator k_a defined by K is compact (or, more generally, if $K^{\square}\in G([a,b],SV_b([a,b],K(X)))$ - hence k_a^2 is compact by Theorem 2.9) then by the theory of compact operators, the unicity of the solution of (K_a^b) that we just proved implies that (K_a^b) has always one and only one solution (see [15], corollary 1 of p. 148, or [9], p. 588) and this completes the proof of III.

REMARKS - a) If the equivalent properties of I of Theorem 3.4 are satisfied, the solution $y_{f,x}\in G([c,d],X)$ of (K_c^d) is a continuous function of $x\in X$ and $f\in G([c,d],X)$.

Indeed, if y and $\hat{y}$ are two solutions of (K_c^d) corresponding to x,f and $\hat{x},\hat{f}$ respectively, then by (ρ) we have

$$\|y-\hat{y}\| \le \|f-\hat{f}\| + \|R_c\|[\|x-\hat{x}\| + \|f-\hat{f}\|] + SV^u[R]\|f-\hat{f}\|$$

b) The equivalent properties of I in Theorem 3.4 depend only on the behaviour of K in an arbitrarily small set of the form

$$\bigcup_{1\le i\le |d|} \{(t,s)\in[a,b]\times[a,b] \mid t_{i-1}\le s\le t\le t_i\}$$

where d is a division of [a,b].

Indeed, by c) and d) of 3.1 the property 4) is equivalent to the property

4_Δ) For any division d of [a,b] the equations $(K_{t_{i-1}}^{t_i})$, $i=1,\dots,|d|$, have always one and only one regulated solution,

hence the result. A result of this type was first proved by Arbex, [1], and, in another context, by Bitzer, [2].

c) If in 4) we replace the conditions $x\in X$ and $f,y\in G([c,d],X)$, respectively by $u\in L(X)$ and $\Phi,U\in G^\sigma([c,d],L(X))$, we denote by 4^σ) the corresponding property. In an analogous way we define 1^σ), 2^σ), 3^σ), 5^σ) and 8^σ). It is immediate that the properties n) and n^σ) are equivalent (see also Theorems 3.2 and 3.3). More generally we may also consider $L(W,X)$ instead of $L(X)$.

d) It is immediate what is the analog of Theorem 3.4 when we give $K\in G_\Delta^\sigma\cdot SV^u(\Gamma_b,L(X))$ or, more generally, $K\in G_\Delta^\sigma\cdot SV^u(\Gamma_{t_0},L(X))$ and consider the equation

$$y(t) - x + \int_{t_0}^{t} \cdot\, d_sK(t,s)\cdot y(s) = f(t) - f(t_0), \quad a\le t\le b$$

In this case the property 9) is replaced by

9_{t_0}) For every $t\in\,]a,t_0]$ the operator $I_X + K(\dot{t-},\dot{t-}) - K(\dot{t-},t)$ is invertible and for every $t\in[t_0,b[$ the operator $I_X+K(\dot{t+},\dot{t+})-K(\dot{t+},t)$ is invertible.

e) If the equivalent properties of part I of Theorem 3.4 are satisfied and if in 4) we have $f(t) = f(c)+\int_c^t g(s)ds$ for some $g\in G([c,d],X)$ then f is continuous and using integration by parts in (ρ) we obtain

$$y(t) = R(t,c)\cdot x + \int_c^t R(t,s)\cdot g(s)ds, \quad c\le t\le d$$

f) If the operator $k_a \in L[G([a,b],X)]$ defined by K is compact or has a compact power then it allows a pseudo-resolvent (see [7], Theorem 17 and the Remark 1 that follows it). If furthermore the equation

$$y(t) + \int_a^t \cdot d_s K(t,s) \cdot y(s) = 0, \qquad a \le t \le b,$$

has $y \equiv 0$ as its unique regulated solution then the equivalent properties of I are verified.

Indeed, if $k_a^m \in K[G([a,b],X)]$ then for every $c \in [a,b]$ we have

$$k_c^m \in K[G([c,b],X)]$$

too. By a) of 3.1 and by the theory of compact operators (see the references given at the end of the proof of III) it follows that (K_c^b) has always one and only one regulated solution, hence the property 3) of Theorem 3.4 is satisfied (compare with the Example 1 that follows 3.1).

g) If the operator k_a defined by K is compact or has a compact power and has a unique quasi-resolvent $R \in G_a^{(\sigma,\cdot)}$ then the equivalent properties of I are true.

Indeed, it follows immediately from the preceding remark by Theorem 3.2 (compare with the Example 2 that follows Theorem 3.3).

h) For the kernels K with finite two-dimensional Vitali variation that Schwabik considers we have $K^{\square} \in BV_a([a,b],BV_a([a,b],K(X))$ (see [19], Theorem 2.7), hence $k_a^2 \in K[G([a,b],X)]$ by Theorem 2.9 and so the Remark f) applies.

It applies too to kernels K with finite two-dimensional Fréchet variation (if $\dim X < \infty$) since then we have $k_a \in K[G([a,b],X)]$ (see [20]).

i) The property 9) is not necessary to assure that (K_c^d) has at most one regulated solution for every $[c,d] \subset [a,b]$ as shows the Example 1 that follows 3.1, since there we have $K(t,t) = 0$ and $K(t+,t) = 1$, hence

$$1 + K(t+,t+) - K(t+,t) = 0 \text{ for every } t \in [0,1[.$$

j) Nor is the property 9) sufficient to assure the unicity of the solution of (K_a^b) as shows the following

EXAMPLE - We take $[a,b] = [0,1]$, $X = \mathbb{C}$ and $K(t,s) = 2\chi_{]0,t/2]}(s)$. Then for every $t \in [0,1[$ we have $K(t,t) = K(t+,t) = 0$ hence $1+K(t+,t+)-K(t+,t)=1$. However the equation

$$y(t) + \int_0^t \cdot d_s K(t,s) \cdot y(s) = 0, \quad 0 \leq t \leq 1,$$

is equivalent to the system

$$y(0) = 0, \; y(t) - 2y^-(\tfrac{t}{2}) + 2y(0+) = 0, \quad 0 < t \leq 1$$

that allows the non-trivial solutions $y(t) = ct$ $(c \in \mathbb{C})$.

k) The resolvent of K is not necessarely given by a Neumann series, as shows an example of Gomes, [3].

ℓ) It is immediate that we have 9) iff for every $t \in [a,b[$ the operator $x \in X \longmapsto x + \lim_{\tau \downarrow t} k_a(\chi_{]t,\tau]} x](\tau) \in X$ is invertible.

m) From the initial part of the proof of III it follows that if we have $K^\square \in G([a,b], SV_b([a,b], L(X)))$ then 9) implies that every (K_c^d) has at most one solution.

n) We denote by $\mathbb{K}$ the class of all kernels $K \in G_0^\sigma \cdot SV^u(\Gamma_a, L(X))$ that satisfy the equivalent properties of I of Theorem 3.4. It follows immediately from the usual results on invertible operators in Banach algebras that $\mathbb{K}$ is an open subset of $G_0^\sigma \cdot SV^u(\Gamma_a, L(X))$ and that the mapping that to each $K \in \mathbb{K}$ associates its resolvent R (where $R - I_X \in \mathbb{K}$) is bicontinous (see the proof of 7) $\Longrightarrow$ 6) in Theorem 3.4).

o) For $K \in G_\Delta^\sigma \cdot SV^u(\Gamma_a, L(X))$ we have $k_a y \in C([a,b],X)$ for every $y \in G([a,b],X)$ iff K is simply continuous in the first variable and along the diagonal (we write $K \in C_\Delta^\sigma \cdot SV^u(\Gamma_a, L(X))$); if furthermore the equivalent properties of I of Theorem 3.4 are satisfied we have $R \in C_\Delta^\sigma \cdot SV^u(\Gamma_a, L(X))$ and for

$f \in C([a,b],X)$ the solution y of (K_c^d) is a continuous function.

p) We recall that functions of bounded variation are particular instances of functions of bounded semi-variation (see 1.8). In particular, if $\dim X < \infty$ we have $G_\Delta^\sigma \cdot SV^u(\Gamma_a, L(X)) = G_\Delta \cdot BV^u(\Gamma_a, L(X))$ (see 1.5 and 1.8).

§4 - APPLICATIONS

1 - Linear delay differential equations.

We give $t_0 < t_1$ and $0 < r < t_1 - t_0$ (the delay). We consider linear delay differential equations

$$\text{(D)} \qquad y'(t) = L(t, y_t) + g(t), \quad t_0 \leq t \leq t_1$$

where $y \in G([t_0 - r, t_1], X)$, $g \in G([t_0, t_1], X)$ and $y_t \in G([-r, 0], X)$, $t_0 \leq t \leq t_1$, is defined in the usual way: $y_t(s) = y(t+s)$, $-r \leq s \leq 0$. We suppose that

$$L: [t_0, t_1] \times G([-r, 0], X) \longrightarrow X$$

is regulated as a function of the first variable and in the second variable it is linear and continuous and depends only on the equivalence class of the elements of $G([-r,0],X)$ (see 1.3); hence by Theorem 2.2 there exists $A \in G^\sigma \cdot SV_0^u([t_0, t_1] \times [-r, 0], L(X))$ such that for every $f \in G([-r,0],X)$ we have

$$L(t,f) = \int_{-r}^{0} \cdot\, d_s A(t,s) \cdot f(s), \qquad t_0 \leq t \leq t_1$$

hence (D) takes the form

$$(\tilde{D}) \qquad y'(t) = \int_{-r}^{0} \cdot\, d_s A(t,s) \cdot y(t+s), \qquad t_0 \leq t \leq t_1$$

We extend A to $[t_0, t_1] \times \mathbb{R}$ defining $A(t,s) = A(t,0)$ for $s \geq 0$ and $A(t,s) = A(t,-r)$ for $s \leq -r$. It is easy to prove that the second member of $(\tilde{D})$ is regulated for every $y \in G([t_0 - r, t_1], X)$ if and only if for every $s \in [-r, 0]$ the function $t \in [t_0, t_1] \longmapsto A(t, s-t) \in L(X)$ is simply regulated and from now on we suppose that this hypothesis is satisfied.

We look for a solution of (D) or $(\tilde{D})$ that satisfies the initial condition $y_{t_0} = \phi$, where $\phi \in G([-r,0],X)$.

If we define

$$K(t,s) = -\int_s^t A(\tau,s-\tau)\,d\tau$$

and

$$f(t) = \int_{t_0}^{t} \left(\int_{t_0-r}^{t_0} d_s A(\tau,s-\tau)\cdot\phi(s-t_0) + g(\tau)\right)d\tau$$

for $t_0 \le s \le t \le t_1$ then our problem takes the form

$$(D_K)\qquad \begin{aligned} y(t) &= \phi(0) + \int_{t_0}^{t} d_s K(t,s)\cdot y(s) = f(t) \quad \text{for } t_0 \le t \le t_1 \\ y(t) &= \phi(t-t_0) \quad \text{for } t_0-r \le t \le t_0 \end{aligned}$$

(see [4], p.91-94).

For $t',t'' \in [t_0,t_1]$ we have

$$SV[K^{t''}-K^{t'}] = \sup\left\{\left\|\int_{t_0}^{t_1} d_s[K(t'',s)-K(t',s)]\cdot h(s)\right\| \,\middle|\, h\in G([t_0,t_1],X), \|h\|\le 1\right\} =$$

$$= \sup\left\{\left\|\int_{t_0}^{t_1} d_s\left(\int_{t'}^{t''} A(\tau,s-\tau)\,d\tau\right)\cdot h(s)\right\| \,\middle|\, h\in G([t_0,t_1],X), \|h\|\le 1\right\} \overset{B}{=}$$

$$\overset{B}{=} \sup\left\{\left\|\int_{t'}^{t''}\left(\int_{t_0}^{t_1} d_s A(\tau,s-\tau)\cdot h(s)\right)d\tau\right\| \,\middle|\, h\in G([t_0,t_1],X), \|h\|\le 1\right\} \le$$

$$\le |t''-t'|\,SV^u[A]$$

where at $\overset{B}{=}$ we applied Theorem 2.12. Hence we proved that

$$K^{\square} \in C([t_0,t_1], SV^u([t_0,t_1],L(X)))$$

and since the kernel K is continuous it follows that the property 9) of

Theorem 3.4 is satisfied for K and for its iterates. Hence, if we have $A(t,s)\in K(X)$ (for instance if $\dim X < \infty$) then it is easy to see that $K(t,s)\in K(X)$ and by the Remark f that follows Theorem 3.4 (and by b of Theorem 2.9) we have the

THEOREM 4.1 - Under the preceding hypothesis, for every $f\in G([t_0,t_1],X)$ and $\phi\in G([-r,0],X)$ the system (D_K) has one and only one regulated solution.

Compare this result, for instance, with [12], p.142. In particular we proved the

THEOREM 4.2 - Let $A\in G^{\sigma}\cdot SV_0^u([t_0,t_1]\times[-r,0],K(X))$ be such that for every $s\in[-r,0]$ the function $t\in[t_0,t_1]\longmapsto A(t,s-t)\in L(X)$ is simply regulated. Then for every $g\in G([t_0,t_1],X)$ and $\phi\in G([-r,0],X)$ the equation $(\tilde{D})$ has one and only one regulated solution $y\in G([t_0-r,t_1],X)$ such that $y_{t_0}=\phi$.

Elsewhere we will extend the results above to linear functional differential equations of neutral type, consider the Green function in the case of generalized boundary conditions (see [4], chapter III), the adjoint system, the resolvent (see [12], p. 144) etc.

2 - Linear integro-differential equations.

Let us sketch the applications of III of Theorem 3.4 to linear integro-differential equations

$$\text{(L)} \qquad y(t) - x + \int_s^t \cdot dA(\sigma)\cdot y(\sigma) = f(t) - f(s), \qquad a\le t\le b$$

This equation is a particular instance of (K): we define $K(t,s) = A(s)$; the condition $K\in G^{\sigma}_{\Delta}\cdot SV^u(\Gamma_a,L(X))$ takes the form $A\in SVG^{\sigma}([a,b],L(X)) = SV([a,b],L(X))\cap G^{\sigma}([a,b],L(X))$ and we have $K^{\square}\in G([a,b],SV([a,b],L(X)))$. Hence in order to apply III of Theorem 3.4 (more precisely, the Remark d that follows it) it is enough to suppose that $A\in SVG^{\sigma}([a,b],K(X))$. Hence by 9) of Theorem 3.4 (and the Remark d that follows it) we have

THEOREM 4.3 - For $A \in SVG^{\sigma}([a,b],K(X))$ the following properties are equivalent:

1) For every $s \in [a,b]$, $x \in X$ and $f \in G([a,b],X)$ the equation (L) has one and only one regulated solution.

2) For every $t \in [a,b[$ $[t \in]a,b]]$ the operator $I_X + A(t\dot{+}) - A(t)$ $[I_X - A(t) + A(t\dot{-})]$ is invertible.

Analogously we have

THEOREM 4.4 - For $A \in SVG^{\sigma}([a,b],K(X))$ the following properties are equivalent

1) For every $s \in [a,b]$, $x \in X$ and $f \in G([s,b],X)$ the equation

$$(L^b) \qquad y(t) - x + \int_s^t \cdot dA(\sigma)\cdot y(\sigma) = f(t) - f(s), \qquad s \le t \le b$$

has one and only one solution $y \in G([s,b],X)$.

2) For every $t \in [a,b[$ the operator $I_X + A(t\dot{+}) - A(t)$ is invertible.

In [10] we consider the equation (L) under more general conditions and we characterize the functions that are resolvents of equations of type (L) and (L^b).

3. Linear Volterra integral equations.

A linear Volterra integral equation

$$(V) \qquad y(t) - x + \int_a^t H(t,s)\cdot y(s)ds = f(t) - f(a), \qquad a \le t \le b$$

may be reduced to an equation (K_a^b): we take

$$K(t,s) = {}_L\int_a^s H(t,\sigma)d\sigma,$$

where ${}_L\int$ denotes the Bochner-Lebesgue integral. *Here we consider only the case where* $\dim X < \infty$. It is easy to prove the

THEOREM 4.5 - We have $K \in G_\Delta \cdot BV^u(\Gamma_a, L(X))$ if and only if H satisfies the following properties:

1) $\sup_{a \le t \le b} \|H^t\|_1 < \infty$ where $\|H^t\|_1 = L\int_a^t \|H(t,s)\| ds$.

2) For every $s \in [a,b]$ the function $t \in [s,b] \longmapsto L\int_a^s H(t,\sigma) d\sigma \in X$ is regulated.

3) The function $t \in [a,b] \longmapsto L\int_a^t H(t,s) ds \in X$ is regulated.

However, if these properties are satisfied the operator defined by H is not necessarely compact and (V) may not have always a solution as is shown by the following

EXAMPLE - We take $X = \mathbb{C}$, $[a,b] = [0,1]$ and $H(t,s) = 0$ if $t = 0$, $H(t,s) = -\frac{1}{t}$ if $0 \le s < t$. Then we have $\|H^t\|_1 = 1$ for $0 < t \le 1$ and $1+K(t+,t+)-K(t+,t) = 1$. It is immediate that every $\lambda \in]0,1]$ is an eigenvalue of the operator defined by (H) and that (K_0^d) does not have always a solution.

THEOREM 4.6 - The operator defined by H is compact if and only if $H^\square \in G([a,b], L_1([a,b], L(X)))$ and then the property 9) of Theorem 3.4 is satisfied and (V) has always (i.e., for every $x \in X$ and $f \in G([a,b],X)$) one and only regulated solution.

In the case where the equation (V) has a resolvent R, we don't know if there exists a function S satisfying properties 1), 2), 3) of Theorem 4.5 and such that

$$R(t,s) = I_X + L\int_t^s S(t,\sigma) d\sigma$$

(and hence the solution of (V) is given by

$$y(t) = f(t) - L\int_a^t S(t,s) \cdot [f(s) - f(a) + x] ds$$

and (R^*), (R_*) are replaced, respectively, by

$$S(t,s) - H(t,s) + L\int_s^t H(t,\tau) \circ S(\tau,s) d\tau = 0, \quad a \le s \le t \le b$$

$$S(t,s) - H(t,s) + L\int_s^t S(t,\tau)\circ H(\tau,s)d\tau = 0, \quad a \le s \le t \le b).$$

4. Associated integrals.

Schwabik in [17], [18] works only with functions of bounded variation and in this case the integral he uses is equivalent to the Young integral, that we denote here by $Y\int$. The interior integral we work with is simpler and much easier to deal with then the Young integral and is also richer in properties (seė Theorems 1.12 and 2.2). However:

THEOREM 4.7 - The interior integral and the Young integral are *associated* in the following sense:

1) For every $\alpha \in SV([a,b],L(X,Y))$ and $f \in G([a,b],X)$ we have

$$\int_a^b \bullet d\alpha(t)\cdot f(t) + Y\int_a^b \alpha(t)\cdot df(t) = \alpha(b)\cdot f(b) - \alpha(a)\cdot f(a)$$

2) For $\alpha \in SVG^{\sigma}([a,b],L(X,Y))$ and $f \in G([a,b],X)$ we have

$$Y\int_a^b d\alpha(t)\cdot f(t) + \int_a^b \bullet\alpha(t)\cdot df(t) = \alpha(b)\cdot f(b) - \alpha(a)\cdot f(a).$$

REMARK - If we suppose that $f \in G^{\sigma}([a,b],L(W,X))$ then 1) and 2) remain valid if we replace $\int\bullet$ and $Y\int$ by $\overset{\sigma}{\int}\bullet$ and $Y\overset{\sigma}{\int}$ respectively.

The proof of Theorem 4.7 follows from [4], Theorem I.4.21, since, for instance,

$$Y\int_a^b \alpha(t)\cdot df(t) = \lim_{d\in D} \sum_{i=1}^{|d|}\Big\{\alpha(t_{i-1})\cdot[f(t_{i-1}+)-f(t_{i-1})] + \alpha(\xi_i^\bullet)\cdot[f(t_i-) - f(t_{i-1}+)] +$$

$$+ \alpha(t_i)\cdot[f(t_i)-f(t_i-)]\Big\} = \alpha(b)\cdot f(b) - \alpha(a)\cdot f(a) - \lim_{d\in D}\sum_{i=1}^{|d|}\Big\{[\alpha(\xi_i^\bullet) -$$

$$- \alpha(t_{i-1})]\cdot f(t_{i-1}+) + [\alpha(t_i)-\alpha(\xi_i^\bullet)]\cdot f(t_i-)\Big\} = \alpha(b)\cdot f(b) - \alpha(a)\cdot f(a) -$$

$$- \int_a^b \bullet d\alpha(t)\cdot f(t) \quad (\text{where } \xi_i^\bullet \in]t_{i-1},t_i[).$$

In the case where $\alpha\epsilon BV([a,b],L(X))$ and $f\epsilon G([a,b],L(X))$ the association was proved by MacNerney, [14], Theorem B. The association allows us to translate the results from one of the integrals into those of the other one (Cf. Remark h after Theorem 3.4).

REFERENCES

[1] - S.E.Arbex, Equações integrais de Volterra-Stieltjes com núcleos descontinuos, Doctor Thesis, Instituto de Matemática e Estatística da Universidade de São Paulo, 1976.

[2] - C.W.Bitzer, Stieltjes-Volterra Integral equations, Illinois J.of Math 14(1970), 434-451.

[3] - J.B.F.Gomes, in preparation.

[4] - C.S.Hönig, Volterra Stieltjes-integral equations, Mathematics Studies 16, North-Holland Publishing Comp., Amsterdam, 1975.

[5] - C.S.Hönig, Volterra-Stieltjes integral equations with linear constraints and discontinuous solutions, Bull. Amer. Math. Soc., 81(1975), 593-598.

[6] - C.S.Hönig, The Dirichlet and substitution formulas for Riemann-Stieltjes integrals in Banach spaces, in "Functional Analysis" edited by Djairo Guedes de Figueiredo, Lectures Notes in Pure and Applied Mathematics, vol. 18, p. 135-189, Marcel Dekker, 1976.

[7] - C.S.Hönig, Fredholm Stieltjes-integral equations, I, in publication.

[8] - C.S.Hönig, The abstract Riemann-Stieltjes integral and its applications to linear differential equations with generalized boundary conditions, Notas do Instituto de Matemática e Estatística da Universidade de São Paulo, Série Matemática nº 1, 1973.

[9] - C.S.Hönig, Análise Funcional e Aplicações, 2 vol., Instituto de Matemática e Estatística da Universidade de São Paulo, 1970.

[10] - C.S.Hönig, The resolvent of a linear Stieltjes integro-differential equation, to appear.

[11] - C.S.Hönig, in preparation.

[12] - J.Hale, Theory of Functional Differential Equations, Applied Mathematical Sciences 3, Springer-Verlag, 1977.

[13] - D.B.Hinton, A Stieltjes-Volterra integral equations theory, Canada J.Math., 18 (1966), 314-331.

[14] - J.S.MacNerney, An integration-by-parts formula, Bull. Amer. Math. Soc., 69, (1963), 803-805.

[15] - A.P.Robertson & W.J.Robertson, Topological Vector Spaces, Cambridge University Press, 1964.

[16] - C.S.Cardassi, Dependência diferenciável das soluções de equações integro-diferenciais em espaços de Banach, Master Thesis, Instituto de Matemática e Estatística da Universidade de São Paulo, 1975.

[17] - St.Schwabik, On Volterra-Stieltjes integral equations, Čas. pro pěst. mat., 99, (1974), 255-278.

[18] - St.Schwabik, Note on Volterra-Stieltjes integral equations, Čas. pro pěst.mat., 102(1977), 275-279.

[19] - J.C.Prandini, Funções de semi-variação limitada de mais de uma variável, Master thesis, Instituto de Matemática e Estatística da Universidade de São Paulo, 1978.

[20] - C.S.Hönig, Compactifying subspaces, in preparation.

- ABV -

RELATIONSHIP IN THE NEIGHBOURHOOD OF INFINITY AND ASYMPTOTIC EQUIVALENCE OF NEUTRAL FUNCTIONAL DIFFERENTIAL EQUATIONS

by A. F. Izé and A. Ventura

0. Introduction.

The study of asymptotic behavior of differential equations is very important to understand the qualitative behavior of the solutions of an ordinary differential equations and several mathematicians, N. Levinson, H. Weyl, P. Hartman, R. Bellman, K. Cooke, J. Hale, L. Cesari and others have done a great deal of work in this area. The theory of funcitonal differential equations is relatively new, has developed mainly in the last twenty years and not many papers appeared on asymptotic behavior of functional differential equations. One early paper was published by Bellman and Cooke, in 1959, [2], followed by several others that consider a nonlinear delay equation as a perturbation of an ordinary differential equation, see for example Cooke [6]. Although it is important in some cases to consider the problem in this setting , the real approach is to consider a nonlinear functional differential equations as a perturbation of a linear functional differential equation, because the linearized equation is still a linear functional differential equation and the difficulties involved in the solution of the problem is connected with the fact that the space of the solutions of the linear system is now an infinite dimensional space. The first paper in this direction was published by J. Hale [14] that is a generalization of Bellman and Cooke paper. A further generalization appeared later, Izé and Molfetta [18] for a more general class of neutral functional differential equations. We should point out that for functional differential equations of the neutral type the methods used to study asymptotic behavior are in some sense more efective that the classical Lyapunov second method since when we use the classical Lyapunov theorems for neutral equations we have always to assume that the operator D is uniformly stable in the sense of definition (1.8).

We should remark that in the proof of Theorem 2.1 we do not assume that the operator D is uniformly stable and then it can be applied even to perturbations of linear equations that have a weird behavior like those given by Gromova and Zverkin [9] and Brumley [4].

1. Preliminary.

Let $r > 0$ be a given real number, $R = (-\infty,\infty)$, $E^n = R^n$ or C^n a complex n-dimensional linear vector space with norm $|.|$ and $C([a,b], E^n)$ the Banach space of continuous functions mapping the interval $[a,b]$ into E^n with the topology of uniform convergence our compact sets. If $[a,b] = [-r,0]$ then $C = C([-r,0], E^n)$ and the norm in C will be given by

$$||\phi|| = \sup_{-r\leq\theta\leq 0} |\phi(\theta)|$$

If $\sigma \in R$, $A \geq 0$ and $x \in C([\sigma-r,\sigma+A], E^n)$, then for every $t \in [\sigma,\sigma+A]$ we let $x_t \in C$ be defined by $x_t(\theta) = x(t+\theta)$, $-r\leq\theta\leq 0$. If $\Omega \subset R \times C$ is open and if $D,f : \Omega \to R^n$ are continuous functions we say that the relation

$$(1.1) \quad \frac{d}{dt}D(t,x_t) = f(t,x_t)$$

is a functional differential equation. A function x is said to be a solution of (1.1) if there are $\sigma \in R$ and $A > 0$ such that $x \in C([\sigma-r,\sigma+A), R^n)$, $(t,x_t) \in \Omega$, $t \in [\sigma,\sigma+A)$ and x satisfies (1.1) on $(\sigma,\sigma+A)$. For a given $\sigma \in R$, $\phi \in C$, $(\sigma,\phi) \in \Omega$, we say that $x(\sigma,\phi)$ is a solution of (1.1) with initial value (σ,ϕ) or a solution of (1.1) through (σ,ϕ) if there is an $A > 0$ such that $x(\sigma,\phi)$ is a solution of (1.1) on $[\sigma-r,\sigma+A)$ and $x_\sigma(\sigma,\phi) = \phi$.

Let X, Y be Banach spaces, $L(X,Y)$ the Banach space of bounded linear mappings from X into Y. If $L \in L(C,R^n)$, then the Riesz representation Theorem implies that there is an $n \times n$ matrix function η on $[-r,0]$ of bounded variation such that

(1.2) $L\phi = \int_{-r}^{0} [d\eta(\theta)]\phi(\theta)$

For any such η we always understand that we have extended the definition to R so that $\eta(\theta) = \eta(-r)$ for $\theta \le -r$, $\eta(\theta) = \eta(0)$ for $\theta \ge 0$.

Let Λ be an open subset of a metric space. We say that $L : \Lambda \to L(C,R^n)$ has smoothness on the measure if for any real β there is a scalar function $\gamma(\lambda,s)$ continuous for $\lambda \in \Lambda$, $s \in R$, $\gamma(\lambda,0) = 0$ such that if

$$L(\lambda)\phi = \int_{-r}^{0} [d\eta(\lambda,\theta)]\phi(\theta), \quad \lambda \in \Lambda, \quad 0 \le s, \text{ then,}$$

(1.3) $$\left|\lim_{h\to 0^+} \int_{\beta+h}^{\beta+s} + \int_{\beta-s}^{\beta-h} [d\eta(\lambda,\theta)]\phi(\theta)\right| \le \gamma(\lambda,s)\,||\phi||$$

If $\beta \in R$ and the matrix $A(\lambda;\beta,L) = \eta(\lambda,\beta^+) - \eta(\lambda,\beta^-)$ is nonsingular on $\lambda = \lambda_o$, we say that $L(\lambda)$ is atomic at β at λ_o. If $A(\lambda;\beta,L)$ is nonsingular on a set $K \subset \Lambda$ we say that $L(\lambda)$ is atomic at β on K.

Let $\Lambda = \Omega \subset R \times C$ and let $L \in C(\Omega, L(C,R^n))$.

If $D : \Omega \to R^n$ has a continuous first derivative with respect to ϕ, then Lemma 5.1 of J. Hale [10], p. 50, implies that D_ϕ has smoothness on the measure.

<u>Definition</u> 1.1. Suppose $\Omega \subset R \times C$ is an open set and $(t,\phi) \in \Omega$. A function $D : \Omega \to R^n$ (not necessarily linear) is said to be atomic at β on Ω if D is continuous together with its first and second Fréchet derivatives with respect to ϕ and D_ϕ, the derivative with respect to ϕ, is atomic at β on Ω.

System (1.1) is called a functional differential equation of neutral type if D is atomic at zero.

We assume also the existence, uniqueness, and continuous dependence with respect to initial condition of the solutions of (1.1).

Assume $\Omega \subset C$ is open, $D : \Omega \to R^n$ is a continuous linear operator given by

(1.4) $D_\phi = \phi(0) - g(\phi)$ with $g(\phi) = \int_{-r}^{0} [d\mu(\theta)]\phi(\theta)$

where $\mu(\theta)$, $-r \le \theta \le 0$, is an $n \times n$ matrix whose elements are of bounded variation and does not have a singular part, that is,

(1.5) $\int_{-r}^{0} [d\mu(\theta)]\phi(\theta) = \sum_{k=1}^{\infty} A_k \phi(-w_k) + \int_{-r}^{0} A(\theta)\phi(\theta)d\theta$

where

$$0 < w_k \le r \quad \text{and} \quad \sum_{k=1}^{\infty} |A_k| + \int_{-r}^{0} |A(\theta)|d\theta < \infty.$$

Our main concern are the systems

(1.6) $\frac{d}{dt}Dx_t = Lx_t$

(1.7) $\frac{d}{dt}[Dx_t - G(t,x_t)] = Lx_t + F(t,x_t)$

where D satisfies (1.4), L satisfies (1.2), $Dx_t - G(t,x_t)$ is atomic at zero and $G(t,\phi)$ does not depend on $\phi(0)$, $\phi = x_\sigma$.

If $\phi \in C$ and $x_t(\sigma,\phi)$ is the unique solution of (1.6) we define the operator $T(t-\sigma) : C \to C$, $t \ge \sigma$ by the relation

(1.8) $x_t(\sigma,\phi) = T(t-\sigma)$, $T(0) = I$

$\{T(t-\sigma)\}_{t \in [\sigma,\infty)}$ is a family of strongly continuous semigroup from C into itself for all $t \ge \sigma$.

In [10] Theorem 10.1, p. 307, is proved that:

<u>Theorem</u> 1.2. i) The infinitesimal generator A of the semigroup $T(t-\sigma)$, $t \ge \sigma$, of Equation (1.6) has domain $\mathcal{D}(A)$ and range $\mathcal{R}(A)$ respectively given by

$$\mathcal{D}(A) = \{\phi \in C : \dot\phi \in C,\ D\dot\phi = L\dot\phi\},\quad A\dot\phi = \dot\phi$$

ii) The spectrum $\sigma(A)$ coincides with the point espectrum (eigenvalues) and $\lambda \in \sigma(A)$ if and only if λ satisfies the characteristic equation

$$\det\Delta(\lambda) = 0,\quad \Delta(\lambda) = \lambda D(e^{\lambda\cdot}I) - L(e^{\lambda\cdot}I)$$

iii) The roots of the characteristic equations have real parts bounded above and if $\lambda \in \sigma(A)$ then the generalized eigenspace $M_\lambda(A)$ is finite dimensional and there is an integer $k = k(\lambda)$ such that $M_\lambda(A) = N(A-\lambda I)^k$ and $C = N(A-\lambda I)^k \oplus R(A-\lambda I)^k$.

iv) Suppose Λ is a finite set $\{\lambda_1, \lambda_2, \ldots, \lambda_p\}$ of elements of $\sigma(A)$, and $\Phi_\Lambda = (\Phi_{\lambda_1}, \ldots, \Phi_{\lambda_p})$, $B_\Lambda = (B_{\lambda_1}, \ldots, B_{\lambda_p})$ where Φ_{λ_j} is a basis for the generalized eigenspace of λ_j and B_{λ_j} is the matrix defined by $A\Phi_{\lambda_j} = \Phi_{\lambda_j} B_{\lambda_j}$, $j = 1,2,\ldots,p$. Then, the only eigenvalue of B_{λ_j} is λ_j and, for any eigenvector a of the same dimension as Φ_Λ, the solution $T(t-\sigma)\Phi_\Lambda a$ of Equation (1.4) with initial value $\Phi_\Lambda a$ at $t = \sigma$ may be defined on $(-\infty,\infty)$ by the relation:

$$
\begin{aligned}
&T(t-\sigma)\Phi_\Lambda a = \Phi_\Lambda e^{B_\Lambda(t-\sigma)} a \\
(1.9)\quad & \\
&\Phi_\Lambda(\theta) = \Phi_\Lambda(0)e^{B_\Lambda \theta} \qquad -r \le \theta \le 0
\end{aligned}
$$

Furthermore, there exists a subspace Q_Λ of C such that $T(t-\sigma)Q_\Lambda \subset Q_\Lambda$, $t \ge \sigma$ and $C = P_\Lambda \oplus Q_\Lambda$, where

$$P_\Lambda = \{\phi \in C : \phi = \Phi_\Lambda a, \text{ for some vector } a\}.$$

We can give an explicit characterization of decomposition of space C via the formal adjoint equation

$$(1.10)\quad \frac{d}{ds}[y(s) - \int_{-r}^{0} y(s-\theta)d\mu(\theta)] = -\int_{-r}^{0} y(s-\theta)d\eta(\theta)$$

where y is an n-dimensional row vector. If $C^* = C([0,r], E^{n*})$ where E^{n*} is the n-dimensional space of row vectors, then, for any $\phi \in C$ the bi-linear forms

$$
\begin{aligned}
&(\alpha,\phi) = \alpha(0)\phi(0) + \int_{-r}^{0} [\frac{d}{du}\int_{0}^{u} \alpha(s-u)d\mu(\theta)\phi(s)ds]_{u=0} \\
(1.11)\quad & - \int_{-r}^{0}\int_{0}^{\theta} \alpha(s-\theta)d\eta(\theta)\phi(s)ds
\end{aligned}
$$

is defined for $\alpha \in C^* = C([0,r], E^{n*})$, $\dot{\alpha} \in C^*$, $\phi \in C$.

The following theorem is proved in [10], p. 309.

Theorem 1.3. If $\Lambda = \{\lambda_1, \ldots, \lambda_p\}$ is a finite set of elements of $\sigma(A)$ and P_Λ is the linear extension of $M_{\lambda_j}(A)$, $j=1,\ldots,p$ with basis Φ_Λ and P^*_Λ is the linear extension of the correspondent generalized eigenspace of the formal adjoint equation with basis Ψ_Λ, then one can choose Ψ_Λ so that $(\Psi_\Lambda, \Phi_\Lambda) = I$, is the identity and

$$C = P_\Lambda \oplus Q_\Lambda$$

$$P_\Lambda = \{\phi \in C : \phi = \phi_\Lambda a, \text{ for some vector } a\} \tag{1.12}$$

$$Q_\Lambda = \{\phi \in C : (\Psi_\Lambda, \phi) = 0\}.$$

In the conditions of theorem 1.3 we say that the space C is decomposed by Λ and therefore if $\phi \in C$, we have $\phi = \phi^{P_\Lambda} + \phi^{Q_\Lambda}$ where $\phi^{P_\Lambda} = \Phi_a$, $a = (\Psi_\Lambda, \phi)$ and $\phi^{Q_\Lambda} = \phi - \phi^{P_\Lambda}$.

Furthermore, if C is decomposed by Λ, then, $\sigma(T(t-\sigma) \mid P_\Lambda) = \sigma(e^{B_\Lambda(t-\sigma)})$ where $A\Phi_\Lambda = \Phi_\Lambda B_\Lambda$ and $A\Psi_\Lambda = B_\Lambda \Psi_\Lambda$.

1.2. Variation of Constants Formula.

Consider $\{T_D(t-\sigma),\ t \geq \sigma\}$ the strongly continuous semigroup of linear mappings associated with the solution of the difference equation $\frac{d}{dt}Dx_t = 0$.

Definition 1.4. The order a_D of the semigroup is defined by $a_D = \inf\{a \in R : \exists\ K = K(a) \text{ with } ||T_D(t-\sigma)\phi|| \leq Ke^{a(t-\sigma)},\ t \geq \sigma\}$.

In [10], we can find some results that can be stated as

Theorem 1.5. If D and L satisfies (1.6), if the matrix μ has no singular part and if $a > a_D$ is fixed, then, the set $\Lambda_a = \{\lambda \in \sigma(A) : \mathrm{Re}\lambda \geq a\}$ is finite and the space C can be decomposed by Λ_a as $C = P \oplus Q$, where P and Q are invariant subspaces under $T(t-\sigma)$ and A, and the space P is finite dimensional and corresponds to the initial data of all those solutions of (1.6) which are of the form $p(t)\exp(\lambda t)$, where $p(t)$ is a polynomial in t and $\lambda \in \Lambda_a$.

If $x_t(\sigma,\phi)$ is a solution of (1.6), then, according to the theorems 1.2 and 1.3 we may write $x_t = x_t^P + x_t^Q$.

Now let $X(t)$, $t \geq \sigma$, be the $n \times n$ matrix function defined for all $t \in [0,\infty)$ of bounded variation in t and continuous in t from the right such that

$$(1.13) \quad D(X_{t-\sigma}) = I + \int_{\sigma}^{t} L(X_{s-\sigma})ds, \quad t \geq \sigma$$

$$X_o(\theta) = \begin{cases} 0 & -r \leq \theta < 0 \\ I & \theta = 0 \end{cases}$$

So according to [10], page 302, a solution of (1.7) with initial value ϕ in σ, satisfies the variation of constants formula

$$(1.14) \quad \begin{aligned} x_t - X_oG(t,x_t) &= T(t-\sigma)[\phi - X_oG(0,\phi)] \\ &+ \int_{\sigma}^{t} \{[-d_sT(t-s)X_o]G(s,x_s) + T(t-s)X_oF(s,x_s)\}ds, \quad t \geq \sigma \end{aligned}$$

where $T(t-\sigma)[\phi - X_oG(0,\phi)]$ is defined as $T(t-\sigma)\phi - X_tG(0,\phi)$ and

$$(1.15) \quad X(t-\sigma)X_o = X_{t-\sigma}$$

The integral in (1.14) are evaluated at each $\theta \in [-r,0]$ as ordinary integrals in E^n. Also, if C is decomposed by Λ as $C = P \oplus Q$, then, Equation (1.14) is equivalent to

$$(1.16) \quad \begin{aligned} x_t^P - X_o^PG(t,x_t) &= T(t-\sigma)[\phi^P - X_o^PG(0,\phi)] \\ &+ \int_{\sigma}^{t} \{[-d_sT(t-s)X_o^P]G(s,x_s) + T(t-s)X_o^PF(s,x_s)ds\} \\ x_t^Q - X_o^QG(t,x_t) &= T(t-\sigma)[\phi^Q - X_o^QG(0,\phi)] \\ &+ \int_{\sigma}^{t} \{[-d_sT(t-s)X_o^Q]G(s,x_s) + T(t-s)X_o^QF(s,x_s)ds\} \end{aligned}$$

where the superscripts P and Q designate the projection of the corresponding function onto the subspaces P and Q, respectively.

However, we must observe that everything is clear in (1.16) except for the meaning of the projections X_o^P, X_o^Q since X_o is not continuous. Projection operators taking C onto P and C are easily determined by means of the adjoint differential equation (1.10) and the bilinear form (1.11). One can show that (Ψ, X_o) is well defined and $(\Psi, X_o) = \Psi(0)$.

Therefore if we put

$$(1.17) \quad X_o^P = \Phi\Psi(0), \quad X_o^Q = X_o - X_o^P$$

the quantities in (1.16) are well defined.

Also in [17], Henry has given some exponential estimates for the solutions of (1.6).

<u>Theorem</u> 1.6. If $\alpha > a_D$ and $\Lambda = \{\lambda \in \sigma(A) : \text{Re}\lambda > \alpha, \det\Delta(\lambda)=0\}$, then C is decomposed by Λ as $C = P \oplus Q$ and there are positive constants M_1, M_2, M_3, M_4 and ε such that for $\sigma \geq 0$

$$(1.18) \quad \begin{aligned} ||T(t-\sigma)\phi^Q|| &\leq M_1 e^{(\alpha-\varepsilon)(t-\sigma)}||\phi^Q||, \quad t \geq \sigma, \ \phi^Q \in Q \\ ||T(t-\sigma)\phi^P|| &\leq M_2 e^{(\alpha+\varepsilon)(t-\sigma)}||\phi^P||, \quad t \geq \sigma, \ \phi^P \in P \end{aligned}$$

and

$$(1.19) \quad \begin{aligned} ||T(t-\sigma)X_o^Q|| &\leq M_3 e^{(\alpha-\varepsilon)(t-\sigma)}, \quad t \geq \sigma \\ ||T(t-\sigma)X_o^P|| &\leq M_4 e^{(\alpha+\varepsilon)(t-\sigma)}, \quad t \leq \sigma \end{aligned}$$

$$(1.20) \quad \begin{aligned} \int^{\infty} ||d_s T(s) X_o^Q|| e^{-(\alpha-\varepsilon)s} &\leq M_3 \\ \int^{\infty} ||d_s T(s) X_o^P|| e^{-(\alpha+\varepsilon)s} &\leq M_4 \end{aligned}$$

The relation (1.14) suggests the possibility of introducing a new variable for the expression on the left-hand side. However, some care must be exercised because the new variable would not be a continuous function on $[-r,0]$.

Let PC be the space of functions $\phi : [-r,0] \to E^n$ which are uniformly continuous on $[-r,0)$ and for which there exists $\phi(0^-)$. If X_o is the matrix defined in (1.15) it is clear that $PC = C + \langle X_o \rangle$, where $\langle X_o \rangle$ is the span of $\{X_o\}$; that is, any $\psi \in PC$ is given by

$$\psi = \phi + X_o b$$

where $\phi \in C$ and $b \in E^n$. We make PC a Banach space by defining

$$||\psi|| = \max(||\phi||, |b|).$$

An element ψ in PC can, therefore, be identified with the pair (ϕ,b) for $\phi \in C$, $b \in E^n$.

For any ψ in PC we can define

$$T(t-\sigma)\psi = T(t-\sigma)\phi + T(t-\sigma)X_0 b, \quad t \geq \sigma$$

The operator $T(t-\sigma)$ takes PC into functions on $[-r,0]$ and is linear, but $T(t-\sigma)$ does not necessarily take PC into PC.

Let PC_G be the closed set in PC defined by

$$PC_G = \{\psi \in PC : \psi(0) = \phi(0^-) - X_0 G(t,\phi)\}$$

Since $G(t,\phi)$ does not depend on $\phi(0)$, the maps

$$h_1 : C \to PC_G, \quad h_1(\phi) = \phi - X_0 G(t,\phi)$$

$$h_2 : PC_G \to C, \quad h_2(\phi) = \phi + X_0 G(t,\phi)$$

are homeomorphisms such that $h_1 o h_2 = h_2 o h_1 = I$, the identity.

For each ψ in PC_G there is a unique function $\phi \in C$ such that

$$\psi = \phi - X_0 G(t,\phi)$$

and conversely. Therefore, the semigroup $T(t-\sigma)$ is well defined on PC_G, since $T(t-\sigma)$ is well defined on C and $\langle X_0 \rangle$.

In the following, we let $||T(t-\sigma)\phi||$, be defined by $\sup\{|T(t-\sigma)\phi(\theta)|, \ -r \leq \theta \leq 0\}$, for $\phi \in PC_G$.

The domain of definition of the infinitesimal generator A of the semigroup $T(t-\sigma)$ can be extended to PC by letting

$$\mathcal{D}(A) = \{\phi \in PC : \dot{\phi} \in PC\}$$

and

$$A\phi(\theta) = \begin{cases} \dot{\phi}(\theta), & -r \leq \theta < 0 \\ g(\dot{\phi}) + L(\phi), & \theta = 0 \end{cases}$$

The variation of constants formula (1.14) for the solutions of (1.7) suggests the change of variables

(1.21) $z_t = x_t - X_o G(t,x_t)$

to obtain a new equation for z_t in PC.

The transformation (1.21) is a well-defined transformation from C to PC since $G(t,\phi)$ depends only upon the values of $\phi(\theta)$ for $-r \le \theta < 0$. Therefore, $G(t,x_t) = G(t,z_t)$ for $-r \le \theta < 0$.

Thus, if

(1.22) $z_t = x_t - X_o G(t,x_t)$, $x_t = z_t + X_o G(t,x_t) \overset{\text{def}}{=} H(t,z_t)$

the relation (1.14) becomes

$$z_t = T(t-\sigma)z_\sigma + \tag{1.23}$$
$$+ \int_\sigma^t \{[-d_s T(t-s)X_o]G(s,z_s) + T(t-s)X_o F(s,H(s,z_s))ds\}$$

Let Φ, Ψ be the matrices defined by the decomposition $C=P\oplus Q$, $(\Psi,\Phi) = I$ and let B be the $p \times p$ matrix such that $T(t)\Phi=\Phi\exp(Bt)$, $t \in (-\infty,\infty)$. The spectrum of B is Λ. For any $\Phi \in PC$ one can define (ψ,ϕ) and, therefore, it is meaningful to put

$$\phi^P = \Phi(\psi,\phi), \quad \phi^Q = \phi - \phi^P, \quad \phi \in PC$$

One can show that (ψ,X_o) is well defined and $(\Psi,X_o) = \Psi(0)$. Therefore, if we put

$$X_o^P = \Phi\Psi(0), \quad X_o^Q = X_o - X_o^P$$

the quantities in (1.14) are well defined. If we apply apropriently the relations (1.22) we can split the equation (1.23) in the following way

$$z_t^P = T(t-\sigma)z_\sigma^P + \int_\sigma^t \{[-d_s T(t-s)X_o^P]G(s,z_s) + T(t-s)X_o^P F(s,H(s,z_s))ds\}$$
(1.24)
$$z_t^Q = T(t-\sigma)z_\sigma^Q + \int_\sigma^t \{[-d_s T(t-s)X_o^Q]G(s,z_s) + T(t-s)X_o^Q F(s,H(s,z_s))ds\}$$

1.3. The Difference Equation.

Consider the difference equation $x(t) = \sum_{k=1}^{\infty} A_k x(t-w_k)$, where $0 < w_k \le r$, $\sum_{k=1}^{\infty} |A_k| < \infty$ and $\sum_{w_k \le \varepsilon} |A_k| \to 0$ as $\varepsilon \to 0$. This equation can be written in the form

(1.25) $D^o x_t = 0, \quad t \geq 0$

where D^o is the difference operator

(1.26) $D^o\phi = \phi(0) - \sum_{k=1}^{\infty} A_k\phi(-w_k), \quad \phi \in C$

We define $C^o = C \cap N(D^o)$, where $N(D^o)$ is the null space of D^o. If $x(\phi)$ is the solution of (1.25) with initial value $\phi \in C^o$, we can write

(1.27) $x_t(\phi) = T^o(t)\phi, \quad t \geq 0$

thus defining a strongly continuous semigroup of operators $\{T^o(t)\}_{t\geq 0}$ on C^o.

If $D\phi = \phi(0) - g(\phi)$ with $g(\phi)$ nonatomic at zero from Lemma 3.2 of [10], we have

Lemma 1.7. If $D : C \to R^n$ is linear and atomic at zero, then there are n functions $\phi_1, \ldots, \phi_n$ such that $D\phi = I$ where $\phi = (\phi_1, \phi_2, \ldots, \phi_n)$.

Definition 1.8. The operator D is said to be uniformly stable if there exist constants $K \geq 1$ and $\alpha > 0$ such that

(1.28) $||T^o(t)\phi|| \leq Ke^{-\alpha(t-\sigma)}||\phi||, \quad \phi \in C^o, \quad t \geq 0$

It is shown in [10] that D uniformly stable implies that there exists an $n \times n$ matrix function $B(t)$ defined and of bounded variation on $[-r,\infty)$, continuous from the left $B(t) = 0$, $-r \leq t \leq 0$, and a constant $M > 0$ such that

$$||T^o(t)\phi|| \leq M||\phi||, \quad t \geq 0, \quad \phi \in C, \quad \sup_{t\geq -r} |B(t)| \leq M.$$

Theorem 1.9. The following conditions are equivalent

i) D is uniformly stable;

ii) $a_D < 0$ where a_D is the order of the semigroup $T^o(t)$;

iii) there are constants $a > 0$ and $b > 0$ such that for any $h \in C([0,\infty), R^n)$, any solution y of the nonhomogeneous equation $Dy_t = h(t)$, $t \geq 0$, satisfies

$$||y_t|| \le be^{-at}||y_o|| + b \sup_{0\le u\le t} |h(u)|, \quad t \ge 0$$

iv) if $D\phi = \phi(0) - \int_{-r}^{o} [d\mu(\theta)]\phi(\theta)$, where μ satisfies the conditions (1.5) of section 1, then, there exists $\delta > 0$ such that all solutions of the characteristic equation $\det[I - \int_{-r}^{o} e^{\lambda\theta} d\mu(\theta)] = 0$ satisfies $Re\lambda \le -\delta$.

2. Main results.

The lemma bellow gives a characterization on the asymptotic behavior of the solutions of the linear neutral differential equation

$$(2.1) \qquad \frac{d}{dt} Dy = Ly_t$$

where D and L are linear.

Lemma 2.1. If $\beta > a_D$ and y_t is a solution of (2.1) such that

$$||y_t||/(\exp \beta t)$$

do not tends to zero as $t \to \infty$, then there exists a non negative integer ℓ and a real number α uniquely determined such that

$$0 < \varliminf_{t\to\infty} ||y_t||/t^{\ell} e^{\alpha t} \le \varlimsup_{t\to\infty} ||y_t||/t^{\ell} e^{\alpha t} < \infty.$$

Proof: If $\beta > a_D$ follows from theorem 1.6, section 1, that the space C is decomposed by the set

$$\Lambda = \{\lambda \in \sigma(A) : Re\lambda \ge \beta, \det \Delta(\lambda) = 0\}$$

as $C = P \oplus Q$, where the subespace P is of finite dimension and there are positive constants M_1, M_2 and δ such that

$$||T(t)\phi^Q|| \le M_2 e^{(\beta-\delta)t} ||\phi^Q||, \quad t \ge 0, \quad \phi^Q \in Q$$

$$||T(t)\phi^P|| \le M_1 e^{(\beta+\delta)t} ||\phi^P||, \quad t \le 0, \quad \phi^P \in P.$$

Whithout loss of generality we can assume $y_t = T(t)\phi$, for some $\phi \in C$ and since $||y_t||/e^{\beta t}$ do not tends to zero as $t \to \infty$, we have

$$||T(t)\phi^Q||/e^{\beta t} \to 0 \quad \text{as} \quad t \to \infty$$

and

$$||T(t)\phi^P||/e^{\beta t} \not\to 0 \quad \text{as} \quad t \to \infty.$$

From (1.9) of theorem 1.2 of section 1 we have

$$T(t)\phi^P = \Phi e^{Bt}b$$

where B is a Jordan matrix with eigenvalues which are the elements of Λ, have real part greater than or equal to β.

Let

$$z(t) \overset{\text{def}}{=} T(t)\phi^P(0) = \Phi(0)e^{Bt}b$$

There are nonnegative integers ℓ and a real number α, $\alpha \geq \beta$, such that

$$(2.2) \qquad 0 < \underline{\lim}_{t\to\infty} |z(t)|/t^{\ell}e^{\alpha t} \leq \overline{\lim}_{t\to\infty} |z(t)|/t^{\ell}e^{\alpha t}$$

We say that

$$(2.3) \qquad 0 < \underline{\lim}_{t\to\infty} ||T(t)\phi^P||/t^{\ell}e^{\alpha t} \leq \overline{\lim}_{t\to\infty} ||T(t)\phi^P||/t^{\ell}e^{\alpha t} < \infty$$

In fact from (2.2) it follows that there are positive numbers c_1, c_2 and T_1 such that

$$c_1 \leq |z(t)|/t^{\ell}e^{\alpha t} \leq c_2 \quad \text{for} \quad t \geq T_1 - r.$$

But for

$$\frac{|z(t+\theta)|}{t^{\ell}e^{\alpha t}} = \frac{|z(t+\theta)|}{(t+\theta)^{\ell}e^{\alpha(t+\theta)}} \cdot e^{\alpha\theta}\left(1 + \frac{\theta}{t}\right)^{\ell}$$

It is casy to see that there is $T_2 \geq T_1$, large enough, such that $t \geq T_2$ and $\theta \in [-r,0]$, we have:

$$\frac{|z(t+\theta)|}{t^{\ell}e^{\alpha t}} \geq \frac{c_1}{2} \cdot e^{\alpha\theta} \geq \frac{c_1}{2} \min\{e^{-\alpha r}, 1\} > 0$$

and

$$\frac{|z(t+\theta)|}{t^{\ell}e^{\alpha t}} \leq c_2 \max\{e^{-\alpha r}, 1\} < \infty.$$

The above equalities imply that (2.3) is true.

Since $\alpha \geq \beta$ it follows that

$$\lim_{t\to\infty} ||T(t)\phi^Q||/t^{\ell}e^{\alpha t} = 0$$

and then

$$0 < \underline{\lim}_{t\to\infty} ||T(t)\phi^P||/t^{\ell}e^{\alpha t} = \underline{\lim}_{t\to\infty} ||T(t)\phi||/t^{\ell}e^{\alpha t}$$

and

$$\overline{\lim}_{t\to\infty} ||T(t)\phi||/t^{\ell}e^{\alpha t} = \overline{\lim}_{t\to\infty} ||T(t)\phi^P||/t^{\ell}e^{\alpha t} < \infty$$

Notation: We say that $||y_t|| \sim t^{\ell}e^{\alpha t}$ if

$$0 < \underline{\lim}_{t\to\infty} ||y_t||/t^{\ell}e^{\alpha t} \leq \overline{\lim}_{t\to\infty} ||y_t||/t^{\ell}e^{\alpha t} < \infty.$$

Consider now $\beta > a_D$ and $y(t)$ a solution of (2.1) such that

$$||y_t|| \sim t^{\ell}e^{\alpha t} \quad \text{with} \quad \ell \in Z_{++},\ \alpha \in R,\ \alpha \geq \beta.$$

Let $\Lambda = \{\lambda \in \sigma(A) : \operatorname{Re} \lambda \geq \sigma,\ \det \Delta(\lambda) = 0\}$.

Let P, Q and B be as in Lemma 2.1 and let N be the order of the larger block of B which has in the diagonal an integer with real part equal to α.

Let

$$P_1 = \{\phi \in P : \lim_{t\to\infty} ||T(t)\phi||/t^{\ell}e^{\alpha t} = 0\}.$$

The next lemma is proved in details in [26].

Lemma 2.2. There exists a subespace P_2 of P and there are projection

$X^{P_i}: P \to P_i, i = 1, 2$ such that $P = P_1 \oplus P_2$, $X^{P_1} + X^{P_2} = I$.

Furthermore there are positive constants M and σ, such that

(2.4) $||T(t)X^{P_1}T(-s)\phi^P|| \leq Mt^{\ell-1}s^{N-\ell}e^{\alpha(t-s)}||\phi^P||, \quad \sigma \leq s \leq t$

(2.5) $||T(t)X^{P_2}T(-s)\phi^P|| \leq Mt\, s^{N-\ell-1}e^{\alpha(t-s)}||\phi^P||, \quad s \geq t \geq \sigma$

Consider now the systems

(L) $\frac{d}{dt} Dy_t = Ly_t$

(P) $\frac{d}{dt}[Dx_t - G(t,x_t)] = Lx_t + f(t,x_t)$

where D, L are linear and $G(t,\phi)$ independent of $\phi(0)$.

Let $\beta > a_D$ and y_t a solution of (L) such that $||y_t|| \sim t^{\ell}e^{\alpha t}$ with $\alpha \geq \beta$. Let $\Lambda = \{\lambda \in \sigma(A) : \text{Re}\,\lambda \geq \alpha\}$, P, Q and B be as in Lemma 2.1 and let N be the larger order of the blocks of B which have integers in the diagonal with real part equal to α.

The results below is the main theorem of this section.

Theorem 2.1. Let $y(t)$ be solution of (L) such that $||y_t|| \sim t^{\ell}e^{\alpha t}$. Let S be the vector subespace of C, defined by

$$S = \{\phi \in C : \lim_{t\to\infty} ||T(t)\phi||/||y_t|| = 0\}.$$

Assume that for $\psi, \Psi \in C$

$$\text{(2.6)} \quad \begin{aligned} &f(t,0) = 0, \ |f(t,\phi) - f(t,\psi)| \leq h_2(t)||\phi-\psi||, \ t \geq 0 \\ &G(t,0) = 0, \ |G(t,\phi) - G(t,\psi)| \leq h_1(t)||\phi-\psi||, \ t \geq 0 \end{aligned}$$

$$\text{(2.7)} \quad \int^{\infty} s^{N-1}h_2(s)ds < \infty, \ \int^{\infty} s^{N-1}h_1(s)ds < \infty, \ \lim_{t\to\infty} h_1(t) = 0$$

Then there is a subset Y_S of C and a real number $\sigma > 0$ such that

a) For every $\sigma \in y + Y_S$ there exists a solution $x(t)$ of (P) such that $x_\sigma = \phi$ and

$$\text{(2.8)} \quad \lim_{t\to\infty} ||y_t - x_t||/||y_t|| = 0$$

b) Y_S is homeomorphic to S, that is, there exists a homeomorphism $W : S \to Y_S$ such that W^{-1} is the restriction to Y_S of a projection X^S from C onto S.

Proof: Let $y(t)$ be a solution of (L) such that $||y_t|| \sim t^{\ell}e^{\alpha t}$.

Let

$x = y + \overline{z}$

then to find solutions $x(t)$ of (P) satisfyings (2.8) is equivalent to find solutions of

$$\text{(2.9)}\quad \begin{aligned} \frac{d}{dt}[D\bar{z}_t - G(t,\bar{z}_t)] &= L\bar{z}_t + f(t,y_t + \bar{z}_t) \\ &\overset{\text{def}}{=} L\bar{z}_t + F(t,\bar{z}_t) \end{aligned}$$

satisfying

$$\text{(2.10)}\quad \lim_{t\to\infty} ||\bar{z}_t||/t^{\ell}e^{\alpha t} = 0$$

Let $P_1 = \{\phi \in P : \lim_{t\to\infty} ||T(t)\phi||/t^{\ell}e^{\alpha t} = 0\}$ and let P_2, X^{P_1} and X^{P_2} be defined as in Lemma 2.2.

Consider now the following integral equation:

$$\begin{aligned} z_t &= T(t-\sigma)\phi^S + \int_\sigma^t \{[-d_sT(t-s)X_o^Q]G(s,z)+T(t-s)X_o^QF(s,H(s,z_s))ds\} \\ &+ \int_\sigma^t \{[-d_sT(t)X^{P_1}T(-s)X_o^P]G(s,z_s) + T(t)X^{P_1}T(-s)X_o^PF(s,H(s,z_s))ds\} \\ &- \int_t^\infty \{[-d_sT(t)X^{P_2}T(-s)X_o^P]G(s,z_s) + T(t)X^{P_2}T(-s)X_o^PF(s,H(s,z_s))ds\} \end{aligned} \quad \text{(2.11)}$$

where ϕ^S is an arbitraryly fixed element of S, $X_o^P = \Phi\Psi(0)$, $X_c^Q = X_o - X_o^P$.

We shall show that if z_t satisfies (2.10) then the last integral above converges.

First of all we need emphasize some facts

From Theorem 1.2 of section 1 we know that the subespace $P = \{\phi \in C : \phi = \Phi a,$ for some vector $a\}$ and also that $T(-s)\Phi a = \Phi e^{-Bs}a$. Thus for $a = B(\Psi,\phi)$, we have

$$-d_sT(t)X^{P_2}\Phi e^{-Bs}(\Psi,\phi) = T(t)X^{P_2}\Phi e^{-Bs}B(\Psi,\phi)ds = T(t)X^{P_2}T(-s)\Phi a\, ds$$

with $\Phi a \in P$.

From (1.22) of section 1 we have

$$H(t,z_t) = z_t + X_oG(t,z_t) = \bar{z}_t$$

therefore

$$|F(t,H(t,z_t))| = |f(t,y_t + H(t,z_t))| \leq h_2(s)||y_t + H(t,z_t)||$$

$$\leq h_2(t)[||y_t|| + [1 + h_1(t)]||z_t||].$$

If z_t satisfies (2.10) with $\lim_{t\to\infty} h_1(t) = 0$, then

$$t^{-\ell}e^{-\alpha t}[||y_t|| + [1 + h_1(t)]||z_t||]$$

is bounded.

It is easy to show from (2.5), (2.6) and (2.7) that if z_t satisfies (1.16) then the last integral of (2.11) is convergent.

It is not difficult to see from variation of constants formula (1.22) of section 1 and condition (2.7) that if z_t is a solution of (2.9) satisfying (2.10) then there exists $\phi^S \in S^1$ such that z_t satisfies (2.11) conversely suppose that z_t is a continuous solution of (2.11). Therefore z_t satisfies

$$z_t = T(t-\sigma)\phi + \int_\sigma^t \{[-d_sT(t-s)X_o]G(s,z_s)$$

$$+ T(t-s)X_oF(s,H(s,z_s))ds\}$$

where

$$\phi = \phi^S - \int_\sigma^\infty \{[-d_sT(\sigma)X^{P_2}T(-s)X_oP]G(s,z_s)$$

(2.12)

$$+ T(\sigma)X^{P_2}T(-s)X_o^PF(s,H(s,z_s))ds$$

Therefore $z(t)$ is a solution of (2.9) and satisfies $z_0 = \phi$.

Now we show that equation (2.11) has continuous solutions which satisfies (2.10) for ϕ^S fixed in S and σ large enough.

Consider the space E of functions g on $C([\sigma,\infty), C)$ such that $\lim_{t\to\infty} \frac{||g(t)||}{t^\ell e^{\alpha t}}$. In E consider the norm

$$||g||_E = \sup||g(t)||/t^\ell e^{\alpha t}$$

one can show the E with this norm is a Banch space if we define

$g(t)$ by (2.12) where we change z_t by $g(t)$ and defining $z_\sigma = g(\sigma)$ and $z(t) = g(t)(0)$ for $t > \sigma$ we say that $z_t = g(t)$, and $g(t)$ has a solution in E. In fact define the operator $(Ug)(t) = g(t)$ if $g \in E$. One can prove now that $U(E) \subset E$, U is an uniform contraction with respect to Φ^S and from a known Theorem see [11] p. 7 there exists a fixed point of U on E and this fixed point is continuous with respect to S.

To prove that the map $\sigma^S \in S \to (U_{\phi^S} g) \in E$ is continuous, is sufficient to prove that the map $\phi^S \in S \to V_{\phi^S} \in E$ where $V_{\phi^S}(t) = T(t-a)\phi^S$ is continuous in S, this follows from the uniform boundedeness principle applied to the set $\{T(t-\sigma)/e^{\alpha t}, t \geq \sigma\} \in L(S,C)$ and from the fact that $\lim_{t\to\infty} T(t-s)\Phi^S/t^\ell e^{\alpha t} = 0$. This implies that

$$||T(\cdot,-\sigma)\phi^S||_E = \sup_{t\geq\sigma} ||T(t-\sigma)\phi^S/t^\ell e^{\alpha t}|| \leq \tilde{K}$$

now we apply the Theorem 3.2, Hale [11], p.7, with $F = X = E$, $G = Y = S$ and U_{ϕ^S} the uniform contraction then for each $\phi^S \in S$ there exists a unique fixed point $g(\phi^S) \in E$ of U_{ϕ^S} and the map $g: S \to E$ is continuous.

Then equation (2.11) has a continuous solution z_t which satisfies $\lim_{t\to\infty} ||z_t||/t^\ell e^{\alpha t} = 0$ and this implies from (1.22), (2.6) and (2.7) that

$$\lim_{t\to\infty} ||x_t - y_t||/t^\ell e^{\alpha t} = 0$$

Consider now the map $W : S \to C$ defined by

$$\begin{aligned} W(\phi^S) &= \phi^S \\ &- \int_\sigma^\infty \{[-dsT(\sigma)X^{P_2}T(-s)X_0^P]G(s,g(\phi^S)(s)) \\ &+ T(\sigma)\, X^{P_2}T(-s)X_0^P F(s,H(s,g(\phi^S)(s))ds\}. \end{aligned}$$

We claim that W is continuous in S.

If ϕ, $\psi \in S$ we have:

$$||W(\phi) - W(\psi)|| \le ||\phi-\psi||$$

$$+ ||\int_\sigma^\infty \{[-d_sT(\sigma)X^{P_2}(-s)X_0^P][G(s,g(\phi)(s)) - G(s,g(\psi)(s))]$$

$$+ T(\sigma)X^{P_2}T(-s)X_0^P[F(s,H(s,g(\phi)(s))) - F(s,H(s,g(\psi)(s)))ds\}||]$$

From Lemma 2.2 and calculations made before we have

$$||W(\phi) - W(\psi)|| \le ||\phi-\psi||$$

$$+ M^{\sigma\ell}e^{\alpha\sigma}\int_\sigma^\infty s^{N-1}h_1(s)s^{-\ell}e^{-\alpha s}||g(\phi)(s) - g(\psi)(s)||ds$$

$$+ M\sigma\ e^{\alpha\sigma}\int_\sigma^\infty s^{N-1}h_2(s)[1+h_1(s)]s^{-\ell}e^{-\alpha s}||g(\phi)(s) - g(\psi)(s)||ds \le$$

$$\le ||\phi-\psi||$$

$$+ M\sigma^\ell e^{\alpha\sigma}[\int_\sigma^\infty s^{N-1}h_1(s)ds + K\int_\sigma^\infty s^{N-1}h_2(s)ds]||g(\phi) - g(\psi)||_E.$$

But we already proved that the map $\phi \in S \to g(\phi) \in E$ is continuous therefore from the above inequality it follows that W is continuous.

Let $Y^S \overset{def}{=} W(S) \subset C$ and consider the map $W : S \to Y_S$.

We shall show afterwards that the W^{-1} is a projection over S.

Let X^P and X^Q be the projections given by the decomposition $C = P \oplus Q$:

We assert that

$$(2.13)\qquad X^QW(\phi^S) = X^Q\phi^S$$

In fact since the subspace P is invariant under $T(t)$, for all $t \in R$, we have

$$\int_\sigma^\infty \{[-d_sT(\sigma)X^{P_2}T(-s)X_0^P]G(s,g(\phi^S)(s))$$

$$+ T(\sigma)X^{P_2}T(-s)X_0^PF(s,g(\phi^S)(s)))ds\}$$

belongs to P. Then (2.13) follows. Using also the invariance of P under $T(t)$ we have

$$X^PW(\phi^S) = X^P\phi^S - \int_\sigma^\infty \{[-d_sT(\sigma)X^{P_2}T(-s)X_0^P]G(s,g(\phi^S)(s))$$

$$+ T(\sigma)X^{P_2}T(-s)X_o^P F(s,H(s,g(\phi^S)(s)))ds\}.$$

Thus

$$T(-\sigma)X^P W(\phi^S) = T(-\sigma)X^P\phi^S - \int_\sigma^\infty\{[-dsX^{P_2}T(-s)X_o^P]G(s,g(\phi^S)(s))$$

$$+ X^{P_2}T(-s)X_o^P F(s,H(s,g(\phi^S)(s)))ds\}.$$

Since $X^P\phi^S \in P_1$, P_1 is invariant under $T(t)$. We note also that it is not true in general that $T(t)X^{P_1} = X^{P_1}T(t)$ and that P_2 is invariant with respect to $T(t)$.

From the remarks above it follows that

$$X^{P_1}T(-\sigma)X^P W(\phi^S) = X^{P_1}T(-\sigma)X^P\phi^S = T(-\sigma)X^P\phi^S \in P_1.$$

Then

$$T(\sigma)X^{P_1}T(-\sigma)X^P W(\phi^S) = X^P\phi^S \tag{2.14}$$

From (2.13) and (2.14) it follows that

$$[T(\sigma)X^{P_1}T(-\sigma)X^P + X^Q]W(\phi^S) = X^P\phi^S + X^Q\phi^S = \phi^S.$$

Therefore we conclude that $W^{-1} : Y_S \to S$ is $[T(\sigma)X^{P_1}T(-\sigma)X^P+X^Q]|_{Y_S}$.

We show now that this map is a projection.

In fact

$$(T(\sigma)X^{P_1}T(-\sigma)X^P)(T(\sigma)X^{P_1}T(-\sigma)X^P) =$$

$$= T(\sigma)X^{P_1}T(-\sigma)T(\sigma)X^{P_1}T(-\sigma)X^P = T(\sigma)X^{P_1}T(-\sigma)X^P.$$

Thus we conclude that $T(\sigma)X^{P_1}T(-\sigma)X^P + X^Q$ is a projection. Furthermore if $\phi \in C$, then $[T(\sigma)X^{P_1}T(-\sigma)X^P + X^Q]\phi$ belongs to S. We show now that it is a projection over S.

In fact if $\phi \in S$ we have $\phi = \phi^Q + \phi^{P_1}$ and then

$$[T(\sigma)X^{P_1}T(-\sigma)X^P + X^Q](\phi^Q+\phi^{P_1}) =$$

$$= T(\sigma)X^{P_1}T(-\sigma)X^P\phi^{P_1} + X^Q\phi^Q = T(\sigma)T(-\sigma)\phi^{P_1} + \phi^Q = \phi.$$

The continuity of this projections follows from the fact that

x^Q, x^P, x^{P_1} and $T(t)$ are continuous. Therefore we proved that W is an homeomorphism.

For ordinary differential equations we can prove the converse of Theorem 2.1, that is, for each solution x_t of (P) there is a solution y_t of (L) such that $||x_t-y_t||/||x_t|| \to 0$ when $t \to \infty$. For delay equations this is no longer true as is shown by the following example given by J. Hale. Consider the equation

$$\dot{y} = 0 \quad \text{and} \quad \dot{x} = -2t \exp(1-2t)x(t-1)$$

which has the solution $x(t) = \exp(-t^2)$ and $[1-\exp(-t^2)]/\exp(-t^2) \not\to 0$ as $t \to \infty$.

However it is possible to give a partial converse of Theorem 2.1, that is, if the Lyapunov number of the solutions of (L) are finite then the converse is true. This was proved for retarded functional differential equations by Rodrigues. We give in the following the extension of these results to neutral equations. We will need the following lemmas:

<u>Lemma</u> 2.3. Let $\rho, g \in L_1((0,\infty),R)$, $\rho, g \geq 0$.

Let $\gamma(t) \geq 0$ be a decreasing smooth function, $\gamma(t) \to 0$ as $t\to\infty$.

Let $u(t) \geq 0$ be a continuous solution of

$$u(t) \leq K + \int_\sigma^t u(s)\rho(s)ds + \frac{1}{\gamma(t)}\int_t^\infty \gamma(s)u(s)g(s)ds, \quad t \geq \sigma$$

such that $\gamma(t)u(t)$ is bounded. Then

$$u(t) \leq \frac{K}{1-\beta} e^{\frac{1}{1-\beta}\int_t^\infty g(s)ds}$$

where $\beta = \int_\sigma^\infty [\rho(s)+g(s)]ds < 1$.

<u>Proof</u>. Let $V(t) \overset{\text{def}}{=} \max_{\sigma\leq s\leq t} u(s)$. Then V is continuous decreasing, $u(t) \leq V(t)$ and $\gamma(t)v(t)$ is bounded.

For a given $t \geq \sigma$ there exists $t_1 \in [\sigma,t]$ such that:

$$V(t) = u(t_1) \leq K + V(t)[\int_\sigma^\infty \rho(s)ds + \int_\sigma^\infty g(s)ds] +$$
$$+ \frac{1}{\gamma(t)}\int_t^\infty \gamma(s)V(s)g(s)ds$$

Let σ be sufficiently large in such a way that $\beta < 1$ then

$$\gamma(t)V(t) \le \frac{1}{1-\beta}[K\gamma(t) + \int_t^\infty \gamma(s)V(s)g(s)ds]$$

and from Gronwall inequality we have

$$\gamma(t)V(t) \le \frac{K}{1-\beta}\gamma(t)e^{\frac{1}{1-\beta}\int_t^\infty g(s)ds}$$

<u>Lemma</u> 2.4. Let $x(t)$ be a solution of (P) such that

$$\overline{\lim_{t\to\infty}} \frac{\log|x_t|}{t} = \mu \in R, \quad \mu > a_D$$

where

(2.15)
$$|f(t,\phi)| \le h_2(t)||\phi||, \quad |G(t,\phi)| \le h_1(t)||\phi||$$
$$\int^\infty h_1(t)dt < \infty, \quad \int^\infty h_2(t)dt < \infty, \quad \lim h_1(t) = 0$$

then there exists $\lambda \in \sigma(A)$ such that $\mathrm{Re}\lambda = \mu$

<u>Proof</u>. Assume that for every $\lambda \in \sigma(A)$, $\mathrm{Re}\lambda \ne \mu$. Let

$$\Lambda = \{\lambda \in \sigma(A) \mid \mathrm{Re}\lambda \ge \mu\}$$

then $C = P \oplus Q$ and there exists $\varepsilon > 0$ such that

$$||T(t-\sigma)\phi^Q|| \le M_1 e^{(\mu-\varepsilon)(t-\sigma)}||\phi^Q||, \quad t \ge \sigma, \quad \phi^Q \in Q$$

$$||T(t-\sigma)\phi^P|| \le M_2 e^{(\mu+\varepsilon)(t-\sigma)}||\phi^F||, \quad t \ge \sigma, \quad \phi^P \in P$$

$$||T(t-\sigma)X_0^P|| \le M_3 e^{(\mu+\varepsilon)(t-\sigma)}, \quad t \le \sigma$$

$$||T(t-\sigma)X_0^Q|| \le M_4 e^{(\mu-\varepsilon)(t-\sigma)}, \quad t \ge \sigma$$

$$\int^\infty ||dT(s)X_0^Q||e^{-(\mu-\varepsilon)s} \le M_3$$

$$\int^\infty ||dT(s)X_0^P||e^{-(\mu+\varepsilon)s} \le M_4$$

Using the variation of constants formula

$$z_t = T(t-\sigma)\phi + \int_\sigma^t \{[-dsT(t-s)X_0^Q]G(s,z_s)+T(t-s)X_0^Q F(s,H(s,z_s))ds\}$$
$$- \int_t^\infty \{[-dsT(t-s)X_0^P]G(s,z_s) + T(t-s)X_0^P F(s,H(s,z_s))ds\}$$

where

$$\Phi = \phi + \int_\sigma^\infty \{[-d_sT(t-s)X_0^P]G(s,z_s) + T(t-s)X_0^PF(s,H(s,z_s))ds\}$$

From the mean value Theorem we have

$$e^{-(\mu+\varepsilon)t}\left|\int_\sigma^t [-d_sT(t-s)X_0^Q]G(s,x_s)ds\right| \leq$$

$$\leq e^{-(\mu+\varepsilon)(t-\sigma)}\left|e^{(\mu-\varepsilon)(t-\xi)}K_1h_1(\xi)\int_0^{t-\sigma} e^{-(\mu-\varepsilon)u}\|d_uT(u)X_0^Q\|\right| \leq$$

$$\leq Ke^{-2\varepsilon t}e^{\varepsilon\sigma+\varepsilon\xi}h_1(\xi) = Ke^{-\varepsilon(t-\sigma)}e^{-\varepsilon(t-\xi)}h_1(\xi), \quad \sigma \leq \xi \leq t$$

if ξ goes to a constant or ξ goes to infinity the expression above goes to zero. Since $d_sT(t-s)X_0^P = Be^{B(t-s)}X_0^P$ and $|F(t,H(t,z_t))| \leq \leq h_2(t)[1+h_1(t)]\|z_t\|$, we can prove that

$$e^{-(\mu+\varepsilon)t}\left|\int_\sigma \{[-d_sT(t-s)X_0^Q]G(s,z_s) + T(t-s)X_0^QF(s,H(s,z_s))ds\}\right|$$

$$+ \int^\infty \{[-d_sT(t-s)X_0^P]G(s,z_s) + T(t-s)X_0^PF(s,H(s,z_s))ds\}$$

goes to zero.

Since μ is the Lyapunov number of x_t implies that $e^{-(\mu+\varepsilon)t}|x_t| \to 0$ as $t \to \infty$, then $e^{-(\mu+\varepsilon)t}T(t-\sigma)\Phi$ also goes to zero when $t \to \infty$. Since we assumed that μ does not belong to $\sigma(A)$ and $\mu > a_D$, there is $\varepsilon > 0$ such that $\mu-\varepsilon$ also does not belong to $\sigma(A)$ and since $e^{-(\mu+\varepsilon)t}T(t-\sigma)\Phi \to 0$, Φ does not belongs to P and from the relations above

$$|T(t-\sigma)\Phi| \leq Me^{(\mu-\varepsilon)(t-\sigma)}|\Phi|,$$

$$e^{-(\mu-\varepsilon)(t-\sigma)}\left|\int_\sigma^t [-d_sT(t-s)X_0^Q]G(s,x_s)ds\right| \leq Ke^{-\varepsilon(t-\xi)}h_2(\xi) \leq K,$$

$$e^{-(\mu-\varepsilon)(t-\sigma)}\left|\int_\sigma^t T(t-s)X_0^QF(s,H(s,z_s))ds\right| \leq$$

$$\leq e^{-(\mu-\varepsilon)(t-\sigma)}\int_\sigma^t M_4e^{(\mu-\varepsilon)(t-s)}F(s,H(s,z_s))ds \leq K\int_\sigma^t h_2(s)(1+h_1(s))|z_s|ds/e^{(\mu-\varepsilon)s},$$

$$e^{-(\mu-\varepsilon)(t-\sigma)}\left|\int_t^\infty [-d_sT(t-s)X_0^P]G(s,z_s)\right| \leq$$

$$\leq Ke^{-(\mu-\varepsilon)(t-\sigma)}\int_t^\infty Be^{B(t-s)}h_1(s)|z_s| \leq Ke^{-(\mu-\varepsilon)(t-\sigma)}\int_t^\infty e^{(\mu+\varepsilon)(t-s)}h_1|z_s| \leq$$

$$\leq K\int_t^\infty e^{2\varepsilon t}e^{-2\varepsilon s}h_1(s)|z_s|/e^{(\mu-\varepsilon)s}$$

Then

$$|z_s|e^{-(\mu-\varepsilon)t} \le K + K_1\int_\sigma^t h_1(s)(1+h_1(s))|z_s|e^{-(\mu-\varepsilon)s}$$

$$+ e^{2\varepsilon t}K_2\int_t^\infty e^{-2\varepsilon s}h_1(s)|z_s|e^{-(\mu-\varepsilon)}ds =$$

$$= K + K_2\int_\sigma^t \rho(s)|z_s|e^{-(\mu-\varepsilon)s} + e^{2\varepsilon t}\int_t^\infty e^{-2\varepsilon s}g(s)|z_s|e^{-(\mu-\varepsilon)s}ds$$

From Lemma 2.3, $z_t e^{-(\mu-\varepsilon)s}$ is bounded, a contradiction, because μ is the Lyapunov number of z_t and then the Theorem is proved.

<u>Lemma</u> 2.5. Let $\mu > a_D$ and let $x(t)$ be a solution of (P) such that

$$\overline{\lim} \frac{\log|x_t|}{t} = \mu \in R$$

and assume that condition (2.21) is satisfied.

Then there exists a nonnegative integer ℓ such that

$$0 < C_1 \le \frac{|x_t|}{t^\ell e^{\mu t}} \le C_2 < \infty$$

<u>Proof</u>. Let

$$\Lambda_P = \{\lambda \in \sigma(A) \mid \mathrm{Re}\lambda = \mu\}$$

$$\Lambda_S = \{\lambda \in \sigma(A) \mid \mathrm{Re}\lambda > \mu\}$$

$$\Lambda_Q = \{\lambda \in \sigma(A) \mid \mathrm{Re}\lambda < \mu\}$$

Then $C = P \oplus S \oplus Q$. As before let $P_1 = \{\phi \in P \mid T(t)\phi/t^\ell e^{\mu t} \to 0\}$ where $\ell \ge 0$ is fixed. We can choose P_2, $\sigma > 0$ in such a way that $P = P_1 \oplus P_2$

$$|T(t)X^{P_1}T(-s)\phi^P| \le Kt^{\ell-1}s^{N-\ell}e^{\mu(t-s)}|\phi^P|, \quad \sigma \le s \le t$$

$$|T(t)X^{P_2}T(-s)\phi^P| \le Kt^{\ell}s^{N-\ell-1}e^{\mu(t-s)}|\phi^P|, \quad \sigma \le t \le s$$

Moreover there exists $\varepsilon > 0$ such that

$$|T(t-s)\phi^Q| \le Ke^{(\mu-\varepsilon)(t-s)}|\phi^Q| \qquad \sigma \le s \le t$$

$$|T(t-s)\phi^S| \le Ke^{(\mu+\varepsilon)(t-s)}|\phi^Q| \qquad \sigma \le t \le s$$

Let $\tilde{Q} \stackrel{\text{def}}{=} Q + P$. The following estimates hold, provided that $\varepsilon > 0$ is small enough

$$|T(t)\phi^{\tilde{Q}}| \le Ke^{(\mu-\varepsilon/2)t}|\phi^{\tilde{Q}}|, \qquad t \ge 0$$

$$|T(t)\phi^{S}| \le Ke^{(\mu+\varepsilon)t}|\phi^{S}|, \qquad t \le 0$$

Since

$$z_t = T(t-\sigma)\phi + \int_\sigma \{[-dsT(t-s)X_0^{\tilde{Q}}]G(s,x_s) + T(t-s)X_0^{\tilde{Q}}F(s,H(s,z_s))ds\}$$

$$\int_t^\infty \{[-dsT(t-s)X_0^{S}]G(s,z_s) + T(t-s)X_0^{S}F(s,H(s,z_s))ds\}$$

From the estimates above it is easy to show that $\lim_{t\to\infty} e^{-(\mu+\varepsilon)t}T(t-\sigma)\phi \to 0$ and then it follows that $\phi^S = 0$. We can prove also using Lemma 2.3 and the estimates above that $|T(t-\sigma)\phi|/t^{N-1}e^{\mu t}$ is bounded, where N = largest order of the blocks of B where B is in the Jordan canonical form and $T(t)\phi^S = e^{Bt}(\Phi,\Psi)$.

Let $\ell = \min\{n \ge 0 \mid |x_t|/t^{\mu}e^{\mu t}$ is bounded for $t \ge \sigma\}$.

We can prove, as we did before, that z_t can be written in the following form

$$z_t = T(t-\sigma)\Psi + \int_\sigma^t \{[-dsT(t-s)X_0^{Q}]G(s,z_s) + T(t-s)X_0^{Q}F(s,H(s,z_s))ds\}$$

$$+ \int_\sigma^\infty \{[-dsT(t)X^{P_1}T(-s)X_0^{P}]G(s,z_s) + T(t)X^{P_1}T(-s)X_0^{P}F(s,H(s,z_s))ds\}$$

$$- \int^\infty \{[-dsT(t)X^{P_2}T(-s)X_0^{P}]G(s,z_s) + T(t)X^{P_2}T(-s)X_0^{P}F(s,H(s,z_s))ds\}$$

Where $\Psi = \phi^Q + \int_\sigma^\infty \{[-dsT(\sigma)X^{P_2}T(-s)X_0^{P}]G(s,z_s) + T(\sigma)X^{P_2}T(-s)X_0^{P}F(s,z_s))ds\}$

The only thing is to prove that the integrals above are convergent and also it follows that

$$t^{-\ell}e^{-\mu t}T(t-\sigma)\psi$$

is bounded. Now it is not difficult, using Lemma 2.3 and the known estimates that $t^{-\ell}e^{-\mu t}T(t-\sigma)\psi \not\longrightarrow 0$ as $t \to \infty$ and then there exists C_1 such that $|T(t-\sigma)\psi|/t^{\ell}e^{\mu t} \ge C_1 > 0$ for every t large enough

Our final conclusion is that there exists constants C_1, C_2 such that $0 < C_1 \le \frac{x_t}{t^{\ell}e^{\mu t}} \le C_2 < \infty$. We have then the following Theorem.

Theorem 2.2. Assume that f and G are continuous. Let $x(t) \neq 0$ for all sufficiently large t be a solution of (P). Let

$$\overline{\lim_{t\to\infty}} \frac{\log|x_t|}{t} = \mu \in R.$$

Let $N = \max\{m \mid (A - \lambda I)^{m+1} = (A - \lambda I)^m, \ \operatorname{Re}\lambda = \mu\}$.

Let us assume that

$$|f(t,\phi)| \le h_2(t)||\phi||, \quad |G(t,\phi)| \le h_1(t)||\phi||$$

$$\int^{\infty} t^{N-1}h_1(t)dt < \infty, \quad \int^{\infty} t^{N-1}h_2(t)dt < \infty, \quad \lim_{t\to\infty} h_1(t) = 0$$

Then there exists a family of solutions y_t of (L) such that

$$\lim_{t\to\infty} \frac{||x_t - y_t||}{||x_t||} = 0$$

The proof follows easily from the estimates before, Lemma 2.3 and the change of variables 1.22.

3. Alekseev's integral formula for nonlinear and non autonomous systems of differential equations of neutral type.

Consider the nonlinear systems of functional differential equations of neutral type

$$\text{(3.1)} \qquad \frac{d}{dt}Dy_t = f(y_t)$$

$$\text{(3.2)} \qquad \frac{d}{dt}Dx_t = f(x_t) + g(t)$$

where D is a linear continuous functional of C into R^n, satisfying

$$D\phi = \phi(0) - \int_{-r}^{0_-} [d\mu(\theta)]\phi(\theta)$$

and the matrix $\mu(\theta)$, $n \times n$ of bounded variation satisfies the same conditions (1.5) of Section 1. The map $f : \Omega \subset C \to R^n$, Ω open, is a continuous nonlinear map with continuous Frechét's derivative with respect to $\phi \in C$, belonging to C and $g : [\sigma,\infty) \to R^n$ is continuous.

To each solution $y_t(\sigma,\phi)$ of (3.1) we can define a linear non autonomous neutral functional differential equations of the form

$$(3.3) \qquad \frac{d}{dt}Dz_t = f_y(t,y_t(s,x_s(\sigma,\phi))z_t, \quad t \geq s \geq \sigma \geq 0$$

with is called *the linear variational equation of (2.1)*, with respect to the solutions $y_t(s,x_s(\sigma,\phi))$.

For the linear equation (3.3) we can associate a family of linear continuous operators $T(t,\sigma:\phi) : C \to C$, $t \geq \sigma \geq 0$, defined for all $\psi \in C$

$$T(t,\sigma:\phi)\psi = z_t(\sigma,\psi), \quad \phi \in C.$$

Since the solutions of (3.3) are continuous in t and σ, for $\sigma \leq t < \infty$, then $T(t,\sigma)$ is strongly continuous and $T(t,t) = I$, $\sigma \leq t < \infty$.

For any piecewise continuous function $\psi : [-r,0] \to R^n$, we can define a solution of (3.3) with initial values ψ and σ. Therefore if the $n \times n$ matrix function X_o, $n \times n$ is defined by

$$X_o(\theta) = \begin{cases} 0 & \text{if } -r \leq \theta < 0 \\ I & \text{if } \theta = 0 \end{cases}$$

then, the operator $T(t,\sigma)$ can also be defined on the colunms of X_o.

<u>Lemma</u> 3.1. If $y_t(s,x_s(\sigma,\phi))$ and $x_s(\sigma,\phi)$ are differentiable with respect to s, then

$$\frac{d}{ds}y_t(s,x_s(\sigma,\phi)) \quad \text{and} \quad T(t,s : x_s(\sigma,\phi))X_o g(t)$$

are solutions of (3.3) such that $t = s$ coincide with $X_o g(t)$.

<u>Proof</u>. It is obvious that $T(t,s : x_s(\sigma,\phi))X_o$ is a solution of (3.3) and then $T(t,s : x_s(\sigma,\phi))X_o g(t)$ is also a solution since g is constant with respect to θ and for $t = s$

$$(3.4) \qquad T(t,s : x_s(\sigma,\phi))X_o g(s) = X_o g(s)$$

Since D is linear, from [23], p. 100,

$$\frac{d}{dt}D(\frac{d}{ds}y_t(s,x_s(\sigma,\phi))) = \frac{d}{dt}\,\frac{d}{ds}D(y_t(s,x_s(\sigma,\phi))) =$$

$$= \frac{d}{ds}\,\frac{d}{dt}D(y_t(s,x_s(\sigma,\phi))) = \frac{d}{ds}f(y_t(s,x_s(\sigma,\phi))) =$$

$$= f_y(y_t(s,x_s(\sigma,\phi)))\frac{d}{ds}y_t(s,x_s(\sigma,\phi))$$

what shows that $\frac{d}{ds}y_t(s,x_s(\sigma,\phi))$ is a solution of (3.3).

Consider now the function h that associates to each s the pair $(s,x_s(\sigma,\phi))$ and $s \to D(y_s(h(s)))$.

This last function may be considered as being the function G, which to every s associates $G(s,h(s))$. Thus we have

$$\frac{d}{ds}G(s,h(s)) = D_1G(s_o,h(s_o)) + D_2G(s,h(s_o))h'(s_o)$$

where D_1 and D_2 are respectively the derivatives with respect to the first variable and to the second variable.

Then we can write

$$\frac{d}{ds}G(s,h(s)) = D_1G(s_o,h(s_o)) + \frac{d}{ds}G(s_o,h(s))|_{s=s_o} =$$

$$= \frac{d}{ds}G(s,h(s))|_{s=s_o} + \frac{d}{ds}G(s_o,h(s))|_{s=s_o}.$$

Thus

$$\frac{d}{ds}D(y_s(s,x_s(\sigma,\phi))) = [\frac{d}{dt}D(y_t(s,x_s(\sigma,\phi)))]_{t=s} =$$

$$= [\frac{d}{dt}D(y_s(t,x_t(\sigma,\phi)))]_{t=s}.$$

But, from (3.1) it follows that

$$\frac{d}{ds}D(y_s(s,x_s(\sigma,\phi))) = f(x_s(\sigma,\phi)) + g(s)$$

and from (3.2), we have

$$[\frac{d}{dt}D(y_t(s,x_s(\sigma,\phi)))]_{t=s} = f(x_s(\sigma,\phi)).$$

Then, $[\frac{d}{dt}D(y_t(t,x_t(\sigma,\phi)))]_{t=s} = g(s)$.

Also from [23] we have

(3.5) $$[D(\frac{d}{ds}y_t(s,x_s(\sigma,\phi)))]_{t=s} = g(s)$$

We show now that

(3.6) $$[\frac{d}{ds}y_t(s,x_s(\sigma,\phi))]_{t=s}(\theta) = X_o(\theta)g(s), \quad \theta \in [-r,0]$$

Consider the function

$$h(\theta) = [\frac{d}{ds}y_t(s,x_s(\sigma,\phi))]_{t=s}(\theta), \quad \theta \in [-r,0].$$

We have that $\theta \in [-r,0)$ fixed and t sufficiently close to s such that $t + \theta < s$ we have

$$y(t+\theta,\ s : x_s(\sigma,\phi)) = x(t+\theta,\ \sigma : \phi)$$

and then

$$\frac{d}{ds}y(t+\theta,\ s : x_s(\sigma,\phi)) = \frac{d}{ds}x(t+\theta,\ \sigma : \phi) = 0$$

for all t and in particular for $t = s$.

Thus $h(\theta) = 0$ for $\theta \in [-r,0)$.

For $\theta = 0$, from

$$D\phi = \phi(0) - \int_{-r}^{0} [d\mu(\theta)]\phi(\theta)$$ and (3.5) we have

$$[\frac{d}{ds}y_t(s,x_s(\sigma,\phi))]_{t=s}(0) = D([\frac{d}{ds}y_t(s,x_s(\sigma,\phi))]_{t=s}) +$$

$$+ \int_{-r}^{0} [d\mu(\theta)][\frac{d}{ds}y_t(s,x_s(\sigma,\phi))]_{t=s}(\theta).$$

We can say that the above function integral is zero since for $\theta<0$, $h(\theta) = 0$ and the function

$$g(\theta) = \int_{-r}^{0} [d\mu(\theta)]\phi(\theta)$$

is non atomic at zero. Then by (3.5) we conclude that

$$[\frac{d}{ds}y_t(s,x_s(\sigma,\phi))]_{t=s}(\theta) = g(s), \quad \theta = 0.$$

Thus for all $\theta \in [-r,0]$, we can write

$$[\frac{d}{ds}y_t(s,x_s(\sigma,\phi))]_{t=s} = X_o g(s).$$

Then from (3.4) and (3.6) and from the unicity of solutions of system (3.3) we finally have

$$\frac{d}{ds}y_t(s,x_s(\sigma,\phi)) = T(t,s : x_s(\sigma,\phi))X_o g(s) \tag{3.7}$$

and the theorem is proved.

<u>Theorem</u> 3.2. If for $(\sigma,\phi) \in [0,\infty) \times \Omega$, Ω open in C, the solution $y_t(\sigma,\phi)$ of (3.1) has continuous Frechét's derivative with respect to t, then

$$x_t(\sigma,\phi) = y_t(\sigma,\phi) + \int_\sigma^t T(t,s : x_s(\sigma,\phi))X_o g(s)ds \tag{3.8}$$

is a solution of (3.2) where $y_t(\sigma,\phi)$ is a solution of (3.1).

<u>Proof</u>. From the hypotheses above we are in the conditions of the Lemma 3.1 and then we can use (3.7).

We show now that $x_t(\sigma,\phi)$, given by (3.8) satisfies system (3.2). In fact

$$\frac{d}{dt}D(x_t(\sigma,\phi)) = \frac{d}{dt}D[y_t(\sigma,\phi) + \int_\sigma^t T(t,s : x_s(\sigma,\phi))X_o g(s)ds].$$

From the linearity of D and the relation (3.1) we have

$$\frac{d}{dt}D(x_t(\sigma,\phi)) = f(y_t(\sigma,\phi)) + \frac{d}{dt}\int_\sigma^t D[T(t,s : x_s(\sigma,\phi))X_o g(s)]ds =$$

$$= f(y_t(\sigma,\phi)) + D[T(t,t : x_t(\sigma,\phi))X_o]g(t) +$$

$$+ \int_\sigma^t \frac{d}{dt}D[T(t,s : x_s(\sigma,\phi))X_o g(s)]ds = f(y_t(\sigma,\phi)) + g(t) +$$

$$+ \int_\sigma^t \frac{d}{dt}D[T(t,s : x_s(\sigma,\phi))X_o g(s)]ds$$

since $D(X_o) = I$.

But from (3.7) and (3.3) and from Lemma 3.1 it follows that

$$\frac{d}{dt}D[T(t,s : x_s(\sigma,\phi))X_o g(s)] =$$

$$= f_y(y_t(s,x_s(\sigma,\phi)))T(t,s : x_s(\sigma,\phi))X_o g(s) =$$

$$= f_y(y_t(s,x_s(\sigma,\phi)))\frac{d}{ds}y_t(s,x_s(\sigma,\phi)) = \frac{d}{ds}f(y_t(s,x_s(\sigma,\phi))).$$

Then

$$\frac{d}{dt}D(x_t(\sigma,\phi)) = f(y_t(\sigma,\phi)) + g(t) + \int_\sigma^t \frac{d}{ds}f(y_t(s,x_s(\sigma,\phi)))ds =$$

$$= f(y_t(\sigma,\phi)) + g(t) + f(x_t(\sigma,\phi)) - f(y_t(\sigma,\phi)) = f(x_t(\sigma,\phi)) + g(t).$$

Corollary. Consider the systems

(3.9) $\frac{d}{dt}Dy_t = f(t,y_t)$

(3.10) $\frac{d}{dt}Dx_t = f(t,x_t) + g(t,x_t)$

where D is a linear continuous functional from C into R^n atomic at zero satisfying conditions (1.4), (1.5) and (2.1), $f : [\sigma,\infty) \times \Omega \to R^n$, Ω open in C continuous function (nonlinear in general) with continuous Frechét's derivative with respect to $\phi \in \Omega$, $g : [\sigma,\infty) \times C \to R^n$ is a continuous function and the solutions $y_t(s,x_s(\sigma,\phi))$, $x_s(\sigma,\phi)$ are differentiable with respect to s. Then we have

(3.11) $\frac{d}{ds}y_t(s,x_s(\sigma,\phi)) = T(t,s : x_s(\sigma,\phi))X_0 g(s,x_s(\sigma,\phi))$

and

(3.12) $x_t(\sigma,\phi) = y_t(\sigma,\phi) + \int_\sigma^t T(t,s : x_s(\sigma,\phi))X_0 g(s,x_s(\sigma,\phi))ds$

is a solution of (3.10) with $y_t(\sigma,\phi)$ solution of (3.10).

Proof. If we consider the system (3.10) we can state a Lemma similar to the Lemma 2.1, and from this follows the Corollary.

3.1. Relative asymptotic equivalence between the solutions of two nonlinear systems of neutral functional differential.

Several results that are true for linear autonomous ordinary and retarded differential equations can not be extended to neutral equations. Zverkin [9] gives an example of a linear autonomous equation of type (3.1) which has all eigenvalues with real part less or equal to zero and have unbounded solutions, what can not happen to ordinary or delay equations. The reason for that, is that the operator $T(t,\sigma)$ associated to an autonomous linear ordinary or delay equation is compact while the operator $T(t,\sigma)$ associated with equation (3.1) is not compact and therefore may have a continuous spectrum with the real part of the eigenvalues densely packed at zero. To avoid this situation Hale and Cruz [13] introduced the following definition.

Definition 3.1. We say that the operator D is uniformly stable if the solution of the difference equation $\frac{d}{dt}D(t,x_t) = 0$, $x_\sigma = 0$ is uniformly asymptotically stable.

We should remark that the operator D associated to the equation in the example given by Zverkin is not uniformly stable. For linear autonomous equations the definition above implies that the continuous spectrum of (3.1) is contained in the unitary circle. Furthermore it can be proves that if D is stable the operator $T(t,\sigma)$ associated with (3.1) is an α-contraction. As an application of Alekseev formula we give theorems that extendes to neutral equations well known theorems on asymptotic equivalence of ordinary differential equations.

Definition 3.2. For each $\phi \in C = C([-r,0],E^n)$ we define the *Euclidean norm* of ϕ by $|\phi|_1(\theta) = |\phi(\theta)|$, $-r \le \theta \le 0$, $|\phi|_1 \in C_+ = C([-r,0],R_+)$, $R_+ = [0,\infty)$.

Definition 3.3. $w(t,\phi) \in R$ for $(t,\phi) \in R \times C([-r,0],R)$ is nondecreasing in ϕ for fixed t if $|\phi|_1 \le |\psi|_1$ implies $w(t,\phi) \le w(t,\psi)$.

Theorem 3.3. Assume that D is uniformly stable, and let $\Lambda(t)$ be a $n \times n$ continuous nonsingular matrix defined on $[\sigma,\infty)$ with values in R^n. Let $\Omega \subset C$ be an open set, $D_1 \subset \Omega$ open and such that $\bar{D}_1 \subset \Omega$ and $D_2 \subset D_1$ such that:

I) $|\Delta_t \gamma_t|_1 \le \rho_t$ implies $\gamma_t \in D_1$ for $t \ge \sigma$ where $\Delta_t(\theta) = \Delta(t+\theta)$, $\rho(t,\sigma,\rho_\sigma)$ is the maximal solution of $\rho = w(t,\rho_t)$ with $||\rho_\sigma|| < \rho_\infty$, and $\rho(t)$ is a positive maximal solution of $\rho = w(t,\rho_t)$ such that $\lim_{t\to\infty} \rho(t) = \rho_\infty$.

II) $||\Delta_t T(t,s,\gamma_s)X_o G(s,\gamma_s)|| \le w(s,|\Delta_s\gamma_s|_1)$ for $t, s \in [\sigma,\infty)$ and $\gamma_s \in \Omega$.

III) $||\Delta_t y_t(\sigma,\psi)|| \le |\phi(0)|$, $t \ge \sigma$, $\psi \in D_2$ and ϕ initial condition of the solution $x_t(\sigma,\phi)$ of (3.10).

Then for every $\phi \in \Omega$ such that $||\phi|| < \rho$, $\rho < \rho_\infty$ and $|\phi(0)| \le \rho_\sigma(0)$ there exists a solution $x_t(\sigma,\phi) \in D_1$ of (3.10) for $t \ge \sigma$ and for such solution there exist a solution $y_t(\sigma,\phi)$ of (3.9), $t \ge 0$, satisfying the condition

(3.13) $$\lim_{t\to\infty} ||\Delta_t(x_t-y_t)|| = 0$$

and conversely for each solution $y_t(\sigma,\phi)$ of (3.9) with $\phi \in D_2$ there exists $t_1 \ge \sigma$ and a solution $x_t(\sigma,\phi)$ of (3.10) such that condition (3.13) is satisfied.

Since $\int^\infty w(s,M)ds < \infty$ for each $M > 0$ if only if for each real number ρ_∞ there exists a solution of $\dot\rho = w(t,\rho_t)$ on some interval $[t_1,\infty)$ such that $\lim \rho(t) = \rho_\infty$, the results proved in [8], [3], [5] are contained in the above Theorem.

REFERENCES

[1] - BACHMAN, G. and NARICI, L., Functional Analysis, Academic Press, New York, (1966).

[2] - BELLMAN, R. and COOKE, K., Asymptotic behavior of solutions of differential-difference equations, Memoirs of the Amer. Math. Soc, 35(1959).

[3] - BRAUER, F. and WONG, J.S.W., On the asymptotics relationships between solutions of two systems of ordinary differential equations, J. Diff. Equations, Vol. 6, nº 3, November (1969), 527-543.

[4] - BRUMLEY, W.F., On the asymptotic behavior of solutions of differential-difference equations of the neutral type, J. Diff. Equations, 7(1970), 175-188.

[5] - CASSAGO Jr., H., Functional differential inequalities, Boletim da Soc. Bras. de Matem., Vol. 9, 1(1978).

[6] - COOKE, L., Asymptotic equivalence of an ordinary and a functional differential equation, J. Math. An. Appl., 51 (1975), 187-207.

[7] - COPPEL, W.A., Stability and asymptotic behavior of differential equations, Heath Mathematical Monographs.

[8] - EVANS, R.B., Asymptotic equivalence of linear functional differential equations, J. Math. An. Appl., 51(1975), 223-228.

[9] - GROMOVA, P.C. and ZVERKIN, A.M., On trigonometric series whose sum is a continuous unbounded function on real line and is the solution of an equation with retarded argument, Differentialniya Uraneniya, 4(1968), 1774-1784.

[10] - HALE, J.K., Theory of Functional Differential Equations, Springer-Verlag, New York, Heidelberg, Berlin, (1977).

[11] - HALE, J.K., Ordinary Differential Equations, Wiley-Interscience, (1969).

[12] - HALE, J.K. and LOPES, O.F., Fixed point theorems and dissipative processes, J. Diff. Equations, Vol. 13, nº 2, March (1973).

[13] - HALE, J.K. and CRUZ, M.A., Asymptotic behavior of neutral functional differential equations, Arch. Rat. Mech. Anal., Vol. 34.

[14] - HALE, J.K., Linear asymptotically autonomous functional differential equations, Rend. Circ. Mat. Palermo (2), 15 (1966), 331-351.

[15] - HARTMAN, P., Ordinary differential equations, New York, (1964), MR 30 = 1270.

[16] - HARTMAN, P. and ONUCHIC, N., On the asymptotic integration of ordinary differential equations, Pacific J. Math., 13 (1963), 1193-1207.

[17] - HENRY, D., Linear autonomous neutral functional differential equations, J. Diff. Equations, 15(1974), 106-128.

[18] - IZÉ, A.F. and MOLFETTA, N.A., Asymptotically autonomous neutral functional differential equations with time-dependent lag, J. Math. An. Appl., Vol. 51, 2(1975).

[19] - IZÉ, A.F. and FREIRIA, A.A., Asymptotic behavior and nonoscillation of Volterra integral equations and functional differential equations, Proc. Amer. Math. Soc., Vol. 52, (1975).

[20] - IZÉ, A.F., Asymptotic integration of a nonhomogeneous singular linear systems of a ordinary differential equations, J. Diff. Equations, Vol. 8, nº 1, July (1970), 1-15.

[21] - IZÉ, A.F., Asymptotic integration of a nonlinear systems of ordinary differential equations, Bol. Soc. Bras. Mat., Vol. 4, Nº 1, (1973), 61-80.

[22] - LAKSHMIKANTHAM, V. and LEELA, S., Differential an Integrals Inequalities, Theory and Applications, Academic Press, N.Y., (1969), Vols. I and II.

[23] - LANG, S., Analysis II, Addison Wesley Publ. Co. Inc.

[24] - ONUCHIC, N., Relationships among the solutions of two systems of ordinary differential equations, Michigan Math. J, 10 (1963), 129-139.

[25] - ONUCHIC, N., Asymptotic relationships at infinity between the solutions of two systems of ordinary differential equations, J. Diff. Equations, 3(1967), 47-58.

[26] - RODRIGUES, H.M., Comportamento assintótico de equações retardadas e aplicações do método alternativo. To appear.

[27] - SMIRNOV, V.I., A Course of Higher Mathematic, Vol. V, Perg. Press, New York (1964).

[28] - STRAUSS, A. and YORKE, J.A., Perturbation on theorem for ordinary differential equations, J. Diff. Equations, 3 (1967), 18-19.

STABILITY IN FUNCTIONAL DIFFERENTIAL EQUATIONS

by Junji Kato

We discuss the stability problem in functional differential equations in connexion with the choice of the phase spaces.

Consider the functional differential equation

(E) $\dot{x}(t) = f(t,x_t)$,

and assume that $f(t,\phi)$ is completely continuous on $I \times X$ for an admissible space X, where $I = [0,\infty)$ and x_t denotes the function on $(-\infty,0]$ defined by $x_t(s) = x(t+s)$ for a given $x(s)$.

1. Admissible phase space.

A linear space X of R^n-valued functions defined on $(-\infty,0]$ with a semi-norm $||\cdot||_X$ is said to be *admissible* if for any $\tau \geq 0$ and any $\phi \in X_\tau$, we have

(h_1) $\phi_t \in X$ for all $t \in [-\tau,0]$ and it is continuous in t,

(h_2) $\mu||\phi(0)|| \leq ||\phi||_X \leq K(\tau) \sup_{-\tau\leq s\leq 0} ||\phi(s)|| + M(\tau)||\phi_{-\tau}||_X$,

where $\mu > 0$ is a constant and $K(\tau)$, $M(\tau)$ are continuous functions on I.

Here, X_τ denotes the set of functions ϕ on $(-\infty,0]$ such that $\phi(s)$ is continuous on $[-\tau,0]$ and $\phi_{-\tau} \in X$. Clearly, X_τ is an admissible subspace of X, too, with the semi-norm $||\phi||_{X_\tau} = \sup\{||\phi_t||_X : -\tau \leq t \leq 0\}$.

Moreover, we say X has *a fading memory* if in (h_2)

(h_3) $K = K(t)$ is a constant and $M(t) = M_o e^{-ct}$ for a $c > 0$.

Let $\gamma \geq 0$ and $\infty \geq h \geq 0$ be given. Let C_h^γ be the space of functions ϕ which are continuous on $[-h,0]$ (or on $(-\infty,0]$ if $h = \infty$) and satisfy $e^{\gamma s}\phi(s) \to 0$ as $s \to -\infty$, and let M_h^γ be the space of functions ϕ which are measurable on $[-h,0]$ and satisfy $\int_{-h}^{0} e^{\gamma s}||\phi(s)||ds < \infty$.

Then, C_h^γ and M_h^γ are typical and useful examples of admissible space, where $||\cdot||_{C_h^\gamma}$ and $||\cdot||_{M_h^\gamma}$ are given by

$$||\phi||_{C_h^\gamma} = \sup_{-h<s\le 0} e^{\gamma s}||\phi(s)||$$

and

$$||\phi||_{M_h^\gamma} = ||\phi(0)|| + \int_{-h}^{0} e^{\gamma s}||\phi(s)||ds,$$

respectively, and they have fading memories if $\gamma > 0$ or $h < \infty$. Especially we can write $C_o^\gamma = R^n$.

It is known that we can establish the fundamental theorems for the solutions of functional differential equations defined on an admissible phase space, see [6].

2. Stability.

The definitions of stabilities will be given as follow: Let Y be an admissible space such that

$$||\phi||_Y \le N||\phi||_X \qquad (\phi \in X \cap Y)$$

for a constant N, and let $x(t)$ represent an arbitrary solutions of (E).

The zero solution of (E) is said to be *stable in* (X,Y) if for any $\varepsilon > 0$ and any $\tau \ge 0$ we can find a $\delta > 0$ so that

$$||x_t||_Y \le \varepsilon \quad \text{whenever} \quad ||x_\tau||_X \le \delta \quad \text{and} \quad t \ge \tau,$$

and it is said to be *asymptotically stable in* (X,Y) if it is stable in (X,Y) and if for any $\tau \ge 0$ there exists a $\delta_o > 0$ such that for any $\varepsilon > 0$ we can make

$$||x_t||_Y \le \varepsilon \quad \text{whenever} \quad ||x_\tau||_X \le \delta_o \quad \text{and} \quad t \ge \tau + T$$

by taking T suitably for each solution.

As usually, we say the stability is *uniform* if δ_o is a constant and δ, T can be chosen as functions of ε alone, while *the exponential stability in* (X,Y) will be defined if

$$||x_t||_Y \le \beta e^{-c(t-\tau)}||x_\tau||_X \quad \text{if} \quad ||x_\tau||_X \le \delta_o, \quad t \ge \tau$$

for positive constants β, c and δ_o.

It is not difficult to prove the following theorems.

Theorem 1. The stability in (X,Y) implies the stability in (X,R^n), and the reverse is true if Y has a fading memory.

Theorem 2. In the case where $f(t,\phi)$ in (E) is linear in ϕ, the zero solution of (E) is exponentially stable in (X,X) if it is uniformly asymptotically stable in (X,X).

3. Liapunov type theorem.

A various kind of Liapunov type theorems have been considered for the stability problem in functional differential equations, for example refer to [1] ~ [11], [16], [17], [19].

Before stating theorems we shall introduce some notations: $a \in K_r$ means that a is continuous, non-decreasing with respect to the argument r, and the set of $a \in K_r$ having the property $a(r) = 0$ for $r = 0$ is denoted by K_r^o and the one having $a(r) > 0$ for $r > 0$ by K_r^+.

For the proof of the following theorems, see [9] also refer to [8].

Theorem 3. Suppose that there exists a set of pairs $\{(\tau,V(t,\phi))\}$ such that $V(t,\phi)$ is continuous on $\{(t,\phi) : \phi \in X_{t-\tau},\ t \geq \tau\}$ and satisfies

(i) $a(||\phi||_Y) \leq V(t,\phi) \leq b(||\phi||_{X_{t-\tau}})$,

(ii) $\dot{V}_{(E)}(t,\phi) \leq -w(t,\phi)$ whenever $p(t,V(t,\phi)) \geq \tau$

and $V(s,\phi_{s-t}) \leq F(V(t,\phi))$ for $s \in [p(t,V(t,\phi)),t]$,

where $a(r) \in K_r^+$, $b(r) \in K_r^o$, $p(t,r) \in K_r$, $F(r) \in K_r$, $p(t,r) \leq t$, $p(t,r) \to \infty$ as $t \to \infty$ for any $r > 0$, $F(r) > r$ for $r > 0$ and $w(t,\phi) \geq 0$. Moreover, we assume that solutions $x(t)$ of (E) satisfy

$$||x_t||_X \leq L(t-\tau,||x_\tau||_X) \quad (t \geq \tau)$$

if $p(t,r) \not\equiv t$, where $L(t,r) \in K_r^o$ is continuous.

Then, the zero solution of (E) is stable in (X,Y), and the stability is uniform if

(1) $q(r) = p(t,r) - t$ is independent of t.

Furthermore, if for any $\varepsilon > 0$

(2) $\int_{\sigma}^{\sigma+T} w(s,x_s)ds \to \infty$ as $T \to \infty$

when $||x_t||_X \geq \varepsilon$ for all $t \geq \sigma$ and any $\sigma \geq \tau$, then the zero solution of (E) is asymptotically stable in (X,Y), and it is uniformly asymptotically stable in (X,Y) whenever (1) holds and the divergence in (2) is uniformly in each solution and T depends on ε alone.

A sufficient condition for the exponential stability in (X,Y) is

(3) $a(r) = ar$, $b(r) = br$, $w(t,\phi) = cV(t,\phi)$, $F(r) = Fr$, $q(r) = q$

for constants a, b, c, F, q.

In the above,

$$\dot{V}_{(E)}(t,\phi) = \sup[\limsup_{s\to t+0} \frac{1}{s-t}\{V(s,x_s) - V(t,\phi)\}]$$

where $x(s)$ is a solution of (E) satisfying $x_t = \phi$ and the "sup" runs over such solutions.

A converse theorem holds for the part of the exponential stability.

Theorem 4. Suppose that $f(t,\phi)$ in (E) satisfies

$$||f(t,\phi) - f(t,\psi)|| \leq B_1||\phi - \psi||_X$$

and that the zero solution of (E) is exponentially stable in (X,Y). Then there exists a continuous function $V(t,\phi)$ satisfying (i) and (ii) in Theorem 3 with (3), especially, $q = 0$ (and, hence, $F > 1$ is arbitrary), and

(iii) $|V(t,\phi) - V(t,\psi)| \leq B||\phi - \psi||_X$.

Corollary. Under the assumptions in Theorem 4, if

$$||X(t,\phi)|| \leq \varepsilon(t)||\phi||_Y, \quad \lim_{t\to\infty} \int_t^{t+1} \varepsilon(s)ds = 0$$

then the zero solution of the perturbed equation of (E)

$$\dot{x}(t) = f(t,x_t) + X(t,x_t)$$

is exponential stable in (X,Y), too.

For the proofs, see [8 : Theorem 7].

4. Applications.

In practical problems, there are some possibilities how we choose the phase space for given functional differential equation. We present several examples.

Example 1. It is well-known that for an autonomous linear equation with finite delay, the distribution of the roots of the characteristic equation $\Delta(\lambda) = 0$ presents a sufficient information about the stability problem. For example, if all roots of $\Delta(\lambda) = 0$ have negative real parts, that is, if

(4) $\Delta(\lambda) \neq 0$ for any λ with $\mathrm{Re}\lambda \leq 0$

then the zero solution of the equation with lag h is uniformly asymptotically stable in (C_h^o, C_h^o), cf [5].

For the case of infinite delay this has been not proved. However, Naito [15] has shown that the same assertion holds if we can choose C_∞^γ or M_∞^γ with $\gamma > 0$ as the phase space for the equation.

Example 2. Consider the equation

$$(5) \qquad \dot{x}(t) = Ax(t) + \int_0^t B(t-s)x(s)ds,$$

which can be considered as a perturbed equation of the linear autonomous equation

$$(6) \qquad \dot{x}(t) = Ax(t) + \int_{-\infty}^{0} B(-s)x_t(s)ds$$

with the perturbation term

$$X(t,\phi) = - \int_{-\infty}^{-t} B(-s)\phi(s)ds.$$

Clearly, the characteristic equation related with (6) is given by

$$\Delta(\lambda) = \det[\lambda E - A - \int_{-\infty}^{0} B(-s)e^{\lambda s}ds].$$

In order to show the uniform asymptotic stability in (C_∞^0, R^n), Miller [14] has obtain a sufficient condition concerning $B(s)$, in addition to (4), which is weaker than

(7) $\quad B_0 = \int_{-\infty}^{0} ||B(-s)||e^{-\nu s}ds < \infty \quad$ for a $\quad \nu > 0.$

On the other hand, the condition (7) make it possible to take C_∞^γ, $0 < \gamma < \nu$, as the phase space for (6), which guarantees the exponential stability in $(C_\infty^\gamma, C_\infty^\gamma)$ by Example 1 and Theorem 2. Therefore, by applying the corollary to Theorem 4 we can conclude the exponential stability in $(C_\infty^\gamma, C_\infty^\gamma)$ of the zero solution of (5), where we note that

$$||X(t,\phi)|| \le B_0 e^{(\gamma-\nu)t}||\phi||_{C_\infty^\gamma}.$$

Example 3. The same idea is applicable to the scalar equation

(8) $\quad \dot{x}(t) = -x(t-h) + X(t, x(\varepsilon t))$

with $0 \le h < \frac{\pi}{2}$, $\varepsilon < 1$ and $|X(t,x)| \le B_0 e^{-\nu t}|x|$ for a $\nu > 0$. Here, we recall that under the condition the zero solution of

$$\dot{x}(t) = -x(t-h)$$

is exponential stable in (C_h^0, R^n) (see [18]) and, hence, in $(C_\infty^\gamma, C_\infty^\gamma)$ for any $\gamma > 0$. Therefore, by noting that

$$|X(t, \phi(\varepsilon t - t))| \le B_0 e^{-(\nu-(1-\varepsilon)\gamma)t}||\phi||_{C_\infty^\gamma}$$

we can prove the exponential stability in $(C_\infty^\gamma, C_\infty^\gamma)$ of the zero solution of (8), where $0 < \gamma < \nu/(1-\varepsilon)$.

Example 4. It has been shown that every solution of the scalar equation

(9) $\quad \dot{x}(t) = -\int_0^t A(t-s)g(x(s))ds$

starting at $t = 0$ approaches zero as $t \to \infty$ if $A(t) \in C^3$, $g(x) \in C^1$, $g(x)x > 0 \quad (x \ne 0)$ and

(10) $\quad (-1)^k A^{(k)}(t) \ge 0 \quad (k = 0,1,2,3), \quad A(\infty) > 0, \quad A(0) \ne A(\infty),$

see [12], [13]. In the proof, the function

$$E(t) = G(x(t)) + \frac{1}{2}A(t)[\int_0^t g(x(s))ds]^2 - \frac{1}{2}\int_0^t A'(t-s)[\int_s^t g(x(u))du]^2 ds$$

has played important roles, where $G(x) = \int_0^x g(s)ds$.

Suppose that the condition (10) is strengthened to

$$(-1)^k B^{(k)}(t) \geq 0 \quad (k = 0,1,2) \quad \text{and} \quad B(t) - B'(t) > 0,$$

where $B(t) = A(t)e^{\gamma t}$ for a $\gamma > 0$. Define

$$V(t,\phi) = G(\phi(0)) + \frac{1}{2}B(t)W(t,\phi) - \frac{1}{2}\int_0^t B'(s)W(s,\phi)ds$$

with setting

$$W(t,\phi) = [\int_{-t}^0 e^{\gamma s} g(\phi(s))ds]^2.$$

Then, we can verify that

$$G(\phi(0)) \leq V(t,\phi) \leq G(\phi(0)) + \frac{1}{2}B(0)W_+(\phi),$$

$$\dot{V}_{(9)}(t,\phi) = -\gamma B(t)W(t,\phi) + \gamma\int_0^t B'(s)W(s,\phi)ds$$

$$+ \frac{1}{2}B'(t)W(t,\phi) - \frac{1}{2}\int_0^t B''(s)W(s,\phi)ds,$$

where $W_+(\phi) = [\int_{-\infty}^0 e^{\gamma s}|g(\phi(s))|ds]^2$. Since $\dot{V}_{(9)}(t,\phi) \leq 0$, easily it follows that the zero solution of (9) is uniformly stable in (W, R^1), where W is a subset of C_∞^0 with the metric

$$||\phi||_W = \{G(\phi(0)) + W_+(\phi)\}^{1/2}$$

though this may not be a semi-norm.

Now, we shall consider the asymptotic stability. Since $\dot{V}_{(9)}(t,\phi) \leq 0$, it is sufficient to show that for a given $\varepsilon > 0$ there is a $T = T(\varepsilon) > 0$ such that

$$\inf_{\tau \leq t \leq \tau+T} V(t,x_t) \leq \varepsilon$$

for any solution $x(t)$ starting at $t = \tau$.

Suppose that

(11) $V(t,x_t) \geq \varepsilon$ for all $t \in [\tau,\tau+T]$

and for a solution $x(t)$ of (9) and an $\varepsilon > 0$.

First of all, by the uniform stability, we may consider that the solutions under consideration satisfy $|x(t)| \leq \alpha$ for all t and for some $\alpha > 0$. Hence, there is a $\beta > 0$ for which $V(t,x_t) \leq \beta$, and $x(t)$, hence $G(x(t))$, is uniformly continuous since

$$|\dot{x}(t)| \leq \int_0^t A(t-s)|g(x(s))|ds \leq B(0)\frac{1}{\gamma} \max_{|x|\leq\alpha} |g(x)|.$$

From this, there is a $\delta > 0$ such that

$$G(x(s)) \leq \frac{1}{2}\varepsilon \quad \text{on} \quad [t-\delta,t]$$

if $G(x(t)) \leq \frac{1}{4}\varepsilon$.

Because

$$\dot{V}_{(9)}(t,x_t) \leq -2\gamma\{V(t,x_t) - G(x(t))\},$$

the number of disjoint intervals of length δ, contained in $[\tau,\tau+T]$, on which $G(x(t)) \leq \frac{1}{2}\varepsilon$ should be less than a fixed number $K \leq \beta/\gamma\varepsilon\delta$.

On the other hand, it is not difficult to see that we can find $\eta > 0$ and $\rho(s) > 0$ such that

$$W(s,x_t) \geq \rho(s) \quad \text{if} \quad G(x(u)) \geq \frac{1}{4}\varepsilon \quad \text{on} \quad [t-\eta,t]$$

for any $s \in (0,t]$, and we have

$$\dot{V}_{(9)}(t,x_t) \leq \{-\gamma B(s) + \frac{1}{2}B'(s)\}\rho(s)$$

if $G(x(u)) \geq \frac{1}{4}\varepsilon$ on $[t-\eta,t]$. Choose and fix $s > 0$ so that $B(s) - B'(s) > 0$. Then, it is impossible that

$$G(x(u)) \geq \frac{1}{4}\varepsilon \quad \text{over} \quad [t_1,t_2]$$

if $t_2 - t_1 \geq \eta + T^*$, where

$$T^* \leq 2\beta/\{2\gamma B(s) - B'(s)\}\rho(s).$$

Summing up the above there arises a contradiction by setting $T = K(\eta + T^*)$ in (11). This shows that the zero solution of (9) is uniformly asymptotically stable in (W,R^1).

Furthermore, by the assumption there is a constant L such that $|g(x)| \leq L|x|$ for $|x| \leq \alpha$. Therefore, we can find a $\mu \in K_r^o$ for which

$$||\phi||_W \le \mu(||\phi||_{C^\nu_\infty}), \quad \nu = \frac{\gamma}{2}.$$

Thus, we have the stabilities in (C^ν_∞, R^1) under the assumptions.

Example 5. In order to prove the uniformly asymptotic stability in (C^o_h, R^n), $0 < h < \infty$, for (E), Burton [2] has assumed the existence of a Liapunov function $V(t,\phi)$ which satisfies

$$a(||\phi(0)||) \le V(t,\phi) \le b(||\phi||_{M^o_h})$$

and

$$\dot{V}_{(E)}(t,\phi) \le -c(||\phi(0)||)$$

for $a(r), c(r) \in K^+_r$, $b(r) \in K^o_r$.

It is not difficult to see that for any given $\alpha > 0$ there exists a $d(r) \in K^+_r$ such that every $\phi \in C^o_{2h}$, $||\phi||_{C^o_{2h}} \le \alpha$, satisfies

$$\int_{-2h}^{0} c(||\phi(s)||ds \ge d||\phi_t||_{M^o_h}$$

for some $t \in [-h,0]$, as essentially proved in [2].

Therefore, for any positive numbers ε, α and β we can find a T so that

$$\int_0^T c(||x(s)||)ds \ge \beta \quad \text{whenever} \quad ||x(t)|| \le \alpha, \ ||x_t||_{M^o_h} \ge \varepsilon$$

on $[0,T]$.

Using this fact, we can prove that the zero solution of (E) is uniformly asymptotically stable in (M^o_h, R^n) and, hence, in (C^o_h, R^n), where we note that

$$||\phi||_{M^o_h} \le (1+h)||\phi||_{C^o_k}.$$

Example 6 [8]. Consider the scalar equation

$$(12) \quad \dot{x}(t) = -ax(t) + \int_{-\infty}^{0} g(t,s,x(t+s))ds,$$

and assume that $a > 0$ is a constant and $g(t,s,x)$ is continuous and satisfies $|g(t,s,x)| \le m(s)|x|$, where

$$\int_{-\infty}^{0} m(s)ds < a, \quad \int_{-\infty}^{0} m(s)e^{-\gamma s}ds < \infty$$

for a $\gamma \geq 0$. Then the zero solution of (12) is uniformly asymptotically stable in (C_∞^γ, R^1).

To prove this, put $V(t,\phi) = \phi(0)^2$, and choose a constant $F > 1$ and a non-positive function $q(r) \in K_r$ of $r \in (0,\infty)$ so that

$$\sqrt{F}\int_{-\infty}^{0} m(s)ds < a, \quad 2\int_{-\infty}^{q(r)} m(s)e^{-\gamma s}ds \leq$$

$$\leq \{a - \sqrt{F}\int_{-\infty}^{0} m(s)ds\}\sqrt{r} = c(r).$$

Then we can see that

$$\dot{V}_{(12)}(t,\phi) \leq -c(V(t,\phi)) \quad \text{whenever} \quad ||\phi||_{C_\infty^\gamma} \leq 1, \quad t+q(V(t,\phi)) \geq \tau$$

and $\quad V(s,\phi_{s-t}) \leq FV(t,\phi) \quad$ for $\quad s \in [t + q(V(t,\phi)),t],$

namely, the conditions in Theorem 3 are satisfied. Thus, the conclusion follows immediately.

REFERENCES

[1] - BARNEA, B.I., A method and new results for stability and instability of autonomous functional differential equations, SIAM J. Appl. Math., 17(1969), 681-697.

[2] - BURTON, T.A., Uniform asymptotic stability in functional differential equations, Proc. Amer. Math. Soc., 68(1978), 195-199.

[3] - DIVER, R.D., Existence and stability of solutions of a delay-differential system, Arch. Rational Mech. Anal., 10(1962), 401-426.

[4] - GRIMMER, R. and SEIFERT, G., Stability properties of Volterra integro-differential equations, J. Differential Eqs., 19(1975), 142-166.

[5] - HALE, J.K., Theory of Functional Differential Equations, Appl. Math. Sci., Vol. 3, Springer-Verlag, 1977.

[6] - HALE, J.K. and KATO, J., Phase space for retarded equations with infinite delay, Funkcial. Ekvac., 21(1978), 11-41.

[7] - KATO, J., On Liapunov-Razumikhin type theorems for functional differential equations, Funkcial. Ekvac., 16(1973), 225-239.

[8] - KATO, J., Stability problem in functional differential equations with infinite delay, Funkcial. Ekvac., 21(1978), 63-80.

[9] - KATO, J., Liapunov's second method in functional differential equations, to appear.

[10] - KRASOVSKII, N.N., Stability of Motion, Stanford Univ. Press, 1963.

[11] - LAKSHMIKANTHAM, V. and LEELA, S., Differential and Integral Inequalities, Vol. 2, Academic Press, 1969.

[12] - LEVIN, J.J., The asymptotic behavior of the solution of a Volterra equation, Proc. Amer. Math. Soc., 14(1963), 534-541.

[13] - LEVIN, J.J. and NOHEL, J.A., Perturbations of a nonlinear Volterra equation, Michigan Math. J., 12(1965), 431-447.

[14] - MILLER, R.K., Asymptotic stability properties of linear Volterra integro-differential equations, J. Differential Eqs., 10(1971), 485-506.

[15] - NAITO, T., On autonomous linear functional differential equations with infinite retardations, J. Differential Eqs., 21(1976), 297-315.

[16] - RAZUMIKHIN, B.S., On the stability of systems with a delay, Prikl. Mat. Meh., 20(1956), 500-512.

[17] - SEIFERT, G., Liapunov-Razumikhin conditions for asymptotic stability in functional differential equations of Volterra type, J. Differential Eqs., 16(1974), 289-297.

[18] - WRIGHT, E.M., A nonlinear difference-differential equation, J. Reine Angew. Math., 194(1955), 66-87.

[19] - YOSHIZAWA, T., Stability theory by Liapunov's second method, Japan Math. Soc. Publ., Vol. 9, 1966.

TOPOLOGICAL EQUIVALENCE IN BIFURCATION THEORY

by Robert Magnus

1. The use of equivalence.

One of the remarkable aspects of bifurcation theory is the way it began as a specialized problem, the nonlinear eigenvalue problem, seen as a deviant branch of functional analysis, and was later brought into contact with singularity theory, a branch of differential topology. Many results of singularity theory have thereby become available to bifurcation theory, but are only slowly becoming known to traditional bifurcation theorists. The two subjects have a quite different character, which explains why bifurcation theorists have laboured to produce results which are implicit in the literature of singularity theory.

I shall adopt the following (for some purposes too limited) definition of bifurcation. Let X, Y and A be Banach spaces, of which A plays a special role and is called the parameter space. Let $f : X\times A \to Y$ be a C^∞ mapping. If $(x_o,a_o) \in X \times A$ is such that $f(x_o,a_o) = 0$ there are two possibilities. The first is that the solutions of $f(x,a) = 0$ in a neighbourhood of (x_o,a_o) have the form $(x(a),a)$ where $a \in x(a)$ is a C^∞ mapping. The contrary of this is the other possibility, in which case (x_o,a_o) is called a bifurcation point. It follows that a necessary condition for bifurcation is that the derivative $D_1f(x_o,a_o)$ should not be invertible. It is usual to assume that $D_1f(x_o,a_o)$ is a Fredholm operator of index 0, so that $D_1f(x_o,a_o)$ is invertible if and only if it is surjective. According to the usual definition x_o is a singularity of $f(\cdot,a_o)$ whenever $D_1f(x_o,a_o)$ is not surjective. Thus a bifurcation point at (x_o,a_o) implies a singularity of $f(\cdot,a_o)$ at x_o. The converse is untrue as is well known.

A more general problem than that of finding bifurcation points is obtained by suppressing the special role of the parameter space. We consider a C^∞ mapping f between Banach spaces X and Y and try to determine up to homeomorphism the set $f^{-1}(0)$ in a neighbourhood of a point x_o where $f(x_o) = 0$. The present paper is mainly connected with this problem.

The central method of singularity theory is the use of various kinds of equivalence. Consider two germs at 0 of C^∞ mapping $f : X \to Y$, $g : X \to Y$. They are said to be *C^∞ contact equivalent* if, in a neighbourhood of 0,

$$f(\phi(x)) = L(x)g(x)$$

where ϕ is a C^∞ diffeomorphism of a neighbourhood of 0 in X onto another neighbourhood of 0 in X such that $\phi(0) = 0$, and L is a C^∞ mapping from a neighbourhood of 0 in X into the space B(Y) of bounded linear operators in Y, such that $L(0) = I$. Note that the same symbol is used for a germ as for a representative of it. If $L(x) = I$ identically f and g are said to be C^∞ right equivalent. If ϕ and L are C^r instead of C^∞, we get the notions of C^r (contact or right) equivalence. In case $r = 0$, it is usual to speak of topological equivalence.

To bring the parameter space into the picture, let us consider two germs at (0,0) of C^∞ mappings $f : X \times A \to Y$, $g : X \times A \to Y$. We can adopt contact equivalence, or else a stronger equivalence relation, just like contact equivalence, in which the diffeomorphism $\phi : X \times A \to X \times A$ has the form $(x,a) \to (\phi_1(x,a),\phi_2(a))$. The latter is advantageous if it can be used, since it preserves the property of a pair of points in $X \times A$ of being over the same point in A. We might call this kind of equivalence parametrised equivalence. It is used in catastrophe theory in the C^∞ form. To illustrate the difference, the familiar pitchfork bifurcation

$$x^3 - \lambda x = 0$$

remains a pitchfork under parametrised equivalence, but it can become a cross under right equivalence.

Connected with equivalence is the notion of sufficiency or determinacy. A germ $f : X \to Y$ at 0 is said to be k-determinate (or k-sufficient; it means the same) if, whenever g is another germ such that $D^j(f-g)(0) = 0$ for $j = 0, 1, \ldots, k$, f and g are equivalent for whichever equivalence relation is being considered. This is a crucial idea. If we are given a bifurcation problem $f : X \times R^p \to Y$ with parameter space R^p, where $D_1 f(x,a)$ is a Fredholm operator of index 0 from X into Y, then the question of bifurcation, say at $(0,0)$, is reduced, by means of the Liapunov--Schmidt method to a bifurcation problem $g : R^n \times R^p \to R^n$. Similarly if $f : X \to Y$ and we wish to determine $f^{-1}(0)$ near 0, and $f'(0)$ is a Fredholm operator of non-negative index, the Liapunov-Schmidt method reduces the problem to a similar one $g : R^m \to R^n$. Usually only a few terms of the Taylor expansion of g are known, but if these terms up to order k form a k-determinate mapping h, then we may replace g by h up to equivalence.

In these cases singularity theory is used in finite dimensions, but recent papers discuss singularity theory directly in infinite dimensions. See for example Arkeryd [1], [2], Chillingworth [1], Guimarães [1] and Magnus [1], [2], [3]. The first paper to treat bifurcation theory with something like a singularity theory attitude was Crandall and Rabinowitz [1]. Their result was generalized by Magnus [4], [5] and Shearer [1], [2]. Somewhat related are McLeod and Sattinger [1] and Chow, Hale, and Mallet-Paret [1], [2]. That these results are implicit in the singularity theory literature has been observed by Guimarães [1] who gives further references (e.g. Kuo [1]).

2. A theorem on topological equivalence.

We shall consider here the space E of germs at 0 of C^∞ mappings $F : R^m \to R^n$ where $m \geq n$ are fixed integers. It can happen that a germ is not k-determinate for any k for C^∞ contact equivalence, and yet the topological structure of the set $f^{-1}(0)$ in a neighbourhood of 0 is determined by the k-jet of f alone. (The k-jet of f is the equivalence class of f under the relation $f \sim g \iff D^j(f-g)(0)=0$ for $j = 0, \ldots, k$. It may be thought of as the Taylor expansion of f up to the kth order terms). There is a terminology connected with this. Two germs f and g are said to be v-equivalent if $f^{-1}(0)$ and $g^{-1}(0)$ are homeomorphic set germs. The notion of determinacy for v-equivalence has been implicit in bifurcation theory for some time (beginning with Crandall and Rabinowitz [1]).

I wish to consider another kind of equivalence, weaker than C^∞ contact equivalence, yet stronger than v-equivalence, and prove some theorems.

Let S be the sphere $\{x : ||x|| = 1\}$ in R^m, the latter being given its Euclidean structure. Let E_p be the germs at $\{0\} \times S$ of C^∞ mappings $f : R \times S \to R^n$. For each $f \in E$ there is a germ $f_p \in E_p$ given by $f_p(r,u) = f(ru)$, where $r \in R$, $u \in S$. We shall say that f and g are contact p-equivalent if

$$f_p(\rho(r,u),\theta(r,u)) = L(r,u)g_p(r,u)$$

where, firstly, $(\rho,\theta) : (a,b) \times S \to (a',b') \times S$ is a C^∞ diffeomorphism, $a < 0 < b$, $a' < 0 < b'$, $\rho(0,u) = 0$ for all $u \in S$, and secondly, L is a C^∞ mapping from $(a,b) \times S$ into $B(R^n)$, the linear operators on R^n, such that $L(0,u) = I$. It is clear that if f and g are contact p-equivalent, they are also toplogically contact equivalent, since we may define a homeomorphism $\phi : R^m \to R^m$ by setting $\phi(x) = \rho(r,u)\theta(r,u)$ where $x = ru$ and $r \geq 0$. One may think of contact p-equivalence as just contact equivalence in polar coordinates; hence the "p". If $L(r,u)$ is the identity we shall speak of right p-equivalence.

Let M_k be the set of germs h in E_p with the property that $\lim_{r\to 0} r^{-k}h(r,u)$ exists for each $u \in S$. (This is the same as saying that $D_1^j h(0,u) = 0$ for $j = 0, \ldots, k-1$).

For each $f \in E$ let J_f be the set of germs h in E_p having the form

$$h(r,u) = f'(ru)g(r,u) + L(r,u)f(ru)$$

for some $g \in E_p$ and for some germ L of a C^∞ mapping from $R \times S$ into $B(R^n)$, the germ being localized at $\{0\} \times S$. We denote by I_f the subset of J_f obtained by supposing that L is 0.

Theorem 1. Let $f \in E$. If $M_{k-1} \subset J_f$ (resp. $M_{k-1} \subset I_f$) then f is k-determinate for contact (resp. right) p-equivalence.

Proof. Let $M_{k-1} \subset J_f$ and let $g \in E$ be such that $f_p - g_p \in M_{k+1}$ (in other words $D^j(f-g)(0) = 0$ for $j = 0, \ldots, k$). We must show that f and g are contact p-equivalent. Consider the mapping

$$(1) \quad F(t,r,u) = (1-t)f(ru) + tg(ru).$$

We shall find a one-parameter family of diffeomorphisms (ρ_t, θ_t) on $R \times S$, such that $\rho_t(0,u) = 0$, and a family of linear mappings $L_t(r,u)$ on R^n, parametrized by (t,r,u) such that

$$L_t(r,u)F(t,\rho_t(r,u),\theta_t(r,u))$$

is independent of t. It is easy to see that this gives the required result, and it is the usual approach to this kind of problem. We need to have

$$\frac{d}{dt}\{L_t(r,u)F(t,\rho_t(r,u),\theta_t(r,u))\} = 0$$

that is

$$(2) \quad \begin{aligned} &\frac{dL}{dt}t(r,u)F(t,\rho_t(r,u),\theta_t(r,u)) \\ &+ L_t(r,u)\frac{\partial F}{\partial t}(t,\rho_t(r,u),\theta_t(r,u)) \\ &+ L_t(r,u)\frac{\partial F}{\partial r}(t,\rho_t(r,u),\theta_t(r,u))\frac{d\rho}{dt}t(r,u) \\ &+ L_t(r,u)D_uF(t,\rho_t(r,u),\theta_t(r,u))\frac{d\theta}{dt}t(r,u) = 0 \end{aligned}$$

Now we seek a vector field $(\alpha(t,r,u),\sigma(t,r,u))$ on $R \times S$ in a neighbourhood of $\{0\} \times S$ and defined for $0 \le t \le 1$, such that (ρ_t, θ_t) form the solution of the initial value problem

$$(3)\quad \begin{aligned} \rho' &= \alpha(t,\rho,\theta) \\ \theta' &= \sigma(t,\rho,\theta) \end{aligned}$$

with starting values $(\rho,\theta) = (r,u)$ at $t = 0$. In order to have $\rho_t(0,u) = 0$ we need to have $\alpha(t,0,u) = 0$ for all $t \in [0,1]$ and $u \in S$.

We shall also seek a matrix $A(t,r,u)$ such that

$$(4)\quad \frac{dL}{dt}t(r,u) = L_t(r,u)A(t,\rho_t(r,u),\theta_t(r,u))$$

Then (2) is satisfied if

$$(5)\quad \begin{aligned} &A(t,r,u)F(t,r,u) + \frac{\partial F}{\partial t}(t,r,u) + \frac{\partial F}{\partial r}(t,r,u)\alpha(t,r,u) \\ &+ D_uF(t,r,u)\sigma(t,r,u) = 0 \end{aligned}$$

Now we use (1) to expand (5). Omitting for convenience the expression (t,r,u) after A, α and σ, we have

$$(6)\quad \begin{aligned} &A((1-t)f(ru) + tg(ru)) + g(ru) - f(ru) \\ &+ [(1-t)f'(ru) + tg'(ru)].(\alpha u + r\sigma) = 0. \end{aligned}$$

Now $g_p - f_p \in M_{k+1}$, and $M_{k-1} \subset J_f$.
Hence $g(ru) - f(ru) = r^2 f'(ru)h(r,u) + r^2K(r,u)f(ru)$ for some $h \in E_p$ and matrix-valued K. Furthermore, for each $\omega \in R^m$, the mapping $(r,u) \to (g'(ru) - f'(ru)).\omega$ belongs to M_k, and so $(g'(ru) - f'(ru)).\omega = rf'(ru)H(r,u,\omega) + rN(r,u,\omega)f(ru)$ where H is R^m-valued, N is matrix-valued and both depend *linearly* on ω. (To see this, consider finitely many ω's forming a basis of R^m and fix H and N on each of them; then take linear combinations). Thus (6) may be written

$$(7)\quad \begin{aligned} &Af(ru) + tr^2A(f'(ru)h(r,u) + K(r,u)f(ru)) \\ &+ r^2f'(ru)h(r,u) + r^2K(r,u)f(ru) + f'(ru)(\alpha u + r\sigma) \\ &+ trf'(ru)H(r,u,\alpha u + r\sigma) + trN(r,u,\alpha u + r\sigma)f(ru) = 0 \end{aligned}$$

Now (7) is satisfied if

(8) $A + tr^2AK(r,u) + r^2K(r,u) + trN(r,u,\alpha u + r\sigma) = 0$

and

(9) $tr^2Ah(r,u) + r^2h(r,u) + \alpha u + r\sigma + trH(r,u,\alpha u + r\sigma) = 0$

Now recall that $\alpha(r,0,u)$ should be 0, so that α is divisible by r. Write $x = (\frac{\alpha}{r})u + \sigma$. Then (8) and (9) become

(10) $A + tr^2AK(r,u) + tr^2N(r,u,x) = -r^2K(r,u)$

and

(11) $trAh(r,u) + x + trH(r,u,x) = -rh(r,u)$

These non-homogeneous linear equations may be solved for A and x when (r,u) is in a neighbourhood of $\{0\} \times S$ and $t \in [0,1]$, because the linear function of (A,x) on the left-hand side of (10) and (11) becomes the identity when $r = 0$. From x we can find α/r and σ. The proof is then complete for $M_{k-1} \subset J_f$. If $M_{k-1} \subset I_f$ it is easily seen that $A = 0$. This ends the proof.

<u>Theorem</u> 2. Let $f \in E$ and suppose that $M_{k-1} \subset J_f$. If $g \in E$ and $f_p - g_p \in M_{k+j}$ where $j \geq 1$, then $f(x) = L(\phi(x))g(\phi(x))$ where $\phi(x) = x + O(|x|^{j+1})$ and $L(x) = I + O(|x|^{j+1})$. If $M_{k-1} \subset I_f$ then $L(x)$ may be taken as I.

<u>Proof</u>. In equations (8) and (9), $h(r,u)$, $K(r,u)$, $N(r,u,\cdot)$ and $H(r,u,\cdot)$ are all divisible by r^{j-1} in virtue of $f_p - g_p \in M_{k+j}$. Hence $A(t,r,u)$ and $rx(t,r,u)$ are both divisible by r^{j+1}, and so $\alpha u + r\sigma$ defines on R^m a vector field (by setting $x = ru$ with $r \geq 0$) which is actually C^j and is $O(|x|^{j+1})$ at $x = 0$. Hence it generates a C^j flow which is of the form $x + O(|x|^{j+1})$. The mapping $x \to A(t,r,u)$, where $x = ru$, $r \geq 0$, is also of class C^j and is $O(|x|^{j+1})$. It follows then that $L(r,u) = I + O(r^{j+1})$.

<u>Remark</u>. Theorem 2 shows that when f is k-determinate for contact

(resp. right) p-equivalence in virtue of $M_{k-1} \subset J_f$ (resp. $M_{k-1} \subset I_f$), then f is also k-determinate for C^1 contact (resp. right) equivalence and the diffeomorphism needed to remove terms of order > k has derivative the identity at the origin. This is important for use in bifurcation problems. (See examples in Section 3).

Theorem 3. Let $f \in E$ be homogeneous of degree k and suppose that $f'(x)$ is surjective for all $x \neq 0$. Then $M_{k-1} \subset I_f$.

Proof. Let $g \in M_{k-1}$. We must show that $g(r,u)=f'(ru)h(r,u)$ where $h \in E_p$. Now $f'(ru) = r^{k-1}f'(u)$, and as we shall show, $f'(u)$ has a right inverse $L(u)$ which is a C^∞ operator-valued function of u. Hence we choose $h(r,u) = r^{-k+1}L(u)g(r,u)$.

To construct $L(u)$, let $\{y_1, \ldots, y_n\}$ be an orthonormal basis for R^n and let $x_i = f'(u)^* y_i$ for $i = 1, \ldots, n$. Since $f'(u)$ is surjective, $f'(u)^*$ is injective; hence the set $\{x_1, \ldots, x_n\}$ is independent and so the matrix $(\langle x_i, x_j\rangle)$ is invertible. Let the inverse matrix be (b_{ij}) and set

$$L(u)y = \Sigma \langle y_i, y\rangle b_{ij} x_j$$

the summation being over repeated subscripts. Then

$$\begin{aligned} f'(u)L(u)y &= \Sigma \langle y_i, y\rangle b_{ij} f'(u)x_j \\ &= \Sigma \langle y_i, y\rangle b_{ij} \langle f'(u)x_j, y_k\rangle y_k \\ &= \Sigma \langle y_i, y\rangle b_{ij} \langle x_j, x_k\rangle y_k \\ &= \Sigma \langle y_i, y\rangle y_i = y. \end{aligned}$$

Remarks. (a) On v-equivalence.

Suppose $f \in E$ is homogeneous of degree k, and suppose that $f'(x)$ is surjective whenever $x \neq 0$ and $f(x) = 0$. A slight variation of the preceding arguments shows that f is k-determinate for v--equivalence. Suppose that f is a polynomial and set $J = \{u \in S : f(u)=0\}$. Then J is a compact submanifold of S. If $g \in E$ has k-jet f, it is known (Magnus [4]) that given $\varepsilon > 0$ there is a neighbourhood U of 0 such that all solutions $x \neq 0$ of $g(x) = 0$ in U lie in the cone $D = \{x : x = ru, 0 < r < \varepsilon, d(u,J) < \varepsilon\}$. We may choose ε so that $f'(x)$ is surjective for $x \in D$, and construct a vector field on the set $D_p = \{(r,u) : ru \in D\}$ which carries f_p to g_p on D_p. This gives, modulo Liapunov-Schmidt, a proof of the conjectured theorem in Magnus [4], easier than that given in Magnus [5]. (For another version see Buchner, Marsden and Schecter [1]). Kuo has given a characterization of v-sufficiency for jets which need not be homogeneous.

(b) The germ $(x^2+y^2)^2$ is seen by our methods to be k-determinate for C^{k-3} right equivalence, (for $k \geq 4$), slight improving a known result.

3. <u>Some remarks on normal forms</u>.

As a result of the influence of singularity theory, bifurcation problems have come to be classified according to certain normal forms. The classification is only partial, however, and is most complete for variational problems subject to structural stability. Consider the more general problem of determining up to equivalence the zero set $f^{-1}(0)$ of a mapping $f : X \to Y$. The Liapunov-Schmidt method reduces this in a neighbourhood of a known zero point to a finite dimensional problem $g : R^m \to R^n$ where $g(0) = 0$, $g'(0) = 0$. We may try to put g into a normal form by applying various kinds of equivalence. For example, if we start with the problem known as *bifurcation at a simple eigenvalue* we get the normal form $(\lambda,x) \to \lambda x$, with its cross-shaped bifurcation diagram. This incidentally includes the pitchfork bifurcation $x^3-\lambda x = 0$, as well as $x^4 - \lambda x = 0$ (a section of the *swallowtail*), $x^5 - \lambda x = 0$, (a section of the *butterfly*) etc., which are classified differently in catastrophe theory because of the stronger equivalence relation.

The normal forms which arise in practice seem usually to be

sufficient for C^∞ contact equivalence, and in fact they always seem to belong to the collection found in catastrophe theory, or to be sections of them. Where can we look for other normal forms which are sufficient only for topological (or C^1) contact equivalence, and which might have some importance ?

Let us consider a Hopf bifurcation. For simplicity we suppose it to have a two-dimensional phase space (if you like, you may imagine a reduction to have been made to the centre manifold)

(1) $x' = A(\lambda)x + F(\lambda,x)$

where

$$x = \begin{bmatrix} x_1 \\ x_2 \end{bmatrix} \in R^2$$

$$A(\lambda) = \begin{bmatrix} -\lambda & -1 \\ 1 & -\lambda \end{bmatrix}$$

$F : R \times R^2 \to R^2$ is C^∞ and $F(\lambda,0) = 0$, $D_2F(\lambda,0) = 0$ for all λ. We look for periodic solutions, with period near to 2π, and λ near 0, using the method of Magnus [4]. First, alter the time scale by setting $s = (1+\omega)t$, regard ω as a variable near to 0, and lock for solutions of

(2) $(1+\omega)x' = A(\lambda)x + F(\lambda,x)$

with period exactly 2π, and λ near 0. We set up the mapping $G : R^2 \times X \to Y$

where

$$X = \{x \in C^1[0,2] : x(0) = x(2\pi)\}$$

$$Y = C[0,2\pi]$$

$$G(\lambda,\omega,x)(t) = (1+\omega)x'(t) - A(\lambda)x(t) - F(\lambda,x(t))$$

Then

$$G'(0)(\lambda,\omega,x) = x' - A(0)x$$

The kernel V of $G'(0)$ is 4-dimensional, consisting of all (λ,ω,x) where $x = \mu(e_1c+e_2s) + \nu(-e_1s+e_2c)$, $(\lambda,\omega,\mu,\nu) \in R^4$, $c(t) = \cos t$, $s(t) = \sin t$,

$$e_1 = \begin{bmatrix} 1 \\ 0 \end{bmatrix} \qquad e_2 = \begin{bmatrix} 0 \\ 1 \end{bmatrix}$$

We consider V as identified with R^4.

We have

$$G^{(2)}(0)(\lambda,\omega,x)^2 = \omega x' + \lambda x - D_2^2F(0,0)x^2$$

Let $\overline{Y} = Y/R(G'(0))$ and let $P : Y \to \overline{Y}$ be the canonical projection. We identify $\overline{Y}$ with R^2 by taking as basis the images of e_1c+e_2s and $-e_1s+e_2c$. Let $B(\lambda,\omega,x) = PG^{(2)}(0)(\lambda,\omega,x)^2$ where $(\lambda,\omega,x) \in V$. An easy calculation now gives

$$B(\lambda,\omega,x) = \begin{bmatrix} \lambda & \omega \\ \omega & \lambda \end{bmatrix} \begin{bmatrix} \mu \\ \nu \end{bmatrix}$$

Where $x = \mu(e_1c+e_2s) + \nu(-e_1s+e_2c)$. If we identify x with the complex number $\mu + i\nu$ and (λ,ω) with $\lambda + i\omega$ then P is just the mapping

$$(\lambda + i\omega,\mu + i\nu) \to (\lambda + i\omega)(\mu + i\nu)$$

or

$$(z_1,z_2) \to z_1z_2$$

from C^2 into C.

Now B is the 2-jet of the reduced bifurcation problem, and B is 2-determinate for C^1 right equivalence by Theorem 3, so that it may constitute a normal form. It is not determinate to any order for C^∞ contact equivalence. The Hopf bifurcation may easily be read from the zero set $B^{-1}(0)$, which consists of two planes, $\mu = \nu = 0$ tangent to (and identical with) the trivial solutions, and $\lambda = \omega = 0$, tangent to a two-dimensional manifold of non-trivial periodic solutions.

The preceding calculation is for a rather special Hopf bifurcation, but fairly easy calculations show that any Hopf bifurcation (characterized by a pair of conjugate eigenvalues $\alpha \pm i\beta$ crossing the imaginary axis with non-zero velocity with respect to the parameter, and at the moment of crossing no other eigenvalue may be an integral multiple of $i\beta$) leads to the same normal form, though finally a linear coordinate change may be required.

The normal form $(z_1,z_2) \to z_1 z_2$ is obviously analogous to the normal form $(\lambda,x) \to \lambda x$ for bifurcation at a simple eigenvalue. They are the second and first members of a series of analogous normal forms of which there are two more, namely, multiplication of quaternions, leading to a bilinear mapping $b : R^4 \times R^4 \to R^4$, and multiplication of Cayley numbers, leading to a bilinear mapping $b : R^8 \times R^8 \to R^8$.

It is possible to realize these normal forms by means of differential equations, although I have no general characterization analogous to a Hopf bifurcation.

Consider the equation

(3) $x' = Ax + F(x)$

where $x \in R^4$, $A : R^4 \to R^4$ is linear and $F : R^4 \to R^4$ is C^∞ and $0(|x|^2)$. Let us assume that A has the matrix

$$\begin{bmatrix} 0 & 0 & -1 & 0 \\ 0 & 0 & 0 & -1 \\ 1 & 0 & 0 & 0 \\ 0 & 1 & 0 & 0 \end{bmatrix}$$

Then all solutions of $x' = Ax$ are 2π-periodic. In general the solutions of (3) need not be periodic, but if $B(\alpha,\beta,\gamma,\delta)$ is the matrix

$$\begin{bmatrix} \alpha & \beta & \gamma & \delta \\ -\beta & \alpha & \delta & -\gamma \\ -\gamma & \delta & \alpha & \beta \\ -\delta & \gamma & -\beta & \alpha \end{bmatrix}$$

then the equation

(4) $x' = Ax + B(\alpha,\beta,\gamma,\delta)x + F(x)$

has a 4-dimensional manifold of non-trivial 2π-periodic solutions in the space $R^4 \times R^4$. The first R^4 is phase space, the second is $(\alpha,\beta,\gamma,\delta)$-space, and the manifold is tangent to $R^4 \times \{0\}$ at the origin. This follows because the normal form (which is found by a calculation very similar to the one for Hopf bifurcation) is the one associated with quaternion multiplication, $(q_1,q_2) \to q_1q_2$, where q_1 belongs to the 4-dimensional space of solutions of $x' = Ax$, and $q_2 = (\alpha,\beta,\gamma,\delta)$.

Instead of adding parameters to the equation, one can include them in the boundary conditions. For example, consider

$$(5) \qquad \begin{aligned} x' &= Ax + F(x) \\ x(2\pi) &= (e+y)x(0) \end{aligned}$$

where y is a parameter in R^4, which is identified with the quaternion algebra, and e is the quaternion identity. The conclusion is that there exists a 4-dimensional manifold of solutions of (5) in the space $R^4 \times R^4$ (phase space xy-space), tangent to $R^4 \times \{0\}$ at the origin. In other words if you look at the solutions of (3),

$$\frac{|x(2\pi) - x(0)|}{|x(0)|^2} \to 0$$

as $|x(0)| \to 0$, a surprising conclusion.

Examples of greater interest could probably be found, but time and space prevent their consideration here.

REFERENCES

ARKERYD, L.

1. Catastrophe theory in Hilbert space, Tech. Report, Math. Dept., University of Gothenburg (1977).
2. Thom's theorem for Banach spaces, J. Lon. Math. Soc. (To appear).

CHILLINGWORTH, D.R.J.

1. A global genericity theorem for bifurcation in variational problems, Preprint, Math. Dept., Univ. of Southampton (1978).

CHOW, S.-N., HALE, J.K. and MALLET-PARET, J.

1. Applications of generic bifurcation, I) Arch. Rat. Mech. Anal. 59(1975), 159-188; II) Ibid 62(1976), 209-235.

CRANDALL, H.G. and RABINOWITZ, P.H.

1. Bifurcation from simple eigenvalues, J. Funct. Anal. 8(1971), 321-340.

GUIMARÃES, L.C.

1. Contact equivalence and bifurcation theory, Thesis, University of Southampton (1978).

KUO, T.-C.

1. Characterization of v-sufficiency of jets, Topology, 11(1972), 115-131.

McLEOD, J.B. and SATTINGER, D.H.

1. Loss of stability and bifurcation at a double eigenvalue, J. Funct. Anal. 14(1973), 62-84.

MAGNUS, R.J.

1. On universal unfoldings of certain real functions on a Banach space, Math. Proc. Cam. Phil. Soc. 81(1977), 91-95.
2. Determinacy in a class of germs on a reflexive Banach space, Math. Proc. Cam. Phil. Soc. 84(1978), 293-302.
3. Universal unfoldings in Banach spaces: reduction and stability, Battelle-Geneva Math. Report 107(1977) (To appear in Math. Proc. Cam. Phil. Soc.).
4. On the local structure of the zero set of a Banach space valued mapping, J. Funct. Anal. 22(1976), 58-72.
5. The reduction of a vector-valued function near a critical point, Battelle-Geneva Math. Report 93(1975).

SHEARER, M.

1. Small solutions of a non-linear equation in Banach space for a degenerate case, Proc. Royal Soc. Edinburgh, 79A (1977), 58-73.
2. Bifurcation in the neighbourhood of a non-isolated singular point, Israel J. Math. 30(1978), 363-381.

BUCHNER, M., MARSDEN, J. and SCHECTER, S.

1. Differential topology and singularity theory in the solution of non-linear equations (preliminary version), University of California, Berkeley.

ON A HARTREE TYPE EQUATION: EXISTENCE OF REGULAR SOLUTIONS

by Gustavo Perla Menzala*

1. Introduction.

In the late twenties, D. R. Hartree, [3], introduced the so-called Hartree equation for the Helium atom. Since then this equation has received considerable attention by a number of authors, some of whom are cited in our references. In this work we should like to discuss the existence of non-trivial solutions $u \in H^1(R^n)$ of the related equation

$$(1.1) \quad -\Delta u - K(x)u - 2u\int K(x-y)u^2(y)dy = \lambda u$$

for some real number λ. Here, x runs in R^n $(n \geq 3)$, the integration is considered over all space R^n, Δ denotes the Laplacian operator and K is a given real-valued function satisfying suitable conditions which we shall make precise in the next section. Quite recently, E. Lieb, [5], studied the equation

$$(2.1) \quad -\Delta u - 2u\int \frac{u^2(y)}{|x-y|}dy = \lambda u$$

in the case $n = 3$, showing the existence and uniqueness (modulo translations) of a positive solution of the equation (2.1). Inspired by Lieb's paper, [5], we shall present in what follows, detailed proofs for the general equation (1.1). Let us give some comments on the equation (1.1): First, in the way that the above problem was formulated, we don't have uniqueness. In fact, if u is a solution, then $-u$ is also a solution. It would be better, for physical reason to ask if, for some real number λ there is a unique positive solution u of (1.1) which is "smooth" at least outside the possible singularities of K. Secondly, the minus sign in the third term of the right hand side of the equation (1.1) it makes an important difference if we compared it with most of the work which has been done

* This research was supported by CEPG-IMUFRJ and FNDCT (Brazil).

on related equation by using other tools, such as monotone operator theory, bifurcation analysis, etc. In section 2.) we introduce some notations and we obtain some estimates on a functional J associated with (1.1). In section 3.) we solve a variational problem associated with (1.1) and in 4.) we show that the function u which minimizes the functional J satisfies (1.1) at least in the sense of distributions. Finally, in section 5.) we study the smoothness of the solution found in 3.) and 4.). In this last section we only treat the case $n = 3$.

2. Preliminaries.

We shall consider the functional J defined by

$$(1.2)\quad J(u) = \int |\text{grad } u|^2 dx - \int K(x)u^2(x)dx - \iint K(x-y)u^2(x)u^2(y)dxdy$$

for $u \in H^1(R^n)$, i.e., the usual Sobolev space of order one. From now on, all integral signs in which no domain is attached will be understood to be taken over all R^n $(n \geq 3)$. We shall use frequently Sobolev's inequality which states that, if $u \in H^1(R^n)$ then $u \in L^p(R^n)$ and

$$(2.2)\quad ||u||_{L^p} \leq C||\text{grad } u||_{L^2}$$

where $p = \frac{2n}{n-2}$, for some positive constant C.

We shall assume that the real-valued function K satisfies the following hypothesis:

1) K can be written as $K = K_1 + K_2$, with $K_1 \in L^{n+1}(R^n) \cap$ $\cap L^{\frac{n-1}{2}}(R^n)$ and $K_2 \in L^{\frac{n+1}{2}}(R^n) \cap L^\infty(R^n)$,

2) Each K_j is nonnegative $(j = 1, 2)$,

3) K is spherically symmetric and decreases with $|x|$.

Let $r_1 > 0$, then we consider

$$K_1^1(x) = \begin{cases} K_1(x) & \text{if} \quad |x| \leq r_1 \\ 0 & \text{otherwise.} \end{cases}$$

Thus, we can write $K = K_1^1 + K_2^1$ where

$$K_2^1(x) = K_1(x)\chi_A(x) + K_2(x)$$

and χ_A denotes the characteristic function of the set $A = \{x, |x| \geq r_1\}$. Observe that our assumptions imply that $K_1 \in L^{n/2}$. Let $a = ||K_1||_{L^{n/2}}$, and let us choose $\varepsilon > 0$ such that

(3.2) $\varepsilon > \max\{1, 1/2\sqrt{a}\ C\}$

where C was obtained as in (2.2). We select r_1 in such a way that

(4.2) $||K_1^1||_{L^{n/2}} \leq \dfrac{1}{4C^2\varepsilon^2}$.

Let us call

(5.2) $h(\varepsilon) = ||K_2^1||_{L^\infty}$.

<u>Lemma</u> 1. Let $u \in H^1(R^n)$ with $||u||_{L^2} \leq \varepsilon$ then

1) $J(u) \geq -h(\varepsilon)\ (\varepsilon^2+\varepsilon^4)$

2) $\inf_{\substack{v \in H^1 \\ ||v||_{L^2} \leq \varepsilon}} J(v) < 0$

3) Given $\delta > 0$, if $J(u) \leq \inf_{\substack{v \in H^1 \\ ||v||_{L^2} \leq \varepsilon}} J(v) + \delta$ then

$$Pu < \delta + 4h(\varepsilon)\ (\varepsilon^2+\varepsilon^4)$$

where

$$Pu = \int K(x)u^2(x)dx + \iint K(x-y)u^2(x)u^2(y)dxdy.$$

<u>Proof</u>. By using Young's inequality we obtain

$$\iint K_1^1(x-y)u^2(x)u^2(y)dxdy \leq ||K_1^1||_{L^{n/2}}||u||^2_{L^{2n/n-2}}||u||^2_{L^2}$$

and

(6.2) $$\iint K_2^1(x-y)u^2(x)u^2(y)dxdy \leq ||K_2^1||_{L^\infty}||u||^4_{L^2}.$$

By using Hölder's inequality we obtain

$$\int K_1^1(x)u^2(x)dx \leq ||K_1^1||_{L^{n/2}}||u||^2_{L^{2n/n-2}}$$

and

$$(7.2)\quad \int K_2^1(x)u^2(x)dx \le ||K_2^1||_{L^\infty}||u||^2_{L^2}.$$

Thus, from (4.2), (5.2), (6.2) and (7.2) we get after some simplifications

$$J(u) = ||\text{grad } u||^2_{L^2} - Pu \ge \frac{1}{2C^2}||u||^2_{L^{2n/n-2}} - h(\varepsilon)\ (\varepsilon^2+\varepsilon^4) \ge -h(\varepsilon)\ (\varepsilon^2+\varepsilon^4)$$

This proves part 1).

In order to prove part 2), it is sufficient to show that for some $v \in H^1(R^n)$ with $||v||_{L^2} \le \varepsilon$ we have $J(v) < 0$. Let us define $v(x) = a\exp(-b|x|^2)$ for $a \neq 0$, $b > 0$. Clearly $v \in H^1(R^n)$ and an easy calculation of $J(v)$ shows that, if "a" is chosen sufficiently large then we have $J(v) < 0$.

To prove part 3), we observe that, by using (6.2) and (7.2) we obtain

$$(8.2)\quad \begin{aligned} Pu \le\ & ||K_1^1||_{L^{n/2}}||u||^2_{L^p}||u||^2_{L^2} + ||K_2^1||_{L^\infty}||u||^4_{L^2} \\ & + ||K_1^1||_{L^{n/2}}||u||^2_{L^p} + ||K_2^1||_{L^\infty}||u||^2_{L^2} \end{aligned}$$

where $p = \frac{2n}{n-2}$. From this and (4.2) with (5.2) we obtain

$$Pu \le \frac{1}{2}||\text{grad } u||^2_{L^2} + h(\varepsilon)\ (\varepsilon^2+\varepsilon^4)$$

which implies that

$$\frac{1}{2}Pu - 2h(\varepsilon)(\varepsilon^2+\varepsilon^4) \le \frac{1}{2}J(u) - h(\varepsilon)(\varepsilon^2+\varepsilon^4) \le$$

$$\le \frac{1}{2}\inf_{\substack{v\in H^1 \\ ||v||_{L^2}\le\varepsilon}} J(v) + \frac{1}{2}\delta - h(\varepsilon)(\varepsilon^2+\varepsilon^4) \le \frac{\delta}{2}.$$

Thus,

$$Pu \le \delta + 4h(\varepsilon)(\varepsilon^2+\varepsilon^4)$$

which proves the lemma.

3. The variational problem.

With all of the above hypotheses we shall prove.

Theorem 1. There exists $u \in H^1(R^n)$ with $||u||_{L^2} \le \varepsilon$ such that

$$J(u) = \inf_{\substack{v \in H^1(R^n) \\ ||v||_{L^2} \le \varepsilon}} J(v).$$

<u>Proof</u>. Let $\{u_k\}_{k=1}^{\infty}$ be a minimizing sequence, that is,

$$u_k \in H^1(R^n), \quad ||u_k||_{L^2} \le \varepsilon \quad \text{and}$$

(1.3) $$\lim_{k\to\infty} J(u_k) = \inf_{\substack{v \in H^1 \\ ||v||_{L^2} \le \varepsilon}} J(v).$$

We observe that we could use a Schwarz spherical rearrangment (see [2]) of the sequence $\{u_k\}_{k=1}^{\infty}$, that is, a sequence $\{u_k^*\}_{k=1}^{\infty}$, such that each u_k^* is spherically symmetric, positive and decreasing with $r = |x|$. To simplify our notation we shall write again by $\{u_k\}_{k=1}^{\infty}$ this sequence of radial functions. By using well known facts of the Schwarz spherical rearrangments for two or three functions we see that (1.3) remains true for this new sequence. Because of part 3) of Lemma 1 it follows that given $\delta > 0$, there exists $N_o > 0$ such that for any $k \ge N_o$ we have

$$J(u_k) \le \inf_{\substack{v \in H^1 \\ ||v||_{L^2} \le \varepsilon}} J(v) + \delta.$$

Thus

$$||\text{grad } u_k||_{L^2}^2 = J(u_k) + Pu_k \le \inf_{\substack{v \in H^1 \\ ||v||_{L^2} \le \varepsilon}} J(v) + 2\delta + 4h(\varepsilon)(\varepsilon^2 + \varepsilon^4)$$

which implies that $\{u_k\}_{k=N_o}^{\infty}$ is bounded in $H^1(R^n)$, so we can find a subsequence, which we shall denote by $\{u_k\}_{k=N_o}^{\infty}$ such that $u_k \to u$, weakly in H^1, and almost everywhere in R^n, to some $u \in H^1(R^n)$. It is easy to see that

$$||\text{grad } u||_{L^2}^2 \le \text{Lim inf } ||\text{grad } u_k||_{L^2}^2$$

because of the weak semicontinuity of the norm. Similarly, it is easy to prove that $||u||_{L^2} \le \varepsilon$.

Now, we claim that

$$Pu_k \to Pu \quad \text{as} \quad k \to \infty.$$

In fact, if this is so, then we can conclude the proof fo the theorem: Since

$$J(u_k) = ||\text{grad } u_k||^2_{L^2} - Pu_k$$

then

$$\text{Lim inf } J(u_k) \geq ||\text{grad } u||^2_{L^2} - Pu = J(u)$$

so

$$J(u) = \inf_{\substack{v \in H^1 \\ ||v||_{L^2} \leq \varepsilon}} J(v) = \lim_{k\to\infty} J(u_k).$$

Thus, it remains to be shown that $Pu_k \to Pu$ as $k \to \infty$. Since each u_k^2 decreases with $r = |x|$ and $\{||u_k||_{H^1}\}$ is a bounded sequence, it follows that there exists a constant C_4 such that, for any $r = |x|$ we have

$$(2.3)\quad \text{Vol}(\Omega_n)u_k^p(r) \leq w_n\int_0^r s^{n-1}u_k^p(s)ds \leq ||u_k||^p_{L^p} \leq C_4$$

where Ω_n denotes the ball $\{y, |y| \leq r\}$, w_n denotes the surface area of the unit ball and $p = \frac{2n}{n-2}$. From (2.3) we deduce that there exists a constant C_5 such that

$$(3.3)\quad u_k^2(r) \leq C_5 r^{2-n}.$$

Similarly, because $||u_k||_{L^2} \leq \varepsilon$, we obtain

$$\text{Vol}(\Omega_n)u_k^2(r) \leq w_n\int_0^r s^{n-1}u_k^2(s)ds \leq \varepsilon^2.$$

Thus

$$(4.3)\quad u_k^2(r) \leq C_6 r^{-n}$$

for some constant C_6.

We define

$$f(x) = \begin{cases} C_6 k(x)|x|^{-n} & \text{if} \quad |x| > 1 \\ C_5 K(x)|x|^{2-n} & \text{if} \quad 0 < |x| < 1 \end{cases}$$

It follows from (3.3) and (4.3) that

$$|K(x)(u_k^2(x) - u^2(x))| \le f(x).$$

Because of our assumptions on K_1 and K_2 it is not difficult to show that $f \in L^1(R^n)$. Now, we apply the Lebesgue dominated convergence theorem to obtain

$$\lim_{k\to\infty} \int K(x)u_k^2(x)dx = \int K(x)u^2(x)dx.$$

Similarly, by using the same idea as above, we consider appropriate positive constants C_7 and C_8 and define

$$g(x,y) = \begin{cases} C_7 K(x-y)|x|^{2-n}|y|^{2-n} & \text{for} \quad 0 < |x|,\ |y| \le 1 \\ C_8 K(x-y)|x|^{-n}|y|^{-n} & \text{for} \quad |x|,\ |y| > 1. \end{cases}$$

Thus, we have

$$|K(x-y)[u_k^2(x)u_k^2(y) - u^2(x)u^2(y)]| \le g(x,y).$$

Because of our assumptions on K_1 and K_2 an easy application of Young's inequality shows that $g \in L^1(R^n)$. Thus, we apply the Lebesgue dominated convergence theorem to obtain

$$\lim_{k\to\infty} \iint K(x-y)u_k^2(x)u_k^2(y)dxdy = \iint K(x-y)u^2(x)u^2(y)dxdy.$$

Thus proves the theorem.

4. A weak solution.

In this section we shall prove that the function $u \in H^1(R^n)$ obtained in Theorem 1 is a weak solution of equation (1.1). First, we observe that if $u \in H^1(R^n)$, $||u||_{L^2} \le \varepsilon$ and $J(u) = \inf_{\substack{v\in H^1 \\ ||v||_{L^2}\le\varepsilon}} J(v)$

then $||u||_{L^2} = \varepsilon$. In fact, suppose that $||u||_{L^2} < \varepsilon$ then consider $v = \varepsilon||u||_{L^2}^{-1}u$. From the definition of J and P it follows that

$$J(v) = ||\text{grad } v||^2_{L^2} - Pv =$$

$$= \varepsilon^2||u||^{-4}_{L^2}||\text{grad } u||^2_{L^2} - \varepsilon^2||u||^{-2}_{L^2}\int K(x)u^2(x)dx$$

$$- \varepsilon^4||u||^{-4}_{L^2}\iint K(x-y)u^2(x)u^2(y)dxdy <$$

$$< \varepsilon^2||u||^{-2}_{L^2} \inf_{\substack{w\in H^1 \\ ||w||_{L^2}\le\varepsilon}} J(w) < \inf_{\substack{w\in H^1 \\ ||w||_{L^2}\le\varepsilon}} J(w)$$

because of part 2) of Lemma 1. This contradiction proves our claim.

<u>Theorem</u> 2. Let $u \in H^1(R^n)$ with $||u||_{L^2} = \varepsilon$ and $J(u) = \inf_{\substack{v\in H^1 \\ ||v||_{L^2}\le\varepsilon}} J(v)$, then there exists λ such that u satisfies equation (1.1) in the sense of distributions.

<u>Proof</u>. Let us consider the functional $J_1 : H^1(R^n) \to R$ given by

$$(1.4) \quad J_1(v) = \iint v^2(x)K(x-y)v^2(y)dxdy$$

and let us show that J_1 is Frẽchet differentiable. For let $v \in H^1(R^n)$ be fixed and let us take $\phi \in C_o^\infty(R^n)$ which denotes the space of C^∞-functions with compact support. Let us show that

$$J_1(v+\phi) - J_1(v) = L(\phi) + R(\phi)$$

for some $L : H^1(R^n) \to R$, linear and continuous and $\lim\limits_{||\phi||_{H^1}\to 0} \frac{|R(\phi)|}{||\phi||_{H^1}} = 0$. By a direct calculation we obtain

$$J_1(v+\phi) - J_1(v) = 4\iint v^2(x)K(x-y)v(y)\phi(y)dxdy$$

$$(2.4) \quad + 4\iint \phi^2(y)K(x-y)v(x)\phi(x)dxdy + 2\iint v^2(x)K(x-y)\phi^2(y)dxdy$$

$$+ 4\iint v(x)K(x-y)\phi(x)v(y)\phi(y)dxdy + \iint \phi^2(x)K(x-y)\phi^2(y)dxdy.$$

Thus, if we call

$$L(\phi) = 4\iint v^2(x)K(x-y)v(y)\phi(y)dxdy$$

then it can easily be seem by using (2.4) that L is linear and continuous. By calling $R(\phi)$ the last four terms of the right hand side of (2.4), then a similar discussion as above, shows that there exist positive constants C_1, C_2, C_3 (which depend on v) such that

$$(3.4)\quad |R(\phi)| \le C_1||\phi||^2_{H^1} + C_2||\phi||^3_{H^1} + C_3||\phi||^4_{H^1}.$$

From (2.4) and (3.4) we conclude that J_1 is F-differentiable at v and

$$(4.4)\quad J_1'(v)\phi = 4\int (Av)(y)v(y)\phi(y)dy$$

where $Av(x) = \int K(x-y)v^2(y)dy$, for any $\phi \in C_o^\infty$. Similarly, if we consider the functionals J_2 and J_3 given by

$$J_2(v) = \int |\text{grad } v|^2 dx$$

$$J_3(v) = \int K(x)v^2(x)dx$$

then, we can show that J_2 and J_3 are F-differentiable and

$$J_2'(v)\phi = 2\int \text{grad } v.\text{grad } \phi \, dx$$

$$J_3'(v)\phi = 2\int K(x)v(x)\phi(x)dx$$

for any $\phi \in C_o^\infty(R^n)$.

Since u minimizes the functional $J = J_2 - J_3 - J_1$ under the constraint $V(u) = 0$ where $V(u) = ||u||_{L^2} - \varepsilon$ and since V is continuously F-differentiable in $H^1(R^n)$ and

$$V'(u)\phi = 2\int u(x)\phi(x)dx$$

for any $\phi \in C_o^\infty$, then it follows by the principle of Euler-Lagrange that there exists a "Lagrange multiplier" λ such that

$$J'(u)\phi = \lambda V'(u)\phi \qquad \forall\ \phi \in C_o^\infty(R^n)$$

which implies that

$$\int [-\Delta\phi - K(x)\phi - 2(Au)(x)\phi]u(x)dx = \lambda\int u(x)\phi(x)dx$$

for any $\phi \in C_o^\infty(R^n)$, hence u is a weak solution of equation (1.1).

5. Regularization.

In this section we shall consider only the case when $n = 3$ and let us study the regularity of the solution u obtained above.

Our assumptions in this section will be the following:

1) The real-valued function $K : R^3 \to R^+$ is spherically symmetric and decreases with $r = |x|$.

2) K can be written as $K = K_1 + K_2$ where $K_1 \in L^1 \cap L^4$, $K_2 \in L^2 \cap L^\infty$.

3) K is twice continuously differentiable in R^3.

Observations.

a) The above hypotheses satisfy the conditions on K given before in the particular case when $n = 3$.

b) In order to include the important example in which $K = |x|^{-a}$ for some $a > 1$, we should, of course, assume instead that $K \in C^2$ in $R^3 - \{0\}$. However by a simple modification of the proofs below, we still can get the final result, i.e., that $u \in C^2$ in $R^3 - \{0\}$, for this case.

Lemma 2. Let $u \in H^1(R^3)$ with $||u||_{L^2} = \varepsilon$, be the weak solution of (1.1) obtained as above. Then

a) $Tu(x) = K*u^2 \in L^\infty(R^3)$ and

b) $u \in H^2$.

Proof. Since $Tu = K_1*u^2 + K_2*u^2$, then by Young's inequality it follows that $Tu \in L^\infty(R^3)$. Now, let M be the multiplication operator defined by

$$Mf(x) = (\lambda+K(x)+Tu(x))f(x) = q_u(x)f(x)$$

for $f \in L^2(R^3)$. Observe that $q_u \in L^\infty + L^2$, because of our assumptions on K and part a). It follows by well known results on perturbation theory (see [4]) that $-\Delta + q_u(x)$ is a self-adjoint operator with domain $\mathcal{D}(-\Delta+q_u) = \mathcal{D}(-\Delta) = H^2(R^3)$, i.e., the Sobolev space of order two. Thus, if u is a weak solution of $-\Delta u + q_u(x)u = 0$ then $u \in H^2(R^3)$.

Theorem 3. Let $u \in H^1(R^3)$ be the weak solution of (1.1) obtained as in Section 4.). Suppose that K satisfies all of the hypotheses given above, then u is twice continuously differentiable.

Proof. First, we can show that $Tu(x) = K*u^2$ is twice continuously differentiable. This is done in the standard way: We approximate u be a sequence of ϕ_m's, $\phi_m \in C_o^\infty(R^3)$ and then we show that $T\phi_m \to Tu$ uniformly in R^3, as $m \to \infty$. Thus Tu is a continuous function. Next we consider $h_j(x) = K*(2u \frac{\partial u}{\partial x_j})$ and we show that $\frac{\partial T}{\partial x_j}\phi_m \to h_j(x)$, uniformly in R^3, as $m \to \infty$. Thus $Tu \in C^1$. Similarly $Tu \in C^2$. Now, we observe that $q_u(x) \in C^2$. Thus $q_u(x)u \in H^2$, which implies that $q_u(x)u = -\Delta u \in H^2$, from which it follows that $u \in H^4$. Finally, we use Sobolev's imbedding theorem to obtain $u \in C^2$. This proves the Theorem.

REFERENCES

[1] - BADER, P., Variational method for the Hartree equation of the helium atom, Proc. Royal Soc. Edinburgh, 82 A, (1978), 27-39.

[2] - HARDY, G.H., LITTLEWOOD, J. and POLYA, G., Inequalities, Cambridge, Univ. Press (1952).

[3] - HARTREE, D.R., The Calculations of Atomic Structures, J. Wiley, N.Y. (1957).

[4] - KATO, T., Perturbation Theory for Linear Operators, Springer--Verlag, N.Y. (1966).

[5] - LIEB, E., Existence and uniqueness of the minimizing solution of Choquard's nonlinear equation, Studies Appl. Math., 57, (1977), 93-105.

[6] - MEDEIROS, L.A. and RIVERA, P.H., Espaços de Sobolev, Textos de Met. Mat., IMUFRJ, Rio de Janeiro (Brasil), (1977).

[7] - REEKEN, M., General theorem on bifurcation and its application to the Hartree equation of the Helium atom, J. Math. Phys. 11, 8, (1970), 2505-2512.

[8] - STRAUSS, W.A., Existence of solitary waves in higher dimensions, Comm Math. Phys., 55(1977), 149-162.

APPROXIMATION - SOLVABILITY OF SOME NONLINEAR OPERATOR EQUATIONS WITH APPLICATIONS

by P. S. Milojević +

0. Introduction.

Let H_1 and H be real Hilbert spaces, $A : D(A) = H_1 \subset H \to H$ a K-positive definite and K-symmetric*, H_o the completion of $D(A)$ with respect to $||x||_o = (Ax,Kx)^{1/2}$ and $N : H_1 \to H$ a given nonlinear mapping. Many problems in partial and ordinary differential equations may be described in terms of seeking a solution of an abstract equation

(I) $Ax+Nx = f \qquad (x \in H_1,\ f \in H)$

Here we present some constructive solvability and existence type results for Eq. (I) with nonlinearities not necessarily compact, which in terms of differential equations means that we can allow nonlinear dependence on the highest order derivatives, and also discuss the rate of convergence of approximate solutions of (I).

After introducing various definitions and examples needed in the sequel, we establish our first basic result (Theorem 1) in Section 1 asserting the constructive solvability of Eq. (I) when $A+N : H_1 \to H$ is A-proper with respect to a suitable scheme for (H_1,H), and the surjectivity of $A+N$ when it is pseudo A-proper. The unboundedness of A in H and the lack of a suitable connection between the norms of H_1 and H_o forces us to introduce a system of Banach spaces $\{F_i \mid i = 1,2,\ldots,N\}$ with $H_1 \subset F_i \subset H_o$ such that the norms of each such triple satisfy a certain multiplicative inequality, and consequently, to express growth conditions on N in terms of the norms of H_o and F_i. In the second part of Section 1 we establish the

* Precise definitions of the notions used in Introduction can be found in Section 1.

+ Research partially supported by FINEP

rate of convergence of approximate solutions to (I) in the norm of any member of a given scale of interpolation spaces for the pair (H_1,H_o) under an additional condition on the A-proper mapping A+N. Our constructive results include as special cases some of the results of Zarubin [16] obtained via a moments method in the case when N is compact.

In Section 2 we show that if N is either ball-condensing or of type (KS), then A+N is A-proper and obtain the corresponding constructive-solvability results involving such classes of mappings. We also show that if N is either pseudo K-monotone, or generalized pseudo K-monotone or of (KM) type, then A+N is pseudo A-proper, and derive the corresponding surjectivity results for such mappings.

Section 3 is devoted to illustrating some of the possible applications of the results of Sections 1 and 2 to boundary value problems for nonlinear partial differential equations. We establish the approximation solvability of

$$\sum_{|\alpha|\le 2m} a_\alpha(x)D^\alpha u(x) + F(x,u(x),Du(x),\ldots,D^{2m}u(x)) = f(x), \quad f \in L_2(Q)$$

$$B_i u = 0 \quad \text{on} \quad \partial Q \qquad (i = 1,2,\ldots,m)$$

with the linear part being either regularly elliptic, in which case the rate of convergence of approximate solutions is also obtained, or strongly elliptic of the second order. Our assumptions on F are such that the induced mapping N is either k-ball contractive or is of monotone type. When F depends only on the derivatives up to order 2m-1, then N is compact and, as a special case of our result in the case of regular elliptic equations, we obtain a result of Zarubin [16]. Other types of applications, including also ordinary differential equations, will be given elsewhere.

Since, in addition to the notion of A-properness, the only tool used in proving our abstract results is the Brouwer's degree, the extension of which exists for multivalued finite dimensional mappings (see, e.g., [9]), we see that our results are also valid for multivalued nonlinearities N in (I). We also add that the constructive study of Eq. (I) with A a linear Fredholm mapping of nonnegative index can be found in Milojevič [11].

1. <u>Approximation-solvability and surjectivity results</u>.

Let X and Y be two Banach spaces $\{X_n\}$ and $\{Y_n\}$ two sequences of oriented finite dimensional subspaces of X and Y respectively, and $P_n : X \to X_n$, $Q_n : Y \to Y_n$ continuous linear projections.

<u>Definition</u> 1. A quadruple $\Gamma_a = \{X_n, V_n; Y_n, Q_n\}$ is called an *admissible scheme* for (X,Y) if $\dim X_n = \dim Y_n$ for each n, V_n is the injection of X_n into X, $\mathrm{dist}(x,X_n) \to 0$ as $n \to \infty$ for each x in X and $||Q_n|| \leq M$ for each n. A quadruple $\Gamma_o = \{X_n, P_n; Y_n, Q_n\}$ is called a *projectionally complete scheme* if $P_n(x) \to x$ and $Q_n(y) \to y$ for each x in X and each y in Y.

<u>Definition</u> 2. ([13]) A mapping $T : X \to Y$ is said to *approximation proper* (A-*proper*) with respect to Γ if $T_n = Q_nT : X_n \to Y_n$ is continuous for each n and, whenever $\{x_{n_k} | x_{n_k} \in X_{n_k}\}$ is a bounded sequence such that $||T_{n_k}x_{n_k} - Q_{n_k}f|| \to 0$ for some f in Y, then there exists an x in X such that:

(i) $Tx = f$, and

(ii) x belongs to the closure of $\{x_{n_k}\}$.

T is said to be *pseudo* A-*proper* with respect to Γ if we do not require (ii) in Definition 2 to hold.

Many examples of A-proper and pseudo A-proper mappings (and their uniform limits) can be found in [10, 12, 13]. We shall just state here a few needed ones, while other new ones can be found in Section 2.

<u>Definition</u> 3. $T : X \to Y$ is said to be K-quasi-bounded if for some $K : X \to Y^*$ and any bounded sequence $\{x_n\}$ in X, the inequality $(Tx_n, Kx_n) \leq c||x_n||$ for each n and some $c > 0$, implies

that $\{Tx_n\}$ is bounded.

Example 1. ([4, 13]) If $T : X \to Y$ is K-quasibounded demicontinuous and of type (KS) (i.e., whenever $x_n \rightharpoonup x$ (weakly) in X and $(Tx_n, K(x_n-x)) \to 0$, then $x_n \to x$ in X), then it is A-proper with respect to Γ_a for a suitably chosen K.

Definition 4. The *ball measure of noncompactness* of a bounded subset D of X is defined by $\chi(D) = \inf\{r > 0 \mid D \subset \bigcup_{i=1}^{n} B(x_i, r)$, x_i in X and n a positive integer$\}$. A map $T : X \to Y$ is said to be k-*ball contractive* if $\chi(T(D)) \leq k\chi(D)$ for each $D \subset X$; it is *ball-condensing* if $\chi(T(D)) < \chi(D)$ whenever $\chi(D) \neq 0$.

The class of ball-condensing mappings is rather extensive and includes, among others, the classes of compact, k-contractive, $k < 1$, and of semi-contractive ([3]) mappings. We have the following important

Example 2. ([10]) Let $A : X \to Y$ be continuous, surjective and a-stable with respect to Γ_o, i.e., for some $c > 0$

$$||Q_nAx-Q_nAy|| \geq c||x-y|| \quad \text{for all } x,y \in X_n, \ n \geq 1,$$

$F : X \to Y$ demicontinuous and either k-ball contractive if $k > c$ or, ball condensing if $c = 1$. Then $T = A+F$ is A-proper with respect to Γ_o. In particular, as A we can take a c-strongly accretive mapping (i.e., $Y = X$ and $(Tx-Ty,z) \geq c||x-y||^2$ for each $x,y \in X$ and $z \in J(x-y)$, where J is the normalized duality mapping).

The main object of this section is to study the solvability and the approximation-solvability of equations of the form

(1) $\quad Ax+Nx = f \qquad (x \in H_1,\ f \in H)$

where H_1 and H are suitably chosen spaces, A is a linear and N a nonlinear mapping. Associate with Eq. (1) a sequence of finite dimensional equations induced by a given scheme Γ

(2) $Q_nAx + Q_nNx = Q_nf \qquad (x \in X_n, \quad n = 1, 2, \ldots).$

The second part of this section will be denoted to studying the rate of convergence of approximate solutions of Eq. (1).

Definition 5. We say that Eq. (1) is *feebly approximation-solvable* with respect to Γ if Eq.'s (2) have solutions x_n for infinitely many n and some subsequence $x_{n_k} \to x$ with $Ax+Nx = f$. If the whole sequence $x_n \to k$, we say that Eq. (1) is *strongly approximation-solvable*, while it is *uniquely approximation-solvable* if it is strongly a.s. and Eq.'s (1) and (2) are uniquely solvable.

Thus, a suitable class of mappings to which this constructive procedure of solvability applies is that of A-proper mappings.

Let us continue by introducing the class of linear mappings A to be considered.

Let H_1 be a Banach space, H a Hilbert space which contains H_1 as a dense subset and the embedding of H_1 into H be continuous. Let $A : D(A) = H_1 \subset H \to H$ as a mapping in H be K-positive definite (K.p.d.) and K-symmetric. This means that there exist a closeble mapping $D(A)$ onto a dense subset $KD(A)$ of H with $D(K) \supseteq D(A)$ and positive constants α and β such that

(3)
$$(Ax,Kx) \geq \alpha||Kx||^2, \quad (Ax,Kx) \geq \beta||x||^2 \quad \text{for} \quad x \in D(A);$$
$$(Ax,Ky) = (Kx,Ay) \quad \text{for} \quad x,y \in D(A).$$

Let H_o denote the completion of $D(A)$ in the metric

$$[x,y] = (Ax,Ky), \quad ||x||_o = [x,x]^{1/2} \quad (x,y \in D(A)).$$

Then H_o can be regarded as a subset of H and the mapping $x \to Kx$ of $D(A)$ into H is bounded from the H_o-norm to H and so can be extended uniquely to a bounded linear mapping K_o of H_o into H and $D(K_o) = H_o \subseteq D(\overline{K})$, where $\overline{K}$ is the closure of K in H. The class of K.p.d. and K-symmetric mappings contains, among others, positive definite and symmetric mappings $(K = I)$, invertible mappings

$(K = A)$, certain ordinary and partial differential mappings of odd and even order (with K properly chosen), bounded symmetrizable mappings, etc.

Another standing hypothesis on A in this section is that $A : H_1 \to H$ is *continuous and bijective*. Denote by $\Gamma_o = \{X_n, P_n; Y_n Q_n\}$ a projectionally complete scheme for the pair (H_1, H) such that $Q_n Kx = Kx$ for each x in X_n and each n. To give an example of such a scheme, we note first that the continuity of A from H_1 into H and inequality (3) imply that $K_{|H_1} : H_1 \to H$ is a continuous injection. Now, let X_n be a sequence of finite dimensional subspaces of H_1 with $\mathrm{dist}(x, X_n) = \inf_{y \in H_n} ||x-y|| \to 0$ as $n \to \infty$ for each x in H_1, and $P_n : H_1 \to X_n$ a continuous linear projection. For each n, define $Y_n = K(X_n)$ and $Q_n : H \to Y_n$ to be orthogonal projection. Then $\dim X_n = \dim Y_n$, $Q_n Kx = Kx$ for each x in X_n and $\mathrm{dist}(y, Y_n) \to 0$ as $n \to \infty$ for each y in H provided additionally that $K(H_1)$ is dense in H. Indeed, for each $\varepsilon > 0$ and $y \in H$ fixed there exists $z_o = Ku_o$ for some $u_o \in H_1$ such that $||y - z_o|| < \varepsilon$ and therefore,

$$\mathrm{dist}(y, Y_n) = \inf_{z \in X_n} ||y - Kz|| \le \inf_{z \in X_n} (||y - z_o|| + ||Ku_o - Kz||) \le$$

$$\le \varepsilon + ||K|| \inf_{z \in X_n} ||u_o - z||_1 \to \varepsilon \quad \text{as} \quad n \to \infty.$$

Hence, $\mathrm{dist}(y, Y_n) \to 0$ for each y in H.

For our second example of Γ_o, assume that $K = I$, $Y = A(X_n)$ and $A(X_n) \subseteq X_n$. Then Γ_o is such that $Q_n x = x$ for each $x \in X_n$. This situation arises when, for example; A has a complete orthonormal system in a Hilbert space H_1 consisting of eigenvectors.

In view of our a-stability assumption on A in Theorem 1 below, the following result is useful.

Lemma 1. Suppose that either one of the following conditions holds:

(i) $\theta ||A|| \le ||Kx||$ for each $x \in D(A)$ and some $\theta > 0$;

(ii) K is closed, $D(A) = D(K)$ and $K^{-1} : R(K) \subset H \to H$ is bounded;

(iii) $K = I$ and $Y_n = A(X_n)$ with $A(X_n) \subseteq X_n$.

Then $A : H_1 \to H$ is a-stable with respect to Γ_o, i.e., there exist a constant $c > 0$ such that

$$||Q_n Ax|| \geq c||x||_1 \quad \text{for each } x \in X_n, \ n \geq 1.$$

Proof. To see that (ii) implies (i), observe first that $X = D(K)$ with $[u,v] = [Ku,Kv]$ becomes a Hilbert space since $R(K) = H$ and that A has a closed extension A_o in H (cf. [13]). Thus, it follows easily that A is closed from X into H, and consequently, (i) holds.

Next, since $A : H_1 \to H$ is a continuous bijection, there exists $\beta > 0$ such that $||Ax|| \geq \beta||x||_1$ for each $x \in H_1$. By inequality (3) and the properties of Γ_o, we get that $\alpha||Kx|| \leq ||Q_n Ax||$ for each $x \in X_n$, $n \geq 1$, and therefore,

$$||Q_n Ax|| \geq \alpha\theta||Ax|| \geq \alpha\beta\theta||x||_1 \quad \text{for each } x \in X_n, \ n \geq 1,$$

whenever $D(A) = D(K)$. In the last case we have immediately that $||Q_n Ax|| = ||Ax|| \geq \beta||x||_1$ for all $x \in X_n$, $n \geq 1$. □

Suppose that for the spaces $H_1 \subset H_o$ there exists a finite number of Banach spaces F_i, $i = 1, \ldots, N$, such that each triple $H_1 \subset F_i \subset H_o$ satisfies the multiplicative inequality

$$(4) \qquad ||x||_i \leq d_i ||x||_1^{\tau_i} ||x||_o^{1-\tau_i} \quad (x \in H_1, \ i = 1, 2, \ldots, N)$$

for some constants d_i and τ_i. Set $F_o = H_1$.

We are now in a position to prove our first basic result.

Theorem 1. Let A be a K.p.d. and K-symmetric mapping and a-stable (equivalently, A-proper) with respect to Γ_o and $N : H_1 \to H$ a nonlinear mapping such that for some $R \geq 1$ and $||x||_1 > R$,

$$(5) \qquad (Nx,Kx) \geq 0, \ ||Nx|| \leq \sum_{j=0}^{N_1} \prod_{i=0}^{N} f_{ij}(||x||_o)||x||_i^{r_{ij}},$$

where r_{ij} satisfy:

$$r_{ij} \geq 0, \quad \sum_{i=0}^{N} r_{ij}\tau_i < 1, \quad j = 0, 1, \ldots, N_1,$$

and $f_{ij}(\xi)$ are continuous nonnegative functions. Then,

(1) If $A+N$ is A-proper with respect to Γ_o, the equation

$$Ax+Nx = f$$

is feebly approximation solvable w.r. to Γ_o for each f in H, and strongly approximation-solvable if $A+N$ is injective.

(2) If $A+N$ is pseudo A-proper with respect to Γ_o, it is surjective, i.e., $(A+N)(H_1) = H$.

<u>Proof</u>. Let f in H be fixed and for each n define the homotopy

$$H_n : [0,1] \times X_n \to Y_n \text{ by } H_n(t,x) = Q_nAx + tQ_nNx - tQ_nf.$$

Then there exists an $r \geq R$ such that $H_n(t,x) \neq 0$ for all $t \in [0,1]$ and $x \in \partial B_n = \partial B(0,r) \cap X_n$ with $n \geq 1$. Its existence is guaranteed if we can show that for all solutions (t_n,x_n) of $H_n(t_n,x_n) = 0$, $\{x_n\}$ is a priori bounded in H_1. Thus, if $H_n(t_n,x_n) = 0$ for some n, by the properties of Γ_o we have that $(Q_nNx_n,Kx_n) = (Nx_n,Kx_n) \geq 0$, and consequently,

$$||x_n||_o^2 = (Ax_n,Kx_n) = (Q_nAx_n,Kx_n) = -t_n(Q_nNx_n,Kx_n) +$$

$$+ t_n(Q_nf,Kx_n) \leq ||f||.||Kx_n|| \leq ||K||.||f||.||x_n||_o.$$

Hence $||x_n||_o \leq c_1$ for each such x_n with $c_1 = ||K||.||f||$ independent of n. Moreover, from $H_n(t_n,x_n) = 0$ and the a-stability of A we get

$$c||x_n||_1 \leq ||Q_nAx_n|| = t_n||Q_nNx_n-Q_nf|| \leq ||Nx_n|| +$$

$$+ c_2, \; c_2 = ||f||.$$

To estimate $||Nx_n||$, set $m_{ij} = \max_{\lambda\in[0,c_1]} f_{ij}(\lambda)$ for each i and j, and by our assumption on N we obtain

$$||Nx_n|| \le \sum_{j=0}^{N_1} \prod_{i=0}^{N} f_{ij}(||x_n||_o)||x||_i^{r_{ij}} \le$$

$$\le \sum_{j=0}^{N_1} \prod_{i=0}^{N} f_{ij}(||x_n||_o)\{d_i^{r_{ij}}||x_n||_1^{\tau_i r_{ij}}\ ||x_n||_o^{(1-\tau_i)r_{ij}}\} \le$$

$$\le \sum_{j=0}^{N_1} \prod_{i=0}^{N} m_{ij} d_i^{r_{ij}} c_1^{(1-\tau_i)r_{ij}}||x_n||_1^{\tau_i r_{ij}} \le \sum_{j=0}^{N_1} m_j \prod_{i=0}^{N} ||x_n||^{\tau_i r_{ij}} =$$

$$= \sum_{j=0}^{N_1} m_j ||x_n||_1^{\partial_j} \le m_j \sum_{j=0}^{N_1} ||x_n||_1^{\partial_j},$$

where $m_j = \max_{0\le i\le N} \{m_{ij} d_i^{r_{ij}} c_1^{(1-\tau_i)r_{ij}}\}$, $m = \max_j m_j$ and

$\partial_j = \sum_{i=0}^{N} \tau_i r_{ij}$. Therefore,

$$c||x_n||_1 \le ||Nx_n|| + c_2 \le m_j \sum_{j=0}^{N_1} ||x_n||^{\partial_j} + c_2 \le c_2 + c_3||x_n||_1^q,$$

where $c_3 = m(1+N_1)$ and $q = \max \partial_j < 1$. Finally, this implies that there exists a constant c_4 independent of n such that $||x_n||_1 \le c_4$ for all x_n with $H_n(t_n,x_n) = 0$. Taking $r > c_4$, we conclude that

$$H_n(t,x) \ne 0 \quad \text{for each } x \in \partial B_n(0,r), \quad t \in [0,1] \quad \text{and} \quad n \ge 1,$$

and by the homotopy theorem for Brouwer's degree,

$$\deg(Q_n A + Q_n N - Q_n f, B_n(0,r),0) = \deg(Q_n A, B_n(0,r),0) \ne 0$$

for each n. Hence, there exists $x_n \in B_n(0,r)$ such that

$$Q_n A x_n + Q_n N x_n = Q_n f \quad \text{for each } n.$$

Now, if $A+N$ is A-proper, there exists a subsequence $x_{n_k} \to x_o$ in H_1 with $Ax_o+Nx_o = f$. Arguing by contradiction, it is easily seen that $x_n \to x_o$ with $Ax_o+Nx_o = f$ if $A+N$ is also injective. On the other hand, if $A+N$ is just pseudo A-proper, than the solvability of the above approximate equations in $B(0,r)$ implies that $Ax+Nx = f$ some x in H_1. □

We first note that N in Theorem 1 can be assumed multivalued. Moreover, equation (1) was studied constructively via a moments method by Zarubin [16] using the Leray-Schauder degree theory when A is positively definite and symmetric and N compact and satisfies condition (5) (with $K \equiv I$). Analysing the proof of Theorem 1 we see that it holds when A is not necessarily K.p.d. or linear. Namely, it is easily seen that the following result holds.

Theorem 2. Let A be K.p.d. and K-symmetric, $L : D(L) = D(A) \subset H \to H$ and $N : H_1 \to H$ two nonlinear mappings such that

(i) $\|Q_n Lx\| \geq c\|x\|_1$ for $x \in X_n$, $n \geq 1$,

and either one of the following two conditions holds

(ii) $(Lx, Kx) \geq \eta\|x\|_o^2$ for $x \in D(L)$, $\eta > 0$

and condition (5) of Theorem 1 holds;

(iii) $\|Nx\|/\|x\|_1 \to 0$ as $\|x\|_1 \to \infty$.

Suppose that $\deg(Q_n L, B_n(0,r), 0) \neq 0$ for all large r, $n \geq 1$. Then

(1) If $L+N$ is A-proper with respect to Γ_o, the equation $Lx+Nx = f$ is feebly approximation solvable for each f in H.

(2) If $L+N$ is pseudo A-proper with respect to Γ_o, then $(L+N)(H_1) = H$.

Remark. If $\theta\|Ax\| \leq \|Kx\|$ for $x \in D(A)$ and $\|Lx\| \geq \beta\|x\|_1$ for $x \in H_1$, then it can be shown that (i) of Theorem 2 holds. In the proof of Theorem 2 (iii) one ignores the intermediate subspaces F_i and works directly with H_1 and H.

We continue our exposition by discussing the rate of convergence of approximate solutions of Eq. (1). To that end, we assume that there is a family of Banach spaces $H_\alpha (0 \leq \alpha \leq 1)$ that forms a scale, i.e.,

1) H_β is densily embedded in H_α with $\beta > \alpha$, and $\|x\|_\alpha \leq c(\alpha,\beta)\|x\|_\beta$ for each $x \in H$ and some constant $c(\alpha,\beta)$;

2) For each $0 \le \alpha < \beta < \gamma \le 1$ there exists a constant $c(\alpha,\beta,\gamma)$ such that for each $x \in H_\alpha$,

$$||x||_\beta \le c(\alpha,\beta,\gamma)||x||_\alpha^{(\gamma-\beta)/(\gamma-\alpha)}||x||^{(\beta-\alpha)/(\gamma-\alpha)}$$

Let E denote a Banach space continuously embedded in H with $K(H_1) \subset E$; in particular E can be H. We have now

Theorem 3. Let A be K.p.d., K-symmetric and A-proper (equivalently, a-stable) w.r. to Γ_o, $N : H_1 \to H$ a nonlinear mapping such that $A+K$ is A-proper w.r. to Γ_o. Suppose that the following conditions holds:

(i) $(Nx-Ny,Kx-Ky) \ge \alpha_1||x-y||_o^2 - (Ax-Ay,Kx-Ky)$;

(ii) $||Nx||_E \le \alpha_2(R)$ whenever $||x||_1 \le R$, and

(iii) $||K(A^{-1}-(Q_nA)^{-1}Q_n)g||_E \le \mu(n)||g||_E$ for each $g \in E$, where $\mu(n) \to 0$ as $n \to \infty$. Suppose that for $f \in E$, the equation

(6) $Q_nAx + Q_nNx = Q_nf$

has a solution x_n for each n and $\{x_n\}$ is bounded in H_1. Then $x_n \to x_o$ in H_1 with x_o being the unique solution of Eq. (1) and the rate of convergence is given by

$$||x_n-x_o||_\beta \le M_\beta\mu^{(1-\beta)/2}(n), \quad 0 \le \beta < 1,$$

where the constant M_β does not depend on n.

Proof. Since K is one-to-one and $Q_nKx = Kx$ for each x in X_n, it follows easily from condition (i) that Eqn's (6) are uniquely solvable. The boundedness of $\{x_n\}$ and the A-properness of $A+N$ imply that $x_{n_k} \to x_o$ in H_1 with $Ax_o+Nx_o = f$. Since $A+N$ is also injective, it follows that $x_n \to x_o$, and since A is invertible and A-proper, there exists a constant $c > 0$ such that $||Q_nAx|| \ge c||x||_1$ for each $x \in X_n$ and each n.

Thus, from

$$x_o+A^{-1}Nx_o = A^{-1}f \quad \text{and} \quad x_n+(Q_nA)^{-1}Q_nNx_n = (Q_nA)^{-1}Q_nf,$$

we have that

$$K(x_o-x_n)+KA^{-1}Nx_o-KA^{-1}Nx_n = K((Q_nA)^{-1}Q_n-A^{-1})Nx_n +$$
$$+ K(A^{-1}-(Q_nA)^{-1}Q_n)f,$$

or, taking the scalar product of this iquality with $A(x_o-x_n)$ and using the K-symmetry of A, we get by (i)

$$\alpha_1||x_o-x_n||_o^2 \le (A(x_o-x_n),K(x_o-x_n)) + (N(x_o-x_n),K(x_o-x_n))$$
$$\le |(A(x_o-x_n),K(A^{-1}-(Q_nA)^{-1}Q_n)Nx_n)| + |(A(x_o-x_n),K(A^{-1}-(Q_nA)^{-1}Q_n)f)|.$$

If α_3 is the norm of the embedding of E into H, then in view of $||x_n||_1 \le R$ the above inequality becomes

$$\alpha_1||x_o-x_n||_o^2 \le \alpha_3||A||(||x_o||_1+R)(||K(A^{-1}-Q_nA)^{-1}Q_n)Nx_n||_E +$$

$$+ ||K(A^{-1}+(Q_nA)^{-1}Q_n)f||_E) \le \alpha_3||A||(||x_o||_1+R)(\alpha_2(R)+||f||_E)\mu(n) = M_o\mu(n),$$

or, $\quad ||x_o-x_n||_o \le (M_o/\alpha_1)^{1/2}\mu^{1/2}(n).$

But, by property 2) of the scale $\{H_\alpha\}$, we get for $\alpha = 0$, $\gamma = 1$ and $0 < \beta < 1$,

$$||x_o-x_n||_\beta \le c(0,\beta,1)||x_o-x_n||_o^{1-\beta}||x_o-x_n||_1^\beta \le$$

$$\le c(0,\beta,1)(M_o/\alpha_1)^{(1-\beta)/2}\mu^{(1-\beta)/2}(n)(||x_o||_1 + ||x_n||_1)^\beta,$$

or

$$||x_o-x_n||_\beta \le M_\beta\mu^{(1-\beta)/2}(n),$$

where $M_\beta = c(0,\beta,1)(M_o/\alpha_1)^{(1-\beta)/2}(||x_o||_1 + R)^\beta$. □

Let us now construct a scheme Γ_o for which condition (iii) of Theorem 3 holds. Suppose that the eigenvectors $\{\phi_i\}$ of a positive definite and symmetric mapping A form an orthonormal basis in Hilbert spaces H_1 and H, and that the corresponding eigenvalues satisfy: $0 < \lambda_1 \le \lambda_2 \le \dots \le \lambda_n \le \dots$ with $\lambda_n \to \infty$. Define

$X_n = \text{lin. sp.} \{\phi_1, \ldots, \phi_n\}$, $Y_n = A(X_n)$ and let $P_n : H_1 \to X_n$, $Q_n : H \to Y_n$ be the corresponding orthogonal projections. Then $\Gamma_o = \{X_n, P_n; Y_n, Q_n\}$ is a projectionally complete scheme for (H_1, H) with $Q_n Ax = Ax$ for all x in X_n. Thus, $A^{-1} - (Q_n A)^{-1} Q_n = A^{-1}(I-Q_n)$, A is a-stable and for each $g \in H$ we have that $g = \sum_{i=1}^{\infty} (g,\phi_i)\phi_i$, and $(I-Q_n)g = \sum_{i=n+1}^{\infty} (g,\phi_i)\phi_i$ since $\{\phi_i\}$ is also an orthonormal basis for H. This implies that

$$||A^{-1}(I-Q_n)g||_H^2 = ||\sum_{i=n+1}^{\infty} \frac{(g,\phi_i)\phi_i}{\lambda}||_H^2 \leq$$

$$\leq \sum_{i=n+1}^{\infty} \frac{|(g,\phi_i)|^2}{\lambda_i^2} \leq \frac{||g||_H^2}{\lambda_{n+1}^2},$$

and so (iii) of Theorem 3 holds.

Now, as a consequence of Theorems 1 and 3 we have for such A and Γ_o (with $K = I$) the following

<u>Theorem</u> 4. Suppose that $N : H_1 \to H$ is a nonlinear mapping such that $A+N$ is A-proper with respect to Γ_o, N satisfies condition (5) of Theorem 1 and (i) of Theorem 3. Then Eq. (1) is uniquely approximation solvable with respect to Γ_o for each f in H and the rate of convergence of approximate solutions x_n to the exact solution x_o is given by

$$||x_o - x_n||_\beta \leq M_\beta/\lambda_{n+1}^{1-\beta}, \quad 0 \leq \beta < 1,$$

where the constant M_β is independent of n.

When A is positive definite and symmetric and N is compact, the approximation solvability of Eq. (1) and the rates of convergence in Theorems 3 and 4 were obtain by Zarubin [16] using a moments methods (with $K = I$) (cf. Section 2).

<u>Remark</u>. As in Theorem 1, analysing the proofs of Theorems 3 and 4 we see that they are still valid if we assume that N is multivalued; of course, in this case assumptions (i) and (ii) are of the form:

(i_1) $(u-v,Kx-Ky) \geq \alpha_1||x-y||_o^2 - (Ax-Ay,Kx-Ky)$ for each $u \in Nx$ and each $v \in Ny$, $(x,y \in H_1)$;

(ii_1) $||u||_E \leq \alpha_2(R)$ for each $x \in Nx$ whenever $||x||_1 \leq R$.

2. Applications to condensing and monotone like perturbations.

In this section we first shaw that $A+N$ is A-proper whenever N is ball-condensing or of type (S), and it is of pseudo A-proper type if N is of pseudo-monotone, generalized pseudo-monotone or (M) type. Then, using the results of Section 1, we derive constructive solvability and surjectivity results for these classes of mappings. All these results are also valid when N is a multivalued mapping.

We begin by the following special case of Example 2.

Proposition 1. Let $A : H_1 \to H$ be K.p.d., K-symmetric and a-stable $N : H_1 \to H$ demicontinuous and either k-ball contractive with $k < c$, or ball-condensing if $c = 1$, where $||Ax|| \geq c||x||_1$, $x \in H_1$. Then $A+N : H_1 \to H$ is A-proper with respect to Γ_o.

To treat the case when N is of type (S), we assume that $\Gamma_o = \{X_n,P_n;A(X_n),Q_n\}$ and $A : H_1 \to H$ is a continuous bijection with $A(X_n) \subseteq X_n$. We have then:

Proposition 2. Suppose that $A : D(A) \subset H \to H$ is monotone and $N : H_1 \to H$ is I-quasibounded and demicontinuous, where I is the identify injection of H_1 into H. Then $A+N : H_1 \to H$ is A-proper with respect to Γ_o if either N is of type (IS) and I is compact, or N is of type (IS_+).

Proof. Let us first show that $A+N : H_1 \to H$ is I-quasibounded and of type (IS). Let $\{x_n\} \subset H_1$ be bounded and $(Ax_n+Nx_n,x_n) \leq$ $\leq c||x_n||_1$ for some $c > 0$. Then by the monotonicity of A,

$$(Nx_n,x_n) \leq (Ax_n+Nx_n,x_n) \leq c||x_n||_1,$$

and consequently, $\{Nx_n\}$ is bounded in H. By the continuity of

$A : H_1 \to H$ it follows that $\{Ax_n+Nx_n\}$ is bounded.
Next, let $x_n \to x_o$ in H_1 and $\lim(Ax_n+Nx_n, x_n-x_o) = 0$.

Now, $0 \le (Ax_n-Ax_o, x_n-x_o) \le ||A(x_n-x_o)||.||x_n-x_o|| \le$ const. for all n and $(Ax_n-Ax_o, x_n-x_o) \to 0$ as $n \to \infty$ if I is compact. Then, if N is of type (IS), we have $\lim(Nx_n, x_n-x_o) = \lim(Ax_n+Nx_n, x_n-x_o) - \lim(Ax_n, x_n-x_o) = 0$ and therefore $x_n \to x_o$ in H_1.

Next, suppose just that N is of type (IS_+). Then $\liminf(Ax_n, x_n-x_o) + \limsup(Nx_n, x_n-x_o) \le \lim(Ax_n+Nx_n, x_n-x_o) = 0$, and $\liminf(Ax_n, x_n-x_o) \ge \liminf(Ax_n-Ax_o, x_n-x_o) + \lim(Ax_o, x_n-x_o) \ge 0$. Therefore from $\limsup(Nx_n, x_n-x_o) \le 0$ we have that $x_n \to x_o$ in H_1, and so $A+N : H_1 \to H$ is of type (IS) in either case. Let us show now that $A+N$ is A-proper with respect to Γ_o (cf. also [4]). Let $\{x_{n_k} \mid x_{n_k} \in X_{n_k}\}$ be bounded in H_1 and $||Q_{n_k}(A+N)x_{n_k} - Q_{n_k} f|| \to 0$ for some f in H. Since $((A+N)x_{n_k}, x_{n_k}) = (Q_{n_k}(A+N)x_{n_k} + Q_{n_k} f, x_{n_k}) - (Q_{n_k} f, x_{n_k})$, $\{((A+N)x_{n_k}, x_{n_k})\}$ is bounded as is $\{(A+Nx_{n_k}\}$ by the I-quasiboundedness of $A+N$. We may assume that $x_{n_k} \to x_o$ in H_1, and consequently, $x_{n_k} \to x_o$ in H. Let $u_{n_k} \in H_1^n$ with $u_{n_k} \to x_o$ in H_1, and since $x_{n_k} - u_{n_k} \to 0$ in H we have

$$((A+N)x_{n_k}, x_{n_k}-u_{n_k}) = (Q_{n_k}(A+N)x_{n_k} - Q_{n_k} f, x_{n_k}-u_{n_k}) + (Q_{n_k} f, x_{n_k}-u_{n_k}) \to 0,$$

and consequently, $((A+N)x_{n_k}, x_{n_k}-x_o) \to 0$. Since $A+N$ is of type (IS), $x_{n_k} \to x_o$ in H_1, i.e., $A+N$ is A-proper with respect to Γ_o. Π

Now as consequence of Propositions 1 and 2 and Theorem 1, we have

<u>Theorem</u> 5. Let A be K-p.d., K-symmetric and a-stable with respect to Γ_o and $N : H_1 \to H$ demicontinuous and either k-ball contractive with $k < c$ or ball-condensing if $c = 1$. Suppose that N satisfies condition (5) of Theorem 1. Then the equation $Ax+Nx = f$ is feebly approximation-solvable with respect to Γ_o for each f in H.

<u>Theorem</u> 6. Let $A : H_1 \to H$ be continuous linear bijection with $A(X_n) \subseteq X_n$ and positive definite and symmetric as a mapping in H. Let $N : H_1 \to H$ be I-quasibounded and demicontinuous and either of type (IS_+) or of type (IS) when I is compact. Suppose that N satisfies condition (5) of Theorem 1. Then the equation $Ax+Nx = f$ is feebly approximation-solvable with respect to $\Gamma_o = \{X_n, P_n; A(X_n), Q_n\}$ for each f in H.

If in Theorems 5 and 6 we assume stronger conditions (corresponding to the conditions in Theorems 3 and 4) instead of condition (5), we obtain the rate of convergence of approximate solutions as given in Theorem 3 and 4.

Next we shall discuss some applications of Theorem 1 part (2) to the solvability of $Ax+Nx = f$ with nonlinear perturbation N of monotone like type. We need

<u>Definition</u> 6.

(1) Let $K : K \to Y^*$. Then a mapping $T : X \to Y$ is said to be *pseudo* K-*monotone* if $x_n \to x_o$ in X and $\limsup (Tx_n, K(x_n - x_o)) \le 0$ implies that for each x in X,

$$\liminf (Tx_n, K(x_n - x)) \ge (Tx_o, x_o - x);$$

and finitely continuous, i.e., continuous from each finite dimensional subspace F of X into the weak topology of Y.

(2) T is said to be *generalized pseudo* K-*monotone* if $x_n \to x_o$, $Tx_n \to y$ and $\limsup (Tx_n, x_n - x_o) \le 0$ imply that $Tx_o = y$ and $(Tx_n, x_n - x_o) \to 0$, and T is finitely continuous.

(3) T is said to be of *type* (KM) if $x_n \to x_o$, $Tx_n \to y$ and $\limsup (Tx_n, x_n - x_o) \le 0$ imply $Tx_o = y$; and T is finitely continuous.

Pseudo monotone and of type (M) mappings ($K = I$, $Y = X^*$) were introduced (in a some what different way) by Brezis [2] while generalized-pseudo monotone by Browder and Hess [5]; for the general setting between two different spaces X and Y see Petryshyn [13] and Milojević and Petryshyn [12].

The next result establishes that each of these mappings is of pseudo A-proper type (compare also with [14]).

Proposition 3. Let X and Y be reflexive Banach spaces, $\Gamma_a = \{X_n, V_n; Y_n, Q_n\}$ an admissible scheme and $K : X \to Y^*$ be a bounded mapping such that

(a_1) $Kx = 0$ implies $x = 0$, K is α-positively homogeneous (i.e, $K(tx) = t^{\alpha}K(x)$ for each $t > 0$, x in X and some $\alpha > 0$), and the range of K is dense in Y^*.

(a_2) For each $x \in X_n$ and $g \in Y$, we have that $(Q_n(g), K(x)) = (g, K(x))$;

(a_3) K is weakly continuous and is uniformly continuous on closed balls in X.

Let $T : X \to Y$ be K-quasibounded and either demiclosed pseudo K-monotone or generalized pseudo K-monotone or of type (KM). Then T is pseudo A-proper w.r. to Γ_a.

Proof. Suppose first that T is pseudo K-monotone. Let $\{x_{n_k} \mid x_{n_k} \in X_{n_k}\}$ be a bounded sequence such that for some $f \in Y$, $Q_{n_k}T(x_{n_k}) - Q_{n_k}(f) \to 0$ is in Y as $k \to \infty$. Then, in view of (a_2) and the equality

$$(Tx_{n_k}, Kx_{n_k}) = (Q_{n_k}Tx_{n_k} - Q_{n_k}f, Kx_{n_k}) + (Q_{n_k}f, Kx_{n_k}),$$

the sequence $\{(Tx_{n_k}, Kx_{n_k})\}$ is bounded, and consequently, $\{Tx_{n_k}\}$ is bounded by the K-quasiboundedness of T. By the reflexivity of X, we may assume that $x_{n_k} \rightharpoonup x_o$ and since $d(x_o, X_n) \to 0$ there exists $y_n \in X_n$ such that $y_n \to x_o$. Let $B(0,r)$ be a ball in X that contains x_o, $\{x_{n_k}\}$ and $\{y_{n_k}\}$. Since $x_{n_k} - y_{n_k} \rightharpoonup 0$, by ($a_2$) and ($a_3$) and the weak continuity of K at 0,

$$(Tx_{n_k},K(x_{n_k}-y_{n_k})) = (Q_{n_k}Tx_{n_k}-Q_{n_k}f,K(x_{n_k}-y_{n_k})) +$$

$$+ (Q_{n_k}f,K(x_{n_k}-y_{n_k})) \to 0 \quad \text{as} \quad k \to \infty.$$

Now, since $K(tx) = t^{\alpha}K(x)$ for $x \in X$ and $t > 0$, we have

$$(Tx_{n_k},K(x_{n_k}-x_o)) = (Tx_{n_k},K(x_{n_k}-y_{n_k})) +$$

$$+ 2^{\alpha}(Tx_{n_k},K(\tfrac{1}{2}(x_{n_k}-x_o)) - K(\tfrac{1}{2}(x_{n_k}-y_{n_k})))$$

with $\frac{1}{2}(x_{n_k}-x_o)$ and $\frac{1}{2}(x_{n_k}-y_{n_k})$ lying in $\overline{B}(0,r)$. For each $t > 0$, define the function $\Psi(t)$ as in [7] by

$$\Psi(t) = \sup\{||Kx-Ky|| \mid ||x-y|| \le t, \quad x,y \in \overline{B}(0,r) .$$

Since K is uniformly continuous on $\overline{B}(0,r)$, the function $\Psi(t)$ is nondecreasing in t, $\Psi(t) \to 0$ as $t \to 0$ and

$$||Kx-Ky|| \le \Psi(||x-y||) \quad \text{for} \quad x,y \in \overline{B}(0,r).$$

This inequality implies, in view of the boundedness of $\{Tx_{n_k}\}$, that

$$|(Tx_{n_k},K(\tfrac{1}{2}(x_{n_k}-x_o)) - K(\tfrac{1}{2}(x_{n_k}-y_{n_k})))| \le c\Psi(\tfrac{1}{2}||y_{n_k}-x_o|| \to 0$$

Thus, from the above discussion we get that $(Tx_{n_k},K(x_{n_k}-x_o)) \to 0$ as $k \to \infty$.

Suppose now that T is pseudo K-monotone. Then for each x in X we have

$$\lim\inf(Tx_{n_k},K(x_{n_k}-x)) \ge (Tx_o,K(x_o-x))$$

Let $x \in B(0,d)$ with $d > r$ and $z_n \in X_n$ such that $z_n \to x$. Then $x_{n_k}-z_{n_k} \to x_o-x$ in X and since K is weakly continuous, we get as before that $(Tx_{n_k},K(x_{n_k}-z_{n_k})) \to (f,K(x_o-x))$ and

$$(Tx_{n_k},K(x_{n_k}-x))) = (Tx_{n_k},K(x_{n_k}-z_{n_k})) +$$

$$+ 2^{\alpha}(Tx_{n_k},K(\tfrac{1}{2}(x_{n_k}-x) - K(\tfrac{1}{2}(x_{n_k}-z_{n_k})))$$

with $\frac{1}{2}(x_{n_k}-z_{n_{k_1}}) \in \overline{B}(0,d)$ for all large k. Thus, as before we get that $(Tx_{n_k},K(\frac{1}{2}(x_{n_k}-x) - K(\frac{1}{2}(x_{n_k}-z_{n_k}))) \to 0$ and consequently, for each x in $B(0,d)$

$$(Tx_{n_k},K(x_{n_k}-x)) \to (f,K(x_o-x)) \quad \text{as} \quad k \to \infty.$$

Thus, in view of the above inequality, we have

$$(7) \qquad (Tx_o,K(x_o-x)) \le (f,K(x_o-x)), \quad x \in B(0,d).$$

This implies that $Tx_o = f$. If not then there would exist an element $y \in Y^*$ such that $(Tx_o-f,y) > 0$. Since $R(K)$ is dense in Y^*, there exists $y_n \in R(K)$ with $y_n \to y$ and let $u_n \in X$ be such that $Ku_n = y_n$. Since

$$\lim(Tx_o-f,Ku_n) = (Tx_o-f,y) > 0,$$

for all large n, we have that $(Tx_o-f,Ku_n) > 0$.

Fix such a large n and since $x_o \in B(0,d)$, $u_t = x_o-tu_n \in B(0,d)$ for all $t > 0$ sufficiently small. Hence, by (7) we have

$$t^{\alpha}(Tx_o-f,Ku_n) = (Tx_o-f,K(tu_n)) \le 0,$$

i.e., $(Tx_o-f,Ku_n) \le 0$, in contradiction to our choice of n.

Thus, $Tx_o = f$ and consequently T is pseudo A-proper.

Next, assume that T is either generalized pseudo K-monotone or of type (KM). Then, since $x_{n_k} \to x_o$, $Tx_{n_k} \to y_o$ and $\lim(Tx_{n_k},K(x_{n_k}-x_o)) \le 0$, we have that $Tx_o = y_o$. We claim that $Tx_o = f$. Let $y \in X$ be arbitrary and choose $y_n \in X_n$ such that $y_n \to y$. Then by (a_2) and (a_3),

$$(y_o-f,Ky) = \lim(Tx_{n_k}-f,Ky_{n_k}) =$$

$$= \lim(Q_{n_k}Tx_{n_k}-Q_{n_k}f,Ky_{n_k}) = (0,Ky) = 0.$$

Hence, since $R(K)$ is dense Y^* and $(y_o-f,w) \to 0$ for each $w \in R(K)$, it follows that $y_o = f$ and so $Tx_o = f$. ∎

For our next result we need to introduce:

Condition (P). $A : D(A) = H_1 \subset H \to H$ is closed and such that

$$(Ax,x) \geq 0 \quad \forall\, x \in D(A), \quad (A^*x,x) \geq 0 \quad \forall\, x \in D(A^*)$$

Now, Condition (P) is equivalent to $A : D(A) \to H$ being maximal monotone (cf. [8], p. 313) and A is therefore demicontinuous (cf. [4]).

Clearly, if $A : D(A) \to H$ is as in Section 1, i.e., positive definite, symmetric and bijective, then A is self-adjoint (and A^{-1} is continuous), and therefore Condition (P) holds.

Proposition 4. Suppose that $A : D(A) \to H$ satisfies Condition (P) and $N : H_1 \to H$ is I-quasibounded and either demiclosed pseudo I-monotone or generalized pseudo I-monotone or of type (IM) with I is compact. Then $A+N : H_1 \to H$ is pseudo A-proper with respect to Γ_o

Proof. As in Proposition 2, we see that $A+N$ is I-quasibounded. Since I is continuous and linear it is easy to see, following the arguments of Browder and Hess [5] that A is both pseudo and generalized pseudo I-monotone (being maximal monotone in H), and so $A+N : H_1 \to H$ is (generalized) pseudo I-monotone if N is such.

Next, suppose that N is of type (IM) and I compact. Then from $x_n \to x_o$ in H_1, $(A+N)x_n \to u$ and $\limsup(Ax_n+Nx_n,x_n-x_o) \leq 0$ we have that $\limsup(Nx_n,x_n-x_o) \leq 0$ as in Proposition 2. Since $x_n \to x_o$ in H and A is maximal monotone in H, we have that $Ax_n \to Ax_o$. Thus, $Nx_n \to u-Ax_o$, and since N is of type (IM), $Nx_o = u-Ax_o$. Hence $A+N : H_1 \to H$ is of type (IM).

In view of this discussion and Proposition 3, we have that $A+N : H_1 \to H$ is pseudo A-proper with respect to Γ_o. □

Now Proposition 4 and Theorem 1, (2), imply the following new surjectivity result.

Theorem 7. Suppose that $A : H_1 \to H$ is a continuous bijection and is closed positive definite and symmetric in H with $A(X_n) \subseteq X_n$.

Suppose that $N : H_1 \to H$ is I-quasibounded and either demiclosed pseudo I-monotone or generalized pseudo I-monotone or of type (IM) with I compact. Then, if N satisfies condition (5) of Theorem 1, A+N is a surjection, i.e., $(A+N)(H_1) = H$.

3. Elliptic boundary value problems with condensing and monotone perturbations.

In this section we use Theorems 2 and 4 to establish the constructive solvability of boundary value problems involving k-ball-contractive and monotone perturbations of regular elliptic and strongly elliptic operators and the rate of convergence of approximate solutions of such problems. The nature of our nonlinearities permits us to allow the nonlinear dependence in boundary value problems on the derivatives of the highest order. Our result concerning regular elliptic operators is an extension of the corresponding result of Zarubin [16] that deals with compact perturbations. Other applications of our abstract results from Sections 1 and 2 to boundary value problems for ordinary and partial differential equations involving nonlinear perturbations of types different from the ones considered here will be treated elsewhere. Rather than striving for most general results, here we are just interested in giving a few illustrations of the abstract theory.

Let Q be a bounded domain in R^n with a sufficiently smooth boundary ∂Q. If $\alpha = (\alpha_1, \ldots, \alpha_n)$ is a multi-index of nonnegative integers, we denote by $D^\alpha = \partial^{\alpha_1}/\partial x_1^{\alpha_1} \ldots \partial^{\alpha_n}/ x_n^{\alpha_n}$ a differential operator of order $|\alpha| = \alpha_1 + \ldots + \alpha_n$. If m is nonnegative integer, and $p \geq 1$, $W_p^m \equiv W_p^m(Q)$ denotes the real Sobolev space with the norm $||u||_{p,m} = \sum_{|\alpha| \leq m} ||D^\alpha u||_p$, where $||\cdot||_p$ is the $L_p(Q)$ norm. Let R^{s_m} denote the vector space whose elements are $\xi = \{(\xi_\alpha) \mid |\alpha| \leq m\}$.

I. Suppose that the problem

$$Lu = \sum_{|\alpha| \leq 2m} a_\alpha(x) D^\alpha u(x) = f(x) \quad \text{in } Q$$

$$B_i u = 0 \quad \text{on} \quad \partial Q, \ (i = 1, 2, \ldots, m),$$

where B_i are boundary differential operators, is regularly elliptic (see, e.g., [1]) and that the homogeneous problem has only the trivial solution.

Consider the problem of finding u in W_2^{2m} such that

(8) $$\sum_{|\alpha| \le 2m} a_\alpha(x) D^\alpha u(x) + F(x,u,Du, \ldots, D^{2m}u) = f(x), \quad f \in L_2(Q)$$

(9) $$B_i u = 0 \quad \text{on} \quad \partial Q, \quad (i = 1, 2, \ldots, m)$$

where F is as specified below. Denote by $\overset{\circ}{W}{}_2^{2m}$ the subspace of W_2^{2m} whose functions satisfy conditions (9). Let $A,N : \overset{\circ}{W}{}_2^{2m} \to L_2$ be defined by $Au = Lu$ and $Nu = F(x,u,Du, \ldots, D^{2m}u)$. Then the boundary value problem (8)-(9) is equivalent to the operator equation

(10) $$Au+Nu = f \, (u \in \overset{\circ}{W}{}_2^{2m}, \quad f \in L_2).$$

We impose the following conditions on A and F:

(A1.) A is symmetric and positive definite in L_2 and possesses an orthonormal basis of eigenfunctions $\{\phi_i\} \subset \overset{\circ}{W}{}_2^{2m}$ (and consequently, in L_2) (cf. [1]).

(F1.) For each fixed v in W_2^{2m}, the mapping $u \to F(x,u,Du,\ldots$ $\ldots, D^{2m-1}u, D^{2m}v)$ is continuous and bounded from W_2^{2m-1} into L_2.

Using the characterization of L_1-convergence, it is not hard to show that condition (F1.) holds if, for examples, $F : Q \times R^{s_{2m-1}} \times R^{n_{2m}} \to R$ satisfies generalized Caractheódory conditions (i.e., $F(\cdot,\eta,\xi)$ is measurable in Q for each $(\eta,\xi) \in R^{s_{2m-1}} \times R^{n_{2m}}$ and $F(x,\cdot,\cdot)$ is continuous for a.e. $x \in Q$, where $n_{2m} = \# \{\alpha = (\alpha_1, \ldots, \alpha_n) \mid |\alpha| = 2m\})$ and there are constants $b_\alpha \ge 0$ and a function $b \in L^2$ such that

$$|F(x,\eta,\xi)| \le \sum_{|\alpha| \le 2m-1} b_\alpha |\eta_\alpha| + b(x)$$

for all $\eta \in R^{s_{2m-1}}$, $\xi \in R^{n_{2m}}$ and $x \in Q$(a.e.).

(F2.) $F : Q \times R^{s_{2m-1}} \times R^{n_{2m}} \to R$ satisfies

$$|F(x,\eta,\xi) - F(x,\eta,\xi')| \le \beta \sum_{|\alpha|=2m} |\xi_\alpha - \xi'_\alpha|$$

for all $(x,\eta) \in Q \times R^{s_{2m-1}}$, $\xi, \xi' \in R^{n_{2m}}$ and some small β.

Under the above assumptions on A we see that if $X_n = \text{lin.sp.}\{\phi_1, \ldots, \phi_n\}$, $Y_n = A(X_n)$, $P_n : \overset{\circ}{W}{}_2^{2m} \to X_n$ and $Q_n = L_2 \to Y_n$ orthogonal projections, respectively, then $\Gamma_o = \{X_n, P_n; Y_n, Q_n\}$ is a projectionally complete scheme for $(\overset{\circ}{W}{}_2^{2m}, L_2)$ such that $Q_n x = x$ for each $x \in X_n$ and A is a-stable with respect to Γ_o.

<u>Theorem</u> 8. Suppose that A and N satisfy conditions (A1.) and (F1.)-(F2.) and that

(11) $(Nu-Nv, u-v) \ge \eta ||u-v||^2_{W_2^m} - (Au-Av, u-v)$, for all

$u, v \in \overset{\circ}{W}{}_2^{2m}$, $\eta > 0$, and

(12) $|Nu| \le \phi(x,u,Du,\ldots,D^r u)(1 + \sum_{i=[(2m-n)/2]}^{2m-1} |D^i u|^{r_i})$,

where $\phi(x, \eta_o, \ldots, \eta_r)$ is a nonnegative continuous function, and the numbers r and r_i satisfy

(13) $0 \le r < [(2m-n)/2]$, $0 \le r_i < (4m+n)/(2i+n)$.

Then the BVP (8)-(9) is uniquely approximation-solvable with respect to Γ_o for each f in L_2 and the rate of convergence of the approximate solutions u_k is given by

(14) $||u_k - u_o||_{2,m(1+\beta)} \le \dfrac{M_\beta}{k^{2m(1-\beta)/n}}$, $(0 \le \beta < 1)$.

<u>Proof</u>. Set $H_1 = \overset{\circ}{W}{}_2^{2m}$ and $H = L_2$ and define the mapping $V : H_1 \times H_1 \to L_2$ by $V(u,v) = F(x, D^\eta u, D^\xi v)$, where $D^\eta u = \{(D^\alpha u) \mid |\alpha| \le 2_{m-1}\}$ and $D^\xi v = \{(D^\alpha v) \mid |\alpha| = 2m\}$. Then for each fixed $u \in \overset{\circ}{W}{}_2^{2m}$, the mapping $V(\cdot,u) : \overset{\circ}{W}{}_2^{2m} \to L_2$ is compact by assumption (F1.) and the compactness of the embedding of W_2^{2m} of into W_2^{2m-1}. For each fixed $u \in \overset{\circ}{W}{}_2^{2m}$ the mapping $V(u,\cdot) : \overset{\circ}{W}{}^{2m} \to L_2$ is k-ball-

-contractive by assumption (F2.). Since $Nu = V(u,u)$, it follows that N is k-set contractive (see [15]). But, the same arguments of [15] imply that N is also k-ball contractive. Hence, in view of Proposition 1, $A+N : H_1 \to H$ is A-proper with respect to Γ_o.

Since $H_o = W_2^{2m}(Q)$, choosing $F_i = W_{2r_i}^{i}$ for $i = [\frac{2m-n}{2}],\dots,2m-1$, we see that for each triple $H_1 \subset F_i \subset H_o$ the multiplicative inequality (4) holds ([6]) with $\tau_i = i/2m+n/2m(1/2-1/2r_i)$.

It follows from the embedding theorem that $||\phi(x,u,\dots,D^r u)||_C \le \phi_1(||u||_o)$, and consequently, by inequality (12), we have

$$||Nu|| \le \phi_1(||u||_o)(\eta_1 + \sum_{i=[(2m-n)/2]}^{2m-1} ||u||_i^{r_i}).$$

Hence, all the conditions of Theorem 4 are satisfied with $H_\beta = W_2^{2m(1+\beta)}$ and so we have

$$||u_k-u_o||_{2,m(1+\beta)} \le \eta_2/\lambda_{k+1}^{1-\beta}, \quad 0 \le \beta < 1.$$

But, it is known that the eigenvalues of A satisfy the inequality

$$\eta_3 k^{2m/n} \le \lambda_k \le \eta_4 k^{2m/n},$$

which together with the last inequality imply (14). □

<u>Remark</u>. When $Nu = F(x,u,Du,\dots,D^{2m-1}u)$, the above result has been proven by Zarubin [16]. Our results can be also used to show that the nonlinearities in Theorems 5 and 6 of [16] can depend also on the highest order derivatives.

II. We shall now use a variant of Theorem 2 in studying a Dirichlet boundary value problem involving k-ball-contractive and monotone perturbations of second order elliptic operators. Consider

(15) $Lu + F(x,u,Du,D^2u) = f, \quad u \in W_2^2 \cap \overset{\circ}{W}{}_2^1, \quad f \in L_2$

where L is a strongly elliptic operator defined on $H_1 = W_2^2 \cap \overset{\circ}{W}{}_2^1$ by

$$Lu = \sum_{|\alpha|=2} a_\alpha(x)D^\alpha u(x) + \sum_{|\alpha|=1} a_\alpha(x)D^\alpha u(x) + a_o(x)u(x),$$

where $a_\alpha \in C^1(Q)$ when $|\alpha| = 2$, while $a_\alpha \in C(Q)$, if $|\alpha| \le 1$. Our assumptions on F are:

(F3.) $F(x,u(x),Du(x),D^2u(x)) = F_1(x,\Delta u(x)) + F_2(x,u(x),Du(x))$,

where

(i) $F_1 : Q \times R \to R$ is continuous and there exist $h \in L_2(Q)$ and a constant $b > 0$ such that

$$|F_1(x,t)| \le b(|h(x)| + |t|), \quad \text{for } x \in Q, \quad t \in R$$

(ii) $(F_1(x,t_1) - F_1(x,t_2))(t_1-t_2) \ge 0$, for $x \in Q$, $t_1,t_2 \in R$

(iii) If we define $N_2(u)(x) = F_2(x,u(x),Du(x))$ for $u \in W_2^1$ and $x \in Q$, then N_2 is continuous from W_2^1 into L_2.

or

(F4.)

(i) For each fixed v in W_2^2, the mapping $u \to F(x,u,Du,D^2v)$ is continuous and bounded from W_2^1 into L_2.

(ii) F satisfies

$$|F(x,\eta,\xi) - F(x,\eta,\xi^1)| \le \beta \sum_{|\alpha|=2} |\xi_\alpha - \xi_\alpha^1|$$

for all $(x,\eta) \in Q \times R^{s_1}$, $\xi, \xi^1 \in R^{n_2}$ and some small β.

We define $K : H_1 \to L_2$ by $Ku = -\Delta u$; it is known that K is a symmetric and positive definite bijection and also K p.d. and K-symmetric in L_2. Let $\{X_n\}$ be a sequence of finite dimensional subspaces of H_1 with $\operatorname{dist}(u,X_n) \to 0$ for each $u \in H_1$, $Y_n = K(X_n)$, $P_n : H_1 \to X_n$ and $Q_n : Y_n \to L_2$ orthogonal projections onto X_n and Y_n respectively. Then $\Gamma_o = \{X_n,P_n;Y_n,Q_n\}$ is a projectionally complete scheme for (H_1,L_2) (with $Q_nKx = Kx$ for each $x \in X_n$, a fact that will not be used below). Since the Sobolevskii inequality can be written as

$$(Lu,Ku) \ge \eta_1||u||_{2,2}^2 - \eta_2||u||_{2,1}^2, \quad \text{for } u \in H_1$$

and some $\eta_1 > 0$, η_2, L is of (SK_+) type in the sense that if $u_n \to u_o$ in H_1 and $\limsup(Lu_n - Lu_o, Ku_n - Ku_o) \le 0$, then $u_n \to u_o$ in H_1. Moreover, if (F3.) holds, then $N_1 : H_1 \to L_2$ defined by $N_1 u = F_1(x, \Delta u(x))$ is bounded, continuous and K-monotone, i.e., $(N_1 u - N_1 v, K(u-v)) \ge 0$ for $u, v \in H_1$. Thus $L+N_1$ is of type (KS_+) and by Example 1 it is A-proper with respect to Γ_o, as is $L+N$, $N = N_1 + N_2$, by the compactness of N_2 from H_1 into L_2. If (F4.) holds, then as in Theorem 8 we obtain that $N : H_1 \to L_2$ defined by $Nu = F(x,u,Du,D^2 u)$ is k-ball-condensing. Assuming that L is injective, it follows from the theory of elliptic partial differential equations that L is onto. Being also A-proper with respect to Γ, we have that $||Q_n Lx|| \ge c||x||_1$ for $x \in X_n$ and some $c > 0$. Thus, $L+N$ is A-proper with respect to Γ_o whenever $k < c$. This discussion and Theorem 2, (iii), imply the validity of the following theorem.

<u>Theorem</u> 9. Suppose that either (F3.) or (F4.) holds, L is injective and

$$||Nu||/||u||_1 \to 0 \quad \text{as} \quad ||u||_1 \to \infty.$$

The equation (15) is feebly approximation-solvable with respect to Γ_o for each f in L_2.

REFERENCES

[1] - AGMON, S., On the eigenfunctions and on the eigenvalues of general elliptic boundary value problems, Commun. Pure and Appl. Math., <u>15</u>(1962), 119-147.

[2] - BREZIS, H., Équations et inéquations non-linéaires dans les espaces vectoriels en dualité, Ann. Inst. Fourier (Grenoble), <u>18</u>(1968), 115-175.

[3] - BROWDER, F.E., Fixed point theorems for nonlinear semi--contractive mappings in Banach spaces, Arch. Rational Mech. Anal., <u>21</u>(1965/1966), 259-269.

[4] - BROWDER, F.E., Nonlinear operators and nonlinear equations of evolution in Banach spaces, Proc. Sympos. Pure Math., Vol.18, part II, Amer. Math. Soc., Providence, R.I. (1976).

[5] - BROWDER, F.E. and HESS, P., Nonlinear mappings of monotone type in Banach spaces, J. Functional Anal., 11(1972), 251-294.

[6] - GLUŠKO, V.P. and KREIN, S.G., Inequalities for the norm of derivaties in L_p spaces with weight, Siberian Math. J., 1 (3), (1960), 343-382.

[7] - KATO, T., Demicontinuity, hemicontinuity and monotonicity, II, Bull. Amer. Math. Soc., 73(1967), 886-889.

[8] - LIONS, J.L., Quelques méthodes de résolutions des problèmes aux limites non linéaires, Dunod; Gauthier-Villars, Paris, (1969).

[9] - MA, T.W., Topological degrees for set-valued compact vector fields in locally convex spaces, Dissertationes Math. Rozprawy Mat., 92(1972), 1-43.

[10] - MILOJEVIĆ, P.S., A generalization of Leray-Schauder theorem and surjectivity results for multivalued A-proper and pseudo A-proper mappings, J. Nonlinear Anal., Theory, Methods Appl. 1, (3) (1977), 263-276.

[11] - MILOJEVIĆ, P.S., Approximation solvability results for equations involving nonlinear perturbations of Fredholm mappings with applications to differential equations, Proc. Seminar Functional Analysis, Holomorphy and Approximation Theory, M. Dekker, New York, (to appear).

[12] - MILOJEVIĆ, P.S. and PETRYSHYN, W.V., Continuation and surjectivity theorems for uniform limits of A-proper mappings with applications, J. Math. Anal. Appl., 62 (2) (1978), 368-400.

[13] - PETRYSHYN, W.V., On the approximation-sovability of equations involving A-proper and pseudo A-proper mappings, Bull. Amer. Math.Soc., 81 (2) (1975), 223-312.

[14] - PETRYSHYN, W.V., On nonlinear equations involving pseudo-A--proper mappings and their uniform limits with applications, J. Math. Anal. Appl., 38(1972), 672-720.

[15] - WEBB, J.R.L., Fixed point theorems for nonlinear semicontractive operators in Banach spaces, J. London Math. Soc., 1 (2) (1969), 683-688.

[16] - ZARUBIN, A.G., On a moments method for a class of nonlinear equations, Siberian Math. J., 19 (3) (1978), 577-586.

THE LEVIN-NOHEL EQUATION ON THE TORUS

by W. M. Oliva

A retarded functional differential equation (RFDE) can be defined on a manifold M. Essentially a (RFDE) F is a map such that to each continuous path $\phi : [-r,0] \to M$ associates a tangent vector $F(\phi)$ at the end of the path.

The main reference is the book of J. K. Hale, *Theory of Functional Differential Equations*, 1977, Springer-Verlag, New York Inc. ([1]).

All the dynamics is described into the set of all continuous paths $\phi : [-r,0] \to M$ which is the Banach manifold $C^0(I,M)$, $I = [-r,0]$, that is, the phase space of the equation. More precisely the definition of the (RFDE) F is given by $F : C^0(I,M) \to TM$ such that $\tau . F = \rho$ where $\tau : TM \to M$ is the canonical projection from the tangent space TM of M onto the manifold M and $\rho : C^0(I,M) \to M$ is the evaluation map $\rho(\phi) = \phi(0)$.

For a smooth (RFDE) F on a compact manifold the largest invariant set $A(F)$ is a non-empty, compact and connected set and is the union of all global solutions of F considered in the space $C^0(I,M)$.

The non-wandering set $\Omega(F)$ is also introduced as a subset of $C^0(I,M)$ and when the flow is one-to-one on $A(F)$, $\Omega(F)$ is invariant and contains the α and ω-limit sets of solutions. In the analytic case and for compact M, the flow is one-to-one on $A(F)$.

Hyperbolic critical points, hyperbolic periodic orbits, local stable and unstable manifolds can also be defined as well as Morse--Smale systems.

In [1], pag. 122, Hale described the ω-limit sets of a scalar (RFDE) called the Levin-Nohel equation; the same equation can be considered on R^n:

$$\dot{x} = - \int_{-r}^{0} a(-\theta) \nabla G(x(t+\theta)) d\theta$$

where $G : R^n \to R$ is a smooth function, $\nabla G : R^n \to R^n$ the gradient of G and $a : [0,r] \to R$ is a C^2-function such that $a(s) \geq 0$, $\dot{a}(s) \leq 0$, $\ddot{a}(s) \geq 0$ for $s \in [0,r]$, $a(0) \neq 0$ and $a(r) = 0$. The following result was essentially proved in [1] in the scalar case $(n = 1)$. The proof for $n > 1$ is the same:

<u>Proposition</u> 1. Let $G : R^n \to R$ and $a : [0,r] \to R$ as above, be analytic functions. If there is a $s_o \in [0,r]$ such that $\ddot{a}(s_o) > 0$ then Ω is the set of critical points of ∇G. If a is linear then Ω is the set of critical points of ∇G plus the periodic orbits of period r generated by solutions of the Newtonian equation

$$\ddot{x} + a(0)\nabla G(x(t)) = 0.$$

The above (RFDE) can be passed to the torus T^n when $G : R^n \to R$ is n-periodic with minimum period $(2\pi, \ldots, 2\pi)$. The covering map $p : R^n \to T^n$ given by $p(x_1, \ldots, x_n) = (e^{ix_1}, \ldots, e^{ix_n})$ induces the Levin-Nohel equation on T^n.

The special case $G(x) = 1 - \cos x$, x being an angle, defines a (RFDE) on the circle S^1 using the scalar equation

$$\dot{x} = -\int_{-r}^{o} a(-\theta)\sin(x(t+\theta))d\theta.$$

<u>Remark</u>. In the torus one can prove the same result given by Proposition 1.

<u>Question</u> 1. Is that proposition true in the non-analytic case ?

If c is a critical point of ∇G the linear analysis can be done using the first variation equation

$$\dot{y} = -\int_{-r}^{o} a(-\theta)H(c)y(t+\theta)d\theta$$

where $H(c)$ is the hessian of G at c.

The characteristic equation is in that case

$$(*) \qquad \lambda = -\int_{-r}^{o} a(-\theta)\varepsilon e^{\lambda\theta}d\theta$$

where ε is one of the real eigenvalues of the hessian $H(c)$.

The following two lemmas describe the characteristic values:

Lemma 1. If $\varepsilon < 0$ there exists just one solution λ_o of (*) on $R(\lambda) \geq 0$; λ_o is real and positive.

Remark that the lemma does not exclude solutions on $R(\lambda) < 0$.

Lemma 2. If $\varepsilon > 0$ there is no solutions of (*) on $R(\lambda) \geq 0$.

Assume now G be a Morse function; then at each critical point c, $H(c)$ has eigenvalues $\varepsilon_1 < \ldots < \varepsilon_k < 0 < \varepsilon_{k+1} < \ldots < \varepsilon_n$ and call $c_1, c_2, \ldots, c_p$ all critical points of G.

The dimension of the unstable manifold of a critical point for the Levin-Nohel equation.

Let c be a critical point and $\varepsilon < 0$ an eigenvalue of $H(c)$. Let $x > 0$ be the unique solution considered in Lemma 1 given by

$$x + \varepsilon \int_{-r}^{0} a(-\theta) e^{x\theta} d\theta = 0.$$

Following closely [1], pag. 168 and calling $T = A - xI$ one can prove that $N(A-xI) = N(A-xI)^2$, and by induction, $N(T) = N(T^2) = \ldots$ $\ldots = N(T^k)$, $k > 2$.

Lemma 3. In the case $\varepsilon < 0$, the linearized equation has the generalized eigenspace associated to $x > 0$ with dimension n.

In fact, that generalized eigenspace is spanned by the functions

$$(e^{x\theta} b_1, \ldots, e^{x\theta} b_n)$$

where $(b_1, \ldots, b_n)$ is a basis of R^n.

Proposition 2. If G is a Morse function as above, the unstable manifold W^u of a critical point c has dimension $k.n$.

Example. In the 2-torus T^2, G being a Morse function near the height function one obtains 4 critical points:

$c_1(\varepsilon_1 < \varepsilon_2 < 0)$ $\quad \dim W^u = 4$

$c_2(\varepsilon_1 < 0 < \varepsilon_2)$ $\quad \dim W^u = 2$

$c_3(\varepsilon_1 < 0 < \varepsilon_2) \quad \dim W^u = 2$

$c_4(0 < \varepsilon_1 < \varepsilon_2) \quad \dim W^u = 0$

Corollary 1. Let F be a Levin-Nohel equation on T^n, $n > 1$, with G a Morse-function. Then $A(F)$ is never a manifold without boundary and F is far way from vector fields.

Consider again the analytic equation F on S^1:

$$(**) \quad \dot{x} = -\int_{-r}^{0} a(-\theta)\sin x(t+\theta)\,d\theta$$

with $a(s)$ nonlinear, analytic and satisfying the hypothesis of Proposition 1.

In this case one can prove the following:

Proposition 3. $A(F)$ is a C^1 manifold, homeomorphic to S^1.

Since $a(s)$ is analytic, ϕ_t is one-to-one on $A(F)$ ([1], pag. 331) then α-limit sets and Ω are invariant. In the space $C^0(I,S^1)$, the unstable manifold of the critical point $\phi \equiv \pi$ (maximum of $G = 1 - \cos x$) has dimension 1. The only question is to prove that $A(F)$ is a manifold in the neighborhood of $\phi \equiv 0$ (minimum of G) because the α-limit set of any solution is the point $\phi \equiv \pi$; and since this solution coincides locally with the unstable manifold, the analyticity implies that it coincides globally. But if $x(t)$ is a solution of the equation, $y = -x(t)$ is also a solution. When we pass to the torus we get

$$u(t) = e^{ix(t)} \qquad \dot{u} = i\dot{x}e^{ix(t)}$$

$$v(t) = e^{-ix(t)} \qquad \dot{v} = -i\dot{x}e^{-ix(t)}$$

and the same one can say about u_t and v_t, which proves that

$$\lim_{t\to\infty} \frac{\dot{u}_t}{\dot{v}_t} = -1.$$

Corollary 1. The analytic Levin-Nohel equation on S^1 given by $G = 1 - \cos x$ is a Morse Smale retarded functional differential equation ([1], pag. 334).

The proof is easy since π is a saddle and 0 is a sink, there are no periodic orbits and transversality conditions are trivially satisfied.

Proposition 4. For generic Levin-Nohel equations on S^1, $A(F)$ is a C^1 manifold homeomorphic to S^1 and the flow induced on $A(F)$ has the same behavior as the gradient flow ∇G on S^1.

First of all we can assume that, generically, $a(s)$ satisfies the hypothesis of Proposition 1 and G is an analytic Morse-function. By the local Morse lemma one can also assume that, in the neighborhood of a critical point in S^1, G is equal to x^2. Then, locally, if $y = x(t)$ is a solution of the Levin-Nohel equation, $y = -x(t)$ is also a solution. Then, locally one has

$$\int_{-r}^{0} a(-\theta)\nabla G(-y(t+\theta))d\theta = -\int_{-r}^{0} a(-\theta)\nabla G(y(t+\theta))d\theta .$$

Using analyticity one gets the result.

Corollary 2. The analytic Levin-Nohel equation on S^1 is a Morse-Smale (RFDE) for generic G, and $a(s)$ nonlinear and satisfying the hypothesis of Proposition 1.

REFERENCE

[1] - HALE, J.K., Theory of Functional Differential Equations, Springer-Verlag, New York Inc., 1977.

NON-SINGULAR STRUCTURAL STABLE FLOWS ON THREE-DIMENSIONAL MANIFOLDS

by M. C. de Oliveira

Introduction.

The interest to study stable flows (see [9], [16] or [10] for background material and definitions) has been great, having in mind the possibility of a global description of their space of trajectories and their invariance under small perturbations. Furthermore, a rather recent point of view on bifurcation theory consists in describing how a system, depending on parameters, fails to be stable when we vary these parameters (see [11]).

It is well known that on manifolds with dimension greater than two, stable vector fields are not C^1-dense ([17] and [20]). The following basic question arises in a natural way: Are there stable systems on any manifold ? The answer is affirmative. Stable systems form, in particular, a C^1 open dense set in the universe of gradient fields (Palis-Smale [12]).

In view of the above results, the existence of stable fields without singularities on manifolds with Euler characteristic zero has a special relevance. Asimov [1] has answered this last question affirmatively, showing the existence of non-singular Morse-Smale vector fields on manifolds with dimension greater than three. On three dimensional manifolds the question of existence of stable fields without singularities was left open.

We have shown that C^r-structural stable vector fields are C^o-dense on the space of all C^r-vector fields, $r \geq 1$ (see [3] and [4]). In particular, we show in this work that structural stable fields without singularities do exist on all three-dimensional compact manifolds. Zeeman [21] has shown the C^o-density of fields with stable limit sets, but not necessarily globally stable. On the other hand, Morgan [6] has recently exhibited counter-examples to Asimov's result on three-

-dimensional manifolds implying, by our work, that on these three--dimensional manifolds we can only have non-singular stable fields with an infinite number of closed orbits. Finally, we call to attention that, to exhibit R^k stable actions, $k \geq 2$, we frequently start with non-singular stable fields, see P. Sad [14].

Statement and proof of the main result.

Theorem. Let M^3 be a compact, C^∞, 3-dimensional manifold without boundary. Then M^3 admits a non-singular structural stable C^r-vector field, $r \geq 1$.

The rest of the article is devoted to the proof of the above theorem.

Let M^3 be a compact, C^∞-differentiable manifold without boundary and we assume that M is endowed with a fixed Riemannian metric. Let $\chi^r(M)$ be the space of all C^r-vector fields on M with the C^r--topology, $r \geq 1$. As the Euler characteristic of M^3 is zero we may choose an $X \in \chi^r(M)$ without singularities.

Let ϕ_t be the flow generated by X. We call a 2-dimensional closed disc $D = D^2$ embedded in M^3 a local cross section of time $\zeta > 0$ for ϕ_t, if D is transversal to ϕ_t and

$$D \cap \phi_{[-\zeta,\zeta]}x = x, \quad \text{for every } x \in D.$$

$$(\text{Here } \phi_{[-\zeta,\zeta]}x = \{y \in M \mid y = \phi_t x, \ -\zeta \leq t \leq \zeta\}).$$

Choose $\varepsilon > 0$. The purpose of the next lemma is to embed in M lots of transversal discs to the flow ϕ_t, all of diameter less than ε, such that when they flow over a little time interval of lenght ε they cover the whole of M.

Lemma 1. There is a $\zeta > 0$ so that the following holds. For each $\varepsilon > 0$ there exist a finite family $\Sigma = \{D_1, \ldots, D_k\}$ of pairwise disjoint local cross sections of time ζ and diameter at most ε so that:

$$M = \bigcup_{i=1}^{k} \phi_{[-\varepsilon,0]} \text{ int } D_i = \bigcup_{i=1}^{k} \phi_{[0,\varepsilon]} \text{ int } D_i$$

Each D_i is ϕ_i-embedded in M, where $\phi_i : D^2 \to M$ and int $D_i = \phi_i(\text{int } D^2)$.

The proof of the above lemma comes immediatly from a lemma of Bowen-Walters [2], p. 188.

We now want to define a bijective function f between the elements of Σ, given by flowing from each element to the next one. The problem associated with such a function is that a discontinuity occurs whenever a point flows from the interior of one disc to the boundary of the next one.

We define $f : \Sigma \to \Sigma$ as follows:

If $x \in \Sigma$ then $f(x) = \phi_{t(x)}x$, where $t(x)$ is the first positive time such that $\phi_{t(x)}x \in \Sigma$. f is clearly a bijection.

Let us call $\Sigma_1 = \{x \in \Sigma \mid f(x) \in \partial\Sigma\}$, where $\partial\Sigma = \bigcup_{j=1}^{k} \partial D_j$.

The set Σ_1 is, in a certain sense, the *trouble set* for f and it requires a special attention. f is not continuous in Σ_1, f is differentiable in $\Sigma \setminus \Sigma_1$. We shall see later on that this set can be made wandering, so it does not interfere with the nonwandering set of f.

By a small perturbation of the embeddings $D_i \to M$, $1 \le i \le k$, we may assume that $f/\partial\Sigma$ is transversal to $\partial\Sigma$. Therefore we obtain:

Lemma 2. Σ_1 is a disjoint union of codimension one submanifolds of Σ.

Stratification of Σ.

Let us consider the following families of sets:

$_0\Sigma = \Sigma$; $_1\Sigma = \{x \in {_0\Sigma} \mid f^{-1}(x) \in \partial\Sigma\}$;...; $_i\Sigma = \{x \in {_{i-1}\Sigma} \mid f^{-i}(x) \in \partial\Sigma\}$.

$\Sigma_0 = \Sigma$; $\Sigma_1 = \{x \in \Sigma_0 \mid f(x) \in \partial\Sigma\}$; ...; $\Sigma_j = \{x \in \Sigma_{j-1} \mid f^j(x) \in \partial\Sigma\}$.

For the two families of sets defined above, we would like to change the flow of X and the embeddings which characterize Σ, using transversality theory, such that each family becomes a stratification of Σ.

Definition. Let J be a manifold. A *stratification* of J is a decomposition of J into a finite number of submanifolds Q_i such that:

(1) $\partial Q_i = \overline{Q}_i - Q_i$ = union of Q_j of lower dimension;

(2) If $z \in Q_j \subset \partial Q_i$ and a submanifold S of J is transversal to Q_j at $z (z \in S)$, then S is transversal to Q_i in a neighbourhood of z.

First we notice that the families $\{{}_i\Sigma \setminus {}_{i+1}\Sigma\}_{i\in N}$ and $\{\Sigma_j \setminus \Sigma_{j+1}\}_{j\in N}$ decompose Σ into a disjoint union of subsets.

Assertion. $\Sigma_j \setminus \Sigma_{j+1}$ is open in Σ_j and ${}_i\Sigma \setminus {}_{i+1}\Sigma$ is open in ${}_i\Sigma$ (i.e., each Σ_j, ${}_i\Sigma$ is closed).

In fact: If $x \in \Sigma_j \setminus \Sigma_{j+1}$ then $y = f^j(x) \in \partial\Sigma$. But $f(y) \notin \partial\Sigma$ because $x \notin \Sigma_{j+1}$. Therefore $f(y) \in \operatorname{int} \Sigma$, i.e., $f^{j+1}(x) \in \operatorname{int} \Sigma$. It follows then, immediatly, that $\Sigma_j \setminus \Sigma_{j+1}$ is open in Σ_j. The proof of ${}_i\Sigma \setminus {}_{i+1}\Sigma$ is similar.

$\Sigma_0 \setminus \Sigma_1$ is an open set of $\Sigma_0 = \Sigma$ and therefore is a submanifold of Σ with codimension zero.

$f/\partial\Sigma$ transversal to $\partial\Sigma$ implies that $\Sigma_1 \setminus \Sigma_2$ is a submanifold of Σ with codimension one and $(\Sigma_1 \setminus \Sigma_2) \cap \partial\Sigma$ is a submanifold of $\partial\Sigma$ with codimension one, i.e., it is a finite set of points in $\partial\Sigma$.

Now, we have to perturb the vector field X and the embeddings which characterize Σ such that $f/\partial\Sigma$ is transversal to $(\Sigma_1 \setminus \Sigma_2) \cap \partial\Sigma$, which will imply that $\Sigma_2 \setminus \Sigma_3$ is a submanifold of Σ with codimension two (finite set of points), $\Sigma_2 \cap \partial\Sigma = \phi$ and $\Sigma_3 = \phi$.

Thus, $\{\Sigma_j \setminus \Sigma_{j+1}\}$, $j = 0, 1, 2$, becomes a stratification of Σ and $\{(\Sigma_j \setminus \Sigma_{j+1}) \cap \partial\Sigma\}$, $j = 0, 1$, becomes a stratification of $\partial\Sigma$ if $f/\partial\Sigma$

is transversal to $\partial\Sigma$ and to $(\Sigma_1\backslash\Sigma_2) \cap \partial\Sigma$. By the same reason $\{{}_i\Sigma\backslash{}_{i+1}\Sigma\}$, $i = 0, 1, 2$ and $\{({}_i\Sigma\backslash{}_{i+1}\Sigma) \cap \partial\Sigma\}$ $i = 0, 1,$ becomes a stratification of Σ and $\partial\Sigma$, respectively.

Let ${}_i\Sigma_j$ be the set defined by ${}_i\Sigma_j = {}_i\Sigma \cap \Sigma_j$. It is easy to see that ${}_i\Sigma_j \cap \partial\Sigma = \phi$ if $i+j \geq 2$. Now, we may perturb the flow of X and the embeddings which characterize Σ such that ${}_i\Sigma_j = \phi$ if $i+j > 2$ and $(\Sigma_j\backslash\Sigma_{j+1}) \cap ({}_i\Sigma\backslash{}_{i+1}\Sigma)$ is a submanifold of Σ with codimension $i+j$, when $0 \leq i+j \leq 2$. In other words, the stratifications meet transversally. If ${}_i\Sigma\backslash{}_{i+1}\Sigma$ meets $\Sigma_j \cap \Sigma_{j+1}$ then the intersection is transversal in Σ and in $\partial\Sigma$.

<u>Triangulation of Σ.</u>

First we triangulate the boundary of Σ. $({}_1\Sigma\backslash{}_2\Sigma) \cap \partial\Sigma$ and $(\Sigma_1\backslash\Sigma_2) \cap \partial\Sigma$ are finite set of points in $\partial\Sigma$. We triangulate $\partial\Sigma$ such that $({}_1\Sigma\backslash{}_2\Sigma) \cap \partial\Sigma$ are 0-simplexes and the triangulation is transversal to $(\Sigma_1\backslash\Sigma_2) \cap \partial\Sigma$. Furthermore, between any two points of $(\Sigma_1\backslash\Sigma_2) \cap \partial\Sigma$ there is at least one 0-simplex.

After we have taken care of the boundary of Σ we extend the triangulation to the interior of Σ in the following may:

> $({}_2\Sigma\backslash{}_3\Sigma)$ is a finite set of points in the interior of Σ and they will be 0-simplexes.

Triangulate the 1-dimensional submanifold $({}_1\Sigma\backslash{}_2\Sigma)$ of Σ such that the triangulation is transversal to $(\Sigma_1\backslash\Sigma_2)$ (recall that ${}_1\Sigma_2 = \phi$). After this, ${}_1\Sigma$ is all triangulated and we extend the triangulation to ${}_0\Sigma = \Sigma$, keeping the transversality to $(\Sigma_1\backslash\Sigma_2)$ and to $(\Sigma_2\backslash\Sigma_3)$. For triangulations of differentiable manifolds see [7].

<u>Construction of the handlebody decomposition of Σ.</u>

Let Q^2 be a C^∞ manifold without boundary. A differentiable handlebody decomposition of Q^2 is a sequence of differentiable 2-dimensional submanifolds with boundary,

$$\phi = T_{-1} \subset T_0 \subset T_1 \subset T_2 = Q^2, \quad \text{such that}$$

$$\overline{T_k - T_{k-1}} = \overset{n_k}{\underset{i=1}{\cup}} D_i^k \times D_i^{2-k}, \quad 0 \le k \le 2.$$

We call be embedded product $D_i^k \times D_i^{2-k} \subset Q$ a k-handle in Q.

The 2-handles $D_i^2 \times 0_i$ are allowed to have creased boundaries. We refer to Smale [18] and Mazur [5] for handlebody theory.

From a C^r-thickening (see [5]) of the triangulation of Σ obtained above we get a generalized handlebody decomposition $\phi = T_{-1} \subset T_0 \subset T_1 \subset T_2 = \Sigma^2$ of $(\Sigma, \partial\Sigma)$. We call it generalized because we allow half-handles of the form $D_i^k \times 1/2D_i^{2-k}$, $k = 0, 1, 2,$ where $1/2D_i^{2-k}$ is the half closed disc, in order to cope with the boundary of Σ.

From the conditions imposed on the triangulation of Σ, the handlebody decomposition of Σ is contructed to have the following properties:

(i) $\quad {}_i\Sigma \setminus {}_{i+1}\Sigma \subset \operatorname{int} T_{2-i}$, $\quad i = 0, 1, 2,$

$\quad ({}_i\Sigma \setminus {}_{i+1}\Sigma) \cap \partial\Sigma \subset \operatorname{int} T_{1-i}$, $\quad i = 0, 1;$

(ii) $\quad \Sigma_j \setminus \Sigma_{j+1}$ has codimension j in Σ, $j = 0, 1, 2.$ It does not meet any k-simplex with $k < j$. Also $(\Sigma_j \setminus \Sigma_{j+1}) \cap \partial\Sigma$, $j = 0, 1,$ has codimension j in $\partial\Sigma$. Therefore $\Sigma_j \setminus \Sigma_{j+1} \subset T_2 \setminus T_{j-1}$, $j = 0, 1, 2$ and $(\Sigma_j \setminus \Sigma_{j+1}) \cap \partial\Sigma \subset \operatorname{int} T_1 \setminus T_{j-1}$, $j = 0, 1.$

<u>Proof of step 1</u>.

After the construction of the handlebody decomposition of Σ we have to isotop the flow of X such that the mapping f between the transversal sections induced by the flow preserves the handle decomposition of Σ. In the diffeomorphism case (Smale [19]), $f \in \operatorname{Diff}^r(M)$ preserved the handlebody decomposition of M, in the sense that $f(T_k) \subset \operatorname{int} T_k$, for all k. For flows we need a stronger condition, namely: The image of a k-handle only meets the boundary of Σ in the interior of a lower index handle, and therefore we require that

$$f(T_k) \subset (\text{int } T_k - \partial\Sigma) \cup (\text{int } T_{k-1} \cap \partial\Sigma), \quad \text{for all } k = 0, 1, 2$$

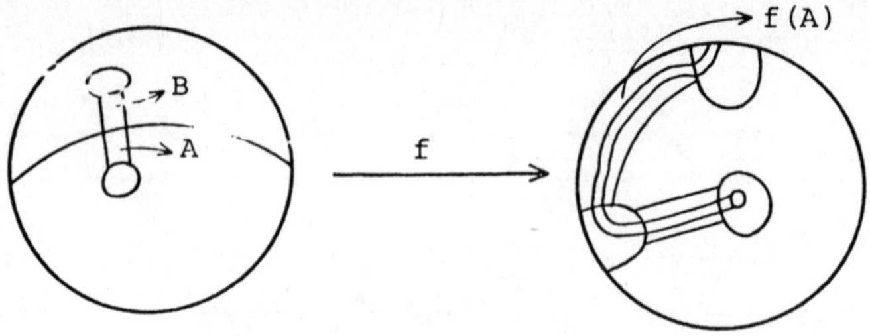

$f(T_o) \subset (\text{int } T_o \setminus \partial\Sigma)$

$h_1 = A \cup B$, $f(T_1) \subset (\text{int } T_1 \setminus \partial\Sigma) \cup (\text{int } T_o \cap \partial\Sigma)$

Figure 1

To produce the condition above we have first to identify some manifolds: Let Γ and Γ_1 be two sections of Σ. If $f(\partial\Gamma_1) \cap \Gamma \neq \phi$, consider the following manifold:

$$S(\Gamma_1,\Gamma) = \{\phi_t x \mid x \in \partial\Gamma_1,\ f(x) \in \Gamma,\ 0 \le t \le t(x)\}.$$

If $\Gamma \cap f^{-1}(\partial\Gamma_1) \neq \phi$, consider the manifold:

$$R(\Gamma_1,\Gamma) = \{\phi_t x \mid x \in \Gamma,\ f(x) \in \partial\Gamma_1,\ 0 \le t \le t(x)\}.$$

In both situations, $t(x)$ is the first positive time such that $\phi_{t(x)}x \in \Gamma$ and $\phi_{t(x)}x \in \partial\Gamma_1$, respectively, i.e., $\phi_{t(x)}x = f(x)$.

Let h_o be a 0-handle of the handlebody decomposition of Σ, then $f(h_o) \subset \text{int } \Sigma$, because $T_o \cap \Sigma_1 = \phi$. It is not difficult to modify the flow of X to produce $f(T_o) \subset (\text{int } T_o \setminus \partial\Sigma)$ by pushing the images of the 0-handles out of the 2-handles and 1-handles. Through all the process of modifying the flow to achieve step 1, the manifolds of type R and S can always be left invariant under the modifications. This means that whenever we change the vector field X, the new vector field is tangent to the manifolds of type R and S.

To obtain $f(T_1) \subset (\text{int } T_1 \setminus \partial\Sigma) \cup (\text{int } T_o \cap \partial\Sigma)$, we treat first the 1-handles which intersect $\partial\Sigma$ and Σ_1.

Let h_1 be a 1-handle such that $h_1 \cap \partial\Sigma \cap \Sigma_1 \neq \phi$, then there exist two sections Γ_2, $\Gamma_3 \in \Sigma$ such that $f(h_1) \cap \Gamma_i \neq \phi$, $i = 2, 3$, and $f(h_1) \subset \Gamma_2 \cup \Gamma_3$, where $h_1 \cap \Sigma_1 \subset f^{-1}(\partial\Gamma_2)$.

Let Γ_1 be the component of Σ containing h_1. Change the flow such that the induced mapping f satisfies $f(h_1 \cap \Sigma_1) \subset \text{int } T_o \cap \partial\Sigma$ and $f(h_1) \subset \text{int } T_1$. This is possible because $f(\partial\Gamma_1) \cap \partial\Sigma \subset \text{int } T_o \cap \partial\Sigma$ and if a 0-handle h_o is attached to the boundary of Σ then $f(h_o) \subset \text{int } T_o \backslash \partial\Sigma$.

The modification leaves the flow fixed on $S(\Gamma_1, \Gamma)$ for all sections $\Gamma \in \Sigma$ and leaves $R(\Gamma_1, \Gamma_2)$ invariant under the modification. On the other manifolds of type R and S the flow is left unchanged. We can proceed as above, separately, for each 1-handle h_1 satisfying $h_1 \cap \partial\Sigma \cap \Sigma_1 \neq \phi$. Therefore, $f(h_1) \subset (\text{int } T_1 \backslash \partial\Sigma) \cup (\text{int } T_o \cap \partial\Sigma)$, for all h_1 such that $h_1 \cap \partial\Sigma \cap \Sigma_1 \neq \phi$.

If $h_1 \cap \partial\Sigma \neq \phi$ and $h_1 \cap \Sigma_1 \neq \phi$ then it is not difficult to obtain $f(h_1) \subset (\text{int } T_1 \backslash \partial\Sigma)$, because $f(\partial\Sigma) \subset \text{int } T_1$.

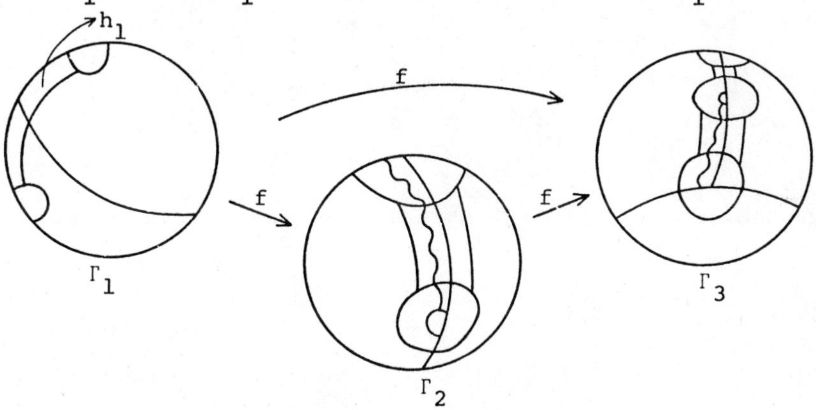

(a) $h_1 \cap \partial\Sigma \cap \Sigma_1 \neq \phi$

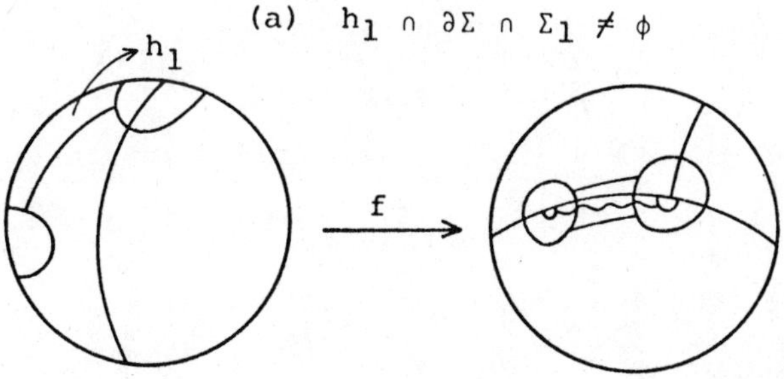

(b) $h_1 \cap \partial\Sigma \neq \phi$, $h_1 \cap \Sigma_1 = \phi$

Figure 2

If $h_1 \cap \partial\Sigma = \phi$ and $h_1 \cap \Sigma_1 \neq \phi$, then we have to modify the flow of X such that $f(h_1 \cap \Sigma_1) \subset \text{int } T_o \cap \partial\Sigma$ and $f(h_1) \subset \text{int } T_1$.

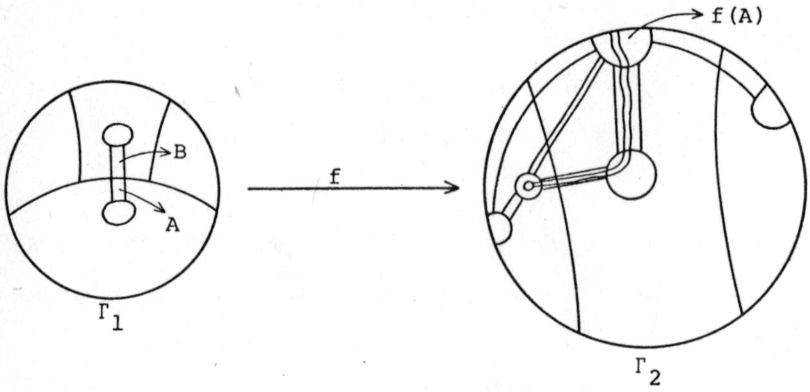

Figure 3

The flow is left fixed on the manifolds of type S and on $R(\Gamma_1,\Gamma')$ for any $\Gamma' \in \Sigma$, $\Gamma' \neq \Gamma_2$. The modification leaves $R(\Gamma_1,\Gamma_2)$ invariant.

The 1-handles which are in int Σ and do not intersect Σ_1 are in the region where f is a diffeomorphism and are easy to treat (see [15]), taking care of leaving the flow fixed on manifolds of type R and S.

Therefore we have obtained

$$f(T_k) \subset (\text{int } T_k - \partial\Sigma) \cup (\text{int } T_{k-1} \cap \partial\Sigma), \quad \text{for all } k = 0, 1, 2.$$

Let $\Omega(f)$ and $\Omega(\phi_t)$ be the nonwandering sets of f and ϕ_t, respectively. From the definitions the nonwandering sets satisfy $\Omega(f) \subset \Omega(\phi_t)$ and $\Omega(\phi_t) \cap \text{int } \Sigma \subset \Omega(f)$ and as a consequence of step 1 $\Omega(\phi_t) \cap \Sigma_1 \cap \text{int } \Sigma = \phi$. Therefore after step 1, we obtain:

<u>Proposition</u> 3. $\Omega(f) = \Omega(\phi_t) \cap \text{int } \Sigma$ and if $x \in \Omega(\phi_t)$ then the orbit of x does not intersect $\partial\Sigma$.

Corollary 4. There exists a neighbourhood U of $\Omega(f)$ in $\text{int}\,\Sigma$ such that $f/U : U \to \text{int}\,\Sigma$ is a diffeomorphism onto an open subset of $\text{int}\,\Sigma$ which contains $\Omega(f)$.

Proof of step 2.

Step 2 for the flow case is exactly the same as the one for diffeomorphisms (see [19]). We have to isotop the flow of X such that the image under f of each k-handle, when it crosses k-handles, crosses them linearly, expanding k-dimensions and contracting (2-k) dimensions, $k = 0, 1, 2$. We do not need to worry about the sets Σ_1 (points which flow to the boundary of Σ), where f is not continuous. In fact: If a k-handle meets Σ_1 then a neighbourhood of the intersection flows to the interior of a lower index handle (by step 1). Therefore we restrict ourselves to the region of Σ where f is a diffeomorphism.

The fact that the manifolds of type R and S can be left invariant under the modifications come from the way we constructed the handlebody decomposition of Σ.

Proposition 5. After step 2, $\Omega(\phi_t)$ has a hyperbolic structure, ϕ_t has no cycles and ϕ_t satisfies the strong transversality condition.

Proof. The hyperbolic structure of $\Omega(\phi_t)$ comes immediatly from Proposition 3 and Corollary 4 together with the fact that $\Omega(f)$ has a hyperbolic structure.

Let $K_i = \bigcap_{n\in Z} f^n(\overline{T_i \setminus T_{i-1}})$, $0 \le i \le 2$. Let $\tilde{K}_i$ be the suspension of K_i, i.e., $\tilde{K}_i$ is the set of all orbits of ϕ_t that pass through K_i. $\Omega(\phi_t) \subset \bigcup_{i=0}^{2} \tilde{K}_i$ and let $\Omega_i(\phi_t) = \Omega(\phi_t) \cap \tilde{K}_i$. If $x \in \Omega_o(\phi_t)$ then x belongs to an isolated closed orbit and the stable manifold of the orbit of x has dimension 3, if $x \in \Omega_2(\phi_t)$ then x belongs to an isolated closed orbit and the unstable manifold of the orbit of x has dimension 3. If $x, y \in \Omega_1(\phi_t)$ then the stable and unstable manifolds of their orbits meet transversally by step 2. Therefore ϕ_t satisfies the strong transversality condition. In particular ϕ_t has no cycles.

As $\Omega(\phi_t)$ is hyperbolic and ϕ_t has no cycles, the flow version of Newhouse's article [8] (see remark on the top of p. 126 of the refered paper) implies that ϕ_t satisfies Axiom A'. This fact together with Proposition 5 implies by [13] that ϕ_t is stable.

REFERENCES

[1] - ASIMOV, D., Round handle and non-singular Morse-Smale flows, Ann. of Math., 102 (1975).

[2] - BOWEN, R. and WALTERS, P., Expansive one-parameter flows, J. Differential Eqns., 12 (1972).

[3] - DE OLIVEIRA, M., C^0-density of structurally stable vector fields, Bull. Amer. Math. Soc., 82 (1976).

[4] - DE OLIVEIRA, M., Construction of stable flows: A density theorem, (to appear).

[5] - MAZUR, B., Differential Topology from the point of view of Simple Homotopy Theory, IHES Publications Mathematiques, 15 (1963).

[6] - MORGAN, J., Non-singular Morse-Smale flows on 3-Dimensional Manifolds, Topology, 18 (1979).

[7] - MUNKRES, J.R., Elementary Differential Topology, Annals of Mathematics Studies 54, Princeton University Press.

[8] - NEWHOUSE, S., On hyperbolic limit sets, Trans. Amer. Math. Soc., 167 (1972).

[9] - NITECKI, Z., Differentiable Dynamics, MIT Press (1971).

[10] - PALIS, J. and DE MELO, Introdução aos Sistemas Dinâmicos, IMPA Projeto Euclides, (1978).

[11] - PALIS, J. and NEWHOUSE, S., Cycles and bifurcation theory, Asterisque, Soc. Math. France, 31 (1976).

[12] - PALIS, J. and SMALE, S., Structural Stability Theorems, Global Anal. Proc. Symp. Pure Math., (1970), 14, Amer. Math. Soc., Providence, R.I., (1970).

[13] - ROBINSON, C., Structural stability for C^1 flows, Proc. Symp. Dynamical Systems (1973/1974) Warwick, Springer Lecture Notes, 468(1975).

[14] - SAD, P., Centralizadores de Campos Vetoriais, Doctoral Thesis, IMPA, (1977).

[15] - SHUB, M. and SULLIVAN, D., Homology theory and dynamical systems, topology, 14(1975).

[16] - SMALE, S., Differentiable dynamical systems, Bull. Amer. Math. Soc., 73(1967).

[17] - SMALE, S., Structural stable systems are not dense, Amer. J. Math., 88(1966).

[18] - SMALE, S., Generalized Poincaré's conjecture in dimensions greater than four, Ann. of Math., 74(1961).

[19] - SMALE, S., Stability and isotopy in discrete dynamical systems, Proc. Symp. Dynamical Systems, Salvador, Brazil, Ed. M. Peixoto, Acad. Press (1973).

[20] - WILLIAMS, R.F., The DA maps of Smale and Structural Stability, Global Anal. Proc. Symp. Pure Math., (1970), 14, Amer. Math. Soc., Providence, R.I., (1970).

[21] - ZEEMAN, E.C., C^0-density of stable diffeomorphisms and flows, Proc. Colloquium on smooth Dynamical Systems, Southampton, (1972). Southampton University reprint.

QUALITATIVE PROPERTIES OF CERTAIN ORDINARY DIFFERENTIAL SYSTEMS

by Nelson Onuchic and Adalberto Spezamiglio

1. Introduction.

Consider the almost linear system of ordinary differential equations

(1) $\dot{x} = A(t)x + f(t,x)$

where $x \in E$ with $E = R^n$ or $E = C^n$, $A(t)$ is an $n \times n$ continuous matrix on $J = [t_o,\infty)$, $t_o \geq 0$ and $f(t,x)$ is an n-dimensional vector function, continuous on $J \times E$. We are concerned with the asymptotic relationships between the solutions of (1) and those of the unperturbed linear system

(2) $\dot{y} = A(t)y$

In this sense, a question frequently posed is that one concerning relative asymptotic equivalence: given a solution $y(t) \neq 0$ of (2), does there exist a solution $x(t)$ of (1) such that $||x(t)-y(t)||/||y(t)|| \to 0$ as $t \to \infty$, and conversely ? Here, $||.||$ represent any convenient norm on E.

In [6], N. Onuchic defined a class of systems of type (1) and applied a result from [3, Th. 1.1] to study these problems. Similar results, by using other approaches may be found in [2], [4] and [10].

We shall consider here the question of relative asymptotic equivalence with weight, between the solutions of the two systems. That is, we define a function $w(t)$ with $w(t) \to 0$ as $t \to \infty$, and we claim that the relative difference $||x(t)-y(t)||/||y(t)|| \to 0$ with larger velocity than $w(t)$ as $t \to \infty$. To be explicit, we shall give conditions to obtain a positive answer to the following problems:

(I) Given a solution $y(t) \neq 0$ of (2), does there exist a solution $x(t)$ of (1) such that $x(t) = y(t) + o(w(t)||y(t)||)$ as $t\to\infty$?

(II) Given a solution $x(t)$ of (1) with $x(t) \neq 0$ for $t \geq t_o$, does there exist a solution $y(t)$ of (2) such that $y(t) = x(t) + o(w(t)||x(t)||)$ as $t \to \infty$?

Following Onuchic [6], H.M. Rodrigues [8] studied problems (I) and (II) with $w(t) = t^{-\mu}$, $t \geq t_o$, where $\mu \geq 0$ is an integer. But, Onuchic and Rodrigues gave no information about the distribuition, on phase space, of the initial values of the solutions under consideration. In fact, they had not any condition implying uniqueness of solutions.

By using a result from [1, Th. 1], we study here problems (I) and (II) with weight $w(t) = t^{-\mu}e^{-\rho t}$, $t \geq t_o$, where $\mu \geq 0$ is an integer and $\rho \geq 0$ a real. Moreover, under suitable conditions, we prove that the sets of initial values of the solutions obtained in problems (I) and (II) are homeomorphic to certain subspaces of the phase space. To this end, we were motivated by a result from [7, Th. 3.2]. We give also information about the number of parameters on which depend the solutions obtained.

2. Preliminaries.

In this section we give definitions, notations and a summary of results to be used in this paper.

The symbol $B = B(J,R)$ denotes a Banach space of real-valued functions defined on J, with the norm of $\varphi \in B$ denoted by $|\varphi|_B$. By $\mathcal{B} = B(J,E)$ we represent the space of measurable functions $x(t)$ defined on J with values in E, such that $||x(t)|| \in B$ and with $|x(t)|_{\mathcal{B}} = |\,||x(t)||\,|_B$. $L(J,E)$ denotes the space of functions defined almost everywhere on J, which are Lebesgue integrable on every compact subinterval of J, with the topology of convergence in the mean of order one on compact subintervals of J. A Banach space $\mathcal{B}$ is *stronger* than $L(J,E)$ if $\mathcal{B}$ is algebraically contained in $L(J,E)$ and convergence in $\mathcal{B}$ implies convergence in $L(J,E)$.

A class of Banach spaces $B = B(J,R)$ to be considered here is the one satisfying the following properties: (i) B is stronger than $L(J,R)$; (ii) If $\varphi \in B$, ψ is measurable and $|\psi(t)| \leq |\varphi(t)|$, then $\psi \in B$ and $|\psi|_B \leq |\varphi|_B$; (iii) If $h_{J'}$ is the characteristic function of the interval $J' \subset J$, then $h_{J'} \in B$ for all intervals $J'=[t_o,T]$, $T > t_o$; (iv) B is lean at infinity, that is, if $\varphi \in B$, then $h_{[t_o,T]}\varphi \to \varphi$ as $T \to \infty$. For example, the spaces $L^p(J,R)$ $(1\le p<\infty)$ and $L_o^\infty(J,R)$ are in this class. If $\psi(t) > 0$ is measurable on J such that ψ and $1/\psi$ are locally bounded on J, then the space $B = L_{\psi,0}(J,R)$ of all measurable functions $\varphi(t)$ such that $\varphi/\psi \in L_o^\infty(J,R)$, with $|\varphi|_B = |\varphi/\psi|_{L^\infty}$ is in this class too.

In the equations

(H) $\dot{y} = A(t)y$,

(NH) $\dot{x} = A(t)x + b(t)$

$A(t)$ is a locally Lebesgue integrable $n \times n$ matrix defined on J, and $b(t) \in L(J,E)$. A pair $(\mathcal{B},\mathcal{D})$ of Banach spaces stronger than $L(J,E)$ is called *admissible for* $A(t)$ if for every $b(t) \in \mathcal{B}$, there is at least one solution $x(t)$ of (NH) in $\mathcal{D}$. Such a solution will be mentioned as a $\mathcal{D}$-*solution*.

From now on, every Banach space used below will be assumed to be stronger than $L(J,E)$.

For a Banach space $\mathcal{D}$, let $E_{o\mathcal{D}}$ denote the set of initial points $y(t_o) \in E$ of $\mathcal{D}$-solutions $y(t)$ of (H). Let E_1 be any subspace of E complementary to $E_{o\mathcal{D}}$ and $P_{o\mathcal{D}}$ the projection of E onto $E_{o\mathcal{D}}$ which annihilates E_1. The following theorem, due to J. L. Massera and J. J. Schäffer [5], p. 295, plays an important role in this work.

Theorem A. Let $(\mathcal{B},\mathcal{D})$ be a pair of Banach spaces admissible for $A(t)$ and $\xi_o \in E_{o\mathcal{D}}$. Then, for each $b(t) \in \mathcal{B}$, equation (NH) has a unique $\mathcal{D}$-solution $x(t)$ satisfying $P_{o\mathcal{D}}x(t_o) = \xi_o$. Furthermore, there exist constants C_o and K depending only on $A(t)$, $\mathcal{B}$, $\mathcal{D}$, and E_1, such that $|x(t)|_{\mathcal{D}} \le C_o||\xi_o|| + K|b(t)|_{\mathcal{B}}$.

From now on, if $(\mathcal{B},\mathcal{D})$ is $A(t)$-admissible, C_o and K always represent the constants of Theorem A. Given $R > 0$, let $S_R = \{f \in \mathcal{D} : |f|_{\mathcal{D}} \le R\}$ and let $V_R = \{\xi \in E_{o\mathcal{D}} : ||\xi|| < R\}$. The main results in this work are applications of Theorems B, stated below, that is a result from, [1, Th. 1]. Let us consider the system

$$(P) \qquad \dot{x} = A(t)x + f(t,x)$$

Theorem B. Let us suppose the following hypotheses: (a) The pair $(\mathcal{B},\mathcal{D})$ is $A(t)$-admissible. (b) There exists $R > 0$ such that $f(t,x(t)) \in \mathcal{B}$, provided $x(t) \in S_R$. (c) There exists λ, $0<\lambda<K^{-1}$ such that $|f(t,x(t)) - f(t,y(t))|_{\mathcal{B}} \le \lambda|x(t)-y(t)|_{\mathcal{D}}$, whenever $x(t)$, $y(t) \in S_R$.

If $\xi_o \in E_{o\mathcal{D}}$, $||\xi_o||$ and $\eta = |f(t,0)|_{\mathcal{B}}$ are sufficiently small in such a way that $C_o||\xi_o|| + K\eta \le (1-\lambda K)R$, then (P) has a unique solution $x(t) \in S_R$ satisfying $P_{o\mathcal{D}}x(t_o) = \xi_o$.

Under the hypotheses of Theorem B, if $\sigma > 0$ is so that $C_o\sigma + K\eta \le (1-\lambda K)R$, we denote by $x(t;\xi)$ the unique $\mathcal{D}$-solution of (P) in S_R such that $P_{o\mathcal{D}}x(t_o;\xi) = \xi$, for each $\xi \in V_\sigma$. Let $\mathcal{F} = \{x(t;\xi) \in \mathcal{D} : \xi \in V_\sigma\}$ equiped with the topology induced by the topology of $\mathcal{D}$. The conection between V_σ, $\mathcal{F}$ and a section $\mathcal{F}(T_o) = \{x(T_o) \in E : x(t) \in \mathcal{F}\}$, $T_o \ge t_o$ stated in Theorem C below, is a result of [7, Th. 3.2].

Theorem C. Suppose that all assumptions of Theorem B are satisfied, with $\eta < (K^{-1}-\lambda)R$. Let $\sigma > 0$ be such that $C_o\sigma + K\eta = (1-\lambda K)R$. If equation (P) satisfies some condition which ensures uniqueness of solutions of $[t_o,\tau] \times E$, $\tau > t_o$, then the mappings

$$T : \xi \in V_\sigma \to x(t;\xi) \in \mathcal{F}$$
$$T_o : x(t;\xi) \in \mathcal{F} \to x(T_o;\xi) \in \mathcal{F}(T_o)$$
$$H : x(T_o;\xi) \in \mathcal{F}(T_o) \to \xi \in V_\sigma$$

are homeomorphisms.

3. Basic Lemmas.

We shall say that a matrix $A(t) = (a_{ij}(t))$, $1 \le i, j \le n$ is in the class $\Delta(n)$, if the following conditions hold:

(Δ1) $A(t)$ is continuous on J and $a_{ij}(t) = 0$ if $i < j$.

(Δ2) $a_{ii}(t) = \alpha + \lambda(t)$ where α is constant and $\int_{t_0}^{t} \mathcal{R}(\lambda(s))ds$ is bounded on J.

(Δ3) $a_{ij}(t)$ is bounded on J, if $i \neq j$.

(Δ4) If $n > 1$, then

$$\varliminf_{t\to\infty} \frac{\left|\int_{t_0}^{t} a_{m,m-1}(t_{m-2})dt_{m-2}\int_{t_0}^{t_{m-2}} a_{m-1,m-2}(t_{m-3})dt_{m-3} \cdots \int_{t_0}^{t_1} a_{21}(s)ds\right|}{t^{m-1}} > 0$$

for $m = 2, \ldots, n$.

A sufficient condition for (Δ4) is

(Δ4') $\varliminf_{t\to\infty} |a_{j+1,j}(t)| > 0, \quad j = 1, 2, \ldots, n-1.$

<u>Lemma</u> 1. If $A(t) \in \Delta(n)$, then for each solution $y(t) \neq 0$ of (H), there is an integer ℓ, $0 \le \ell \le n-1$, such that

$$0 < \varliminf_{t\to\infty} \frac{||y(t)||}{t^{\ell} e^{\mathcal{R}(\alpha)t}} \le \varlimsup_{t\to\infty} \frac{||y(t)||}{t^{\ell} e^{\mathcal{R}(\alpha)t}} < \infty$$

<u>Lemma</u> 2. Let $A(t) = \mathrm{diag}(A_1(t), A_2(t), \ldots, A_N(t))$ where $A_k(t) \in \Delta(n_k)$, $k = 1, 2, \ldots, N$. Let us denote by α_k the constant element in the diagonal of $A_k(t)$. Let $\mathcal{R}(\alpha_1) \ge \mathcal{R}(\alpha_2) \ge \ldots \ge \mathcal{R}(\alpha_N)$. Then, for each solution $y(t) \neq 0$ of (H), there are integers q, ℓ, with $1 \le q \le N$, $0 \le \ell \le n_q - 1$, such that

$$(3) \qquad 0 < \varliminf_{t\to\infty} \frac{||y(t)||}{t^{\ell} e^{\mathcal{R}(\alpha_q)t}} \le \varlimsup_{t\to\infty} \frac{||y(t)||}{t^{\ell} e^{\mathcal{R}(\alpha_q)t}} < \infty$$

For a proof of the above Lemmas, see [6].

<u>Lemma</u> 3. Let $B(t) = A(t) - \frac{\ell}{t}I$ where $A(t) \in \Delta(n)$ and $\ell \ge 0$ is an integer. Let us suppose that $\mathcal{R}(\alpha) \le 0$ and $0 \le \rho \le -\mathcal{R}(\alpha)$, with $\ell \ge \mu + n$ when $\rho = -\mathcal{R}(\alpha)$. Then, every solution $y(t)$ of $\dot{y} = B(t)y$ satisfies $t^{\mu}e^{\rho t}||y(t)|| \to 0$ as $t \to \infty$.

<u>Proof</u>. By making the change of variable $x = t^{\ell}y$ in $\dot{y} = B(t)y$, we obtain $\dot{x} = A(t)x$. As a consequence of Lemma 1, for each solution $x(t)$ of $\dot{x} = A(t)x$, there is an integer m, $0 \le m \le n-1$ and a

constant C, such that $||x(t)|| \le Ct^m e^{R(\alpha)t}$. This implies $||y(t)|| \le Ct^{m-\ell} e^{R(\alpha)t}$ for each solution $y(t)$ of $\dot{y} = B(t)y$. Then, $t^{\mu} e^{\rho t} ||y(t)|| \le Ct^{m-\ell+\mu} e^{[\rho+R(\alpha)]t}$ implies the required result.

We shall give now the meaning of the number "p", which will define the number of parameters mentioned in the Introduction. Let $B(t) = \operatorname{diag}(B_1(t), \ldots, B_N(t))$ where $B_k(t) \in \Delta(n_k)$, $k = 1, 2, \ldots, N$. Let us denote by β_k the constant element in the diagonal of $B_k(t)$, and let $R(\beta_1) \ge \ldots \ge R(\beta_{q-1}) > R(\beta_q) = \ldots = R(\beta_{q+r-1}) = 0 > R(\beta_{q+r}) \ge \ldots \ge R(\beta_N)$. Let $s+1 = \max\{n_k : k = q, \ldots, N\}$ and let ℓ an integer satisfying $0 \le \ell \le s$. For $k = q, \ldots, N$, let $s_k = \min\{\ell-\mu, n_k\}$, if $\ell > \mu$, and $s_k = 0$, if $\ell \le \mu$. For $k = q+r, q+r+1, \ldots, N$, let $s_k' = s_k$ if $\rho = -R(\beta_k)$; $s_k' = n_k$, if $0 \le \rho < -R(\beta_k)$ and $s_k' = 0$ if $-R(\beta_k) < \rho$. Under these conditions, we define

$$p = s_{q+r}' + \ldots + s_N', \quad \text{if} \quad \rho > 0$$

$$p = s_q + \ldots + s_{q+r+1} + s_{q+r}' + \ldots + s_N', \quad \text{if} \quad \rho = 0.$$

From now on we shall be assuming $t_o \ge 1$. We also fix the Banach space $\mathcal{D} = L_{\psi,o}(J,E)$ where $\psi(t) = t^{-\mu} e^{-\rho t}$, $t \ge t_o$.

<u>Lemma</u> 4. Let $B(t)$, s and ℓ be given as above. Let $E_{o\mathcal{D}}$ the space of initial values $y(t_o)$ of $\mathcal{D}$-solutions $y(t)$ of the system $\dot{y} = [B(t) - \frac{\ell}{t}I]y$. Then, $\dim E_{o\mathcal{D}} \ge p$.

<u>Proof</u>. For $k = q, q+1, \ldots, N$ we consider the systems

$$(4) \qquad \dot{y}_k = [B_k(t) - \frac{\ell}{t}I]y_k$$

Let $q+r \le k \le N$. When $0 \le \rho < -R(\beta_k)$, we have by Lemma 3 n_k solutions of (4) in $\mathcal{D}$. If $\rho = -R(\beta_k)$ and $\mu < \ell$, let $y_1^k = \ldots = y_{n_k-s_k}^k = 0$. Then, we have a system in s_k variables $y_{n_k-s_k+1}^k, \ldots, y_{n_k}^k$. Here, $y_k = \operatorname{col}(y_1^k \ldots y_{n_k}^k)$. Since $s_k \le \ell - \mu$, we have $\ell \ge \mu + s_k$, and by Lemma 3 we conclude that every solutions of the new system is in $\mathcal{D}$. Hence, we have s_k solutions of (4) in $\mathcal{D}$.

If $q \le k \le q+r-1$ and $\rho = 0$, we proceed as above for $\rho = -R(\beta_k)$ and we get s_k solutions of (4) in $\mathcal{D}$. By analysing the definition of p, we conclude the proof.

Let $B(t)$ and s as in Lemma 4. Let $h(t)$ be a continuous and positive function on J, satisfying

$$(5)\qquad \int_{t_o}^{\infty} t^{\mu+s}e^{\rho t}h(t)\,dt < \infty$$

One knows that there exists a continuous function $\gamma(t)$ on J, with $\gamma(t) \geq 1$, $\gamma(t) \to \infty$ when $t \to \infty$ and

$$\int_{t_o}^{\infty} t^{\mu+s}e^{\rho t}h(t)\gamma(t)\,dt < \infty.$$

Let us consider the Banach space $\mathcal{B} = L_{\psi,0}(J,E)$, where $\psi(t) = h(t)\gamma(t)$, $t \geq t_o$. Under these conditions, we have now a fundamental result in this work:

<u>Lemma</u> 5. Let $B(t)$, s and ℓ as in Lemma 4, with the restriction $(B_k(t))_{ii} = \beta_k$, $i = 1, 2, \ldots, n_k$; $k = 1, 2, \ldots, N$. Then, the pair $(\mathcal{B},\mathcal{D})$ is $[B(t) - \frac{\ell}{t}I]$-admissible.

<u>Proof</u>. We split the system $\dot{y} = [B(t) - \frac{\ell}{t}I]y + b(t)$ where $b \in \mathcal{B}$, into the N systems

$$(6)\qquad \dot{y}_k = [B_k(t) - \tfrac{\ell}{t}I]y_k + b_k(t).$$

In (6), $b(t) = \mathrm{col}(b_1(t),\ldots,b_N(t))$. We shall consider three cases: (1) $-\mathcal{R}(\beta_k) < \rho$; (2) $\rho = -\mathcal{R}(\beta_k)$; (3) $\rho < -\mathcal{R}(\beta_k)$.

<u>Case</u> (1). There is a solution of (6), $y_k(t) = \mathrm{col}(y_1^k(t), \ldots, y_{n_k}^k(t))$ where $y_i^k(t)$ is a sum of integrals

$$(7)\qquad \sigma_m^i(t) = \int_{\infty}^{t} c_1(t_1)dt_1\int_{\infty}^{t_1} c_2(t_2)dt_2 \cdots \int_{\infty}^{t_{m-1}} d_m(t_m)\exp\left(\int_{t_m}^{t} (\beta_k - \tfrac{\ell}{s})ds\right)dt_m$$

$$(m = 1, \ldots, i)$$

where $c_j(t)$ are bounded and $d_m \in \mathcal{B}$. By using these facts, we have that there is a constant C such that

$$|\sigma_m^i(t)| \leq Ct^{-\ell}e^{\mathcal{R}(\beta_k)t}\int_{t}^{\infty}dt_1\int_{t_1}^{\infty}dt_2 \cdots \int_{t_{m-1}}^{\infty} t_m^{\ell}e^{-\mathcal{R}(\beta_k)t_m}\psi(t_m)dt_m$$

Multiplying the two members by $t^{\mu}e^{\rho t}$, and observing that $\rho + \mathcal{R}(\beta_k) > 0$, we see that there is a $T \geq t_o$ such that the function $t^{\mu-\ell}\exp[\rho+\mathcal{R}(\beta_k)]t$

is increasing, for $t \geq T$. Hence,

(8) $$t^{\mu}e^{\rho t}|\sigma_m^i(t)| \leq C\int_t^{\infty} dt_1 \int_{t_1}^{\infty} dt_2 \cdots \int_{t_{m-1}}^{\infty} t_m^{\mu} e^{\rho t_m}\psi(t_m)dt_m \quad (t \geq T)$$

The integral on the right hand of (8) converges if and only if

$$\int_t^{\infty} \tau^{\mu+m-1}e^{\rho\tau}\psi(\tau)d\tau < \infty$$

(see [10, Lemma 3]). But this is the case, for $\mu+m-1 \leq \mu+n_k-1 \leq \mu+s$. The case (1) is then proved.

<u>Case</u> (2). If $\mu \geq \ell$, we get a solution in $\mathcal{D}$ just as in the above, with obvious adaptations. Let us suppose $\mu < \ell$. We consider two cases: (I) $n_k \leq \ell-\mu$; (II) $n_k > \ell-\mu$.

<u>Case</u> (I). If $n_k \leq \mu+1$, we proceed as in case (1), with obvious modifications. Let us suppose $n_k > \mu+1$. We determine a solution $y_k(t) = \mathrm{col}(y_1^k(t),\dots,y_{n_k}^k(t))$ of (6) as follows: for $i=1,2,\dots,\mu+1$, $y_i^k(t)$ is a sum of integrals of type (7). We have in this case

$$t^{\ell}e^{\rho t}|\sigma_m^i(t)| \leq C\int_t^{\infty} dt_1 \int_{t_1}^{\infty} dt_2 \cdots \int_{t_{m-1}}^{\infty} t_m^{\ell} e^{\rho t_m}\psi(t_m)dt_m.$$

The integral on the right hand converges, for $\ell+m-1 \leq \mu+s$, and since $\mu < \ell$, $t^{\mu}e^{\rho t}|\sigma_m^i(t)| \to 0$ as $t \to \infty$. The other coordinates are determined by induction. We shall detail below only the case $v=\mu+1+1$. The component $y_v^k(t)$ is a solution of the equation

(9) $$\dot{y}_v^k = [\beta_k - \frac{\ell}{t}]y_v^k + \Sigma_1(t) + b_v^k(t)$$

where

$$\Sigma_1(t) = a_{v,1}^k(t)y_1^k(t) + \dots + a_{v,v-1}^k(t)y_{v-1}^k$$

By the conditions on $y_i^k(t)$, $i = 1,2,\dots,v-1$, we have that

(10) $$t^{\ell}e^{\rho t}|\Sigma_1(t)| \to 0 \quad \text{as} \quad t \to \infty.$$

Let the solution of (9) given by

$$y_v^k(t) = \int_{t_0}^{t} [\Sigma_1(u) + b_v^k(u)]\exp\left(\int_u^t [\beta_k - \frac{\ell}{s}]ds\right)du$$

Then, we have

$$t^{\ell-1}e^{\rho t}|y_v^k(t)| \le \frac{1}{t}\int_{t_o}^{t} u^{\ell}e^{\rho u}|\Sigma_1(u)|du + \frac{C}{t}\int_{t_o}^{t} u^{\ell}e^{\rho u}\psi(u)du$$

The second integral on the right converges when $t \to \infty$, for $\ell \le s \le \mu+s$. If the first integral is unbounded, we apply the L'Hospital rule and we conclude, by using (10), that $t^{\ell-1}e^{\rho t}|y_v^k(t)| \to 0$ when $t \to \infty$. We observe now that $\mu+1+1 \le n_k$ implies $\ell-2\mu-1 \ge 1$ and so $\ell-1 \ge 2\mu+1 > \mu$. Hence, $t^{\mu}e^{\rho t}|y_v^k(t)| \to 0$ as $t \to \infty$.

Case (II) is proved by arguments as in case (I). So we proceed ahead.

Case (3). $\rho < -R(\beta_k)$. We consider the solution $y_k(t)$ of (6) in which the coordinate $y_i^k(t)$ is a sum of integrals of the form

$$\sigma_m^i(t) = \int_{t_o}^{t} c_1(t_1)dt_1\int_{t_o}^{t_1} c_2(t_2)dt_2 \ldots \int_{t_o}^{t_{m-1}} d_m(t_m)\exp\left(\int_{t_m}^{t}[\beta_k - \frac{\ell}{s}]ds\right)dt_m$$

$(m = 1,\ldots,i)$

There is a constant C such that

$$(11) \qquad |\sigma_m^i(t)| \le Ct^{-\ell}e^{-R(\beta_k)t}\int_{t_o}^{t} dt_1\int_{t_o}^{t} dt_2 \ldots \int_{t_o}^{t} t_m^{\ell}e^{-R(\beta_k)t_m}\psi(t_m)dt_m$$

Here, we have used the fact that $t_j \le t$, $j = 1,2,\ldots,m-1$. By integrating (11) and using $t - t_o \le t$, it follows

$$|\sigma_m^i(t)| \le Ct^{-\ell}e^{R(\beta_k)t}t^{m-1}\int_{t_o}^{t} t_m^{\ell}e^{-R(\beta_k)t_m}\psi(t_m)dt_m$$

Let $\delta > 0$ be such that $\rho + R(\beta_k) + \delta = \tau < 0$. Thus,

$$t^{\mu}e^{\rho t}|\sigma_m^i(t)| \le Ct^{\mu-\ell+m-1}e^{\tau t}e^{-\delta t}\int_{t_o}^{t} t_m^{\ell}e^{-R(\beta_k)t_m}\psi(t_m)dt_m.$$

There is a $T \ge t_o$ such that the function $t^{\mu-\ell+m-1}e^{\tau t}$ is decreasing for $t \ge T$. Therefore,

$$t^{\mu}e^{\rho t}|\sigma_m^i(t)| \le Ce^{-\delta t}\int_{t_o}^{t} e^{\delta t_m}t_m^{\mu+m-1}e^{\rho t_m}\psi(t_m)dt_m \qquad (t \ge T)$$

Since $\mu+m-1 \le \mu+n_k-1 \le \mu+s$, we apply [9, Lemma 3.6] and we conclude that $t^{\mu}e^{\rho t}|\sigma_m^i(t)| \to 0$ when $t \to \infty$. The proof of the lemma is complete.

4. Qualitative Properties.

Theorem 1 and 2 given below, state asymptotic relations between the solutions of systems (H) and (P).

Theorem 1. Let $A(t)$ be an $n \times n$ matrix satisfying the hypotheses of Lemma 2, and $s+1$ the maximum of n_k, $k = 1,2, \ldots, N$. Let $f(t,x)$ be continuous and satisfying

$$||f(t,x)-f(t,y)|| \leq h(t)||x-y||, \quad f(t,0) = 0$$

on $J \times E$, where $h(t)$ satisfies (5). If $y(t) \neq 0$ is a solution of (H), there exists a p-parameter family of solutions $x(t)$ of (P) defined for $t \geq$ some $T_o \geq t_o$, satisfying

$$x(t) = y(t) + o(t^{-\mu}e^{-\rho t}||y(t)||) \quad \text{as} \quad t \to \infty$$

Moreover, the set of initial points $x(T_o)$ is homeomorphic to a subspace of E, of dimension $\geq p$.

Remark. The number p is defined in the proof. When $p = 0$, by "p-parameter family of solutions" we mean "at least one solution".

Proof of Theorem 1. We can suppose, without loss of generality, that $(A_k(t))_{ii} = \alpha_k$, for $i = 1,2,\ldots,n_k$; $k = 1,2,\ldots,N$. One can see this by making the change of variables

$$y_j^k = \exp\left(\int_{t_o}^t \lambda_k(\tau)d\tau\right)v_j^k; \quad x_j^k = \exp\left(\int_{t_o}^t \lambda_k(\tau)d\tau\right)u_j^k$$

respectively in the systems (H) and (P).

Given the solution $y(t) \neq 0$ of (H), we apply Lemma 2 to obtain integers q and ℓ, with $1 \leq q \leq N$, $0 \leq \ell \leq n_q-1$, such that (3) is satisfied. The change of variable

$$z = \frac{x - y(t)}{t^{\ell}e^{at}}$$

where $a = \mathcal{R}(\alpha_q)$, takes system (P) to system

$$\text{(12)} \qquad \dot{z} = [B(t) - \frac{\ell}{t}I]z + g(t,z)$$

where $B(t) = A(t) - aI = \text{diag}(B_1(t),\ldots,B_N(t))$ and $g(t,z) = f(t,t^{\ell}e^{at}z + y(t))t^{-\ell}e^{-at}$. First, we shall show that there is a p-parameter family of solutions $z(t)$ of (12), satisfying $t^{\mu}e^{\rho t}||z(t)||\to 0$

as $t \to \infty$.

From (3), we get a constant C such that $||y(t)|| \le Ct^{\ell}e^{at}$, $t \ge t_o$. Since $||f(t,x)|| \le h(t)||x||$ on $J \times E$, we have

$$(13) \quad ||g(t,z)|| \le h(t)(||z|| + C)$$

and

$$(14) \quad ||g(t,z_1)-g(t,z_2)|| \le h(t)||z_1-z_2||$$

for $t \ge t_o$, z, z_1 and z_2 in E.

Let us fix $R > 0$ and consider the Banach spaces $\mathcal{B} = L_{\psi,0}(J,E)$, $\mathcal{D} = L_{\varphi,0}(J,E)$ where $\psi(t) = (R+C)h(t)\gamma(t)$, $\varphi(t) = t^{-\mu}e^{-\rho t}$, $t \ge t_o$, with $h(t)$ satisfying (5) and $\gamma(t)$ satisfying conditions following (5).

Let the system

$$(15) \quad \dot{y} = [B(t) - \frac{\ell}{t}I]y$$

and let $E_{o\mathcal{D}}$ be the space of initial values $y(t_o)$ of $\mathcal{D}$-solutions $y(t)$ of (15). We fix a subspace E_1 of E, complementary to $E_{o\mathcal{D}}$. Since $(\mathcal{B},\mathcal{D})$ is admissible relatively to (15), by Lemma 5, we take the constants C_o and K of Theorem A.

There is $T_o \ge t_o$ such that

$$(16) \quad \frac{1}{\gamma(t)} \le \frac{R}{4K} \qquad (t \ge T_o)$$

If $\tilde{g}(t,z) = \chi_{[T_o,\infty)}(t)g(t,z)$, $t \ge t_o$, $z \in E$, we shall apply Theorem B to system

$$(17) \quad \dot{z} = [B(t) - \frac{\ell}{t}I]z + \tilde{g}(t,z)$$

Let us verify the hypotheses of Theorem B: (a) is just Lemma 5. For (b), let $z(t) \in S_R$. Clearly $||z(t)|| \le |z|_{\mathcal{D}}$, and by (13), $||\tilde{g}(t,z(t))||/(R+C)h(t)\gamma(t) \le 1/\gamma(t)$. From the hypotheses about $\gamma(t)$, we conclude that $\tilde{g}(t,z(t)) \in \mathcal{B}$. For (c), let $z_1(t)$, $z_2(t) \in S_R$. Then, $||\tilde{g}(t,z_1(t))-\tilde{g}(t,z_2(t))|| \le \chi_{[T_o,\infty)}(t)h(t)|z_1-z_2|_{\mathcal{D}}$. Since $\chi_{[T_o,\infty)}(t)h(t) \in B = L_{\psi,0}(J,R)$, we have

$$|\tilde{g}(t,z_1(t))-\tilde{g}(t,z_2(t))|_{\mathcal{B}} \le |\chi_{[T_o,\infty)}(t)h(t)|_B |z_1-z_2|_{\mathcal{D}}$$

Taking $\lambda = |\chi_{[T_o,\infty)}(t)h(t)|_B$ and using (16), we have

$$\lambda = \sup_{t\ge T_o} \frac{1}{(R+C)\gamma(t)} \le \frac{R}{4K(R+C)} < \frac{1}{4K} < \frac{1}{K}.$$

Now, let $\xi \in E_{o\mathcal{D}}$ be such that $||\xi|| \le R/2C_o$ and let $\eta=|g(t,0)|_{\mathcal{B}}$. Since $||\tilde{g}(t,0)|| \le \chi_{[T_o,\infty)}(t)h(t)C$, it follows that $\eta \le \lambda C$ and so $C_o||\xi|| + K\eta \le R/2 + KCR/4K(R+C) < 3R/4$. By using $\lambda < 1/4K$, we have $3R/4 < (1-\lambda K)R$. By Theorem B, there is a unique solution $\tilde{z}(t;\xi) \in S_R$ of (17) satisfying $P_{o\mathcal{D}}\tilde{z}(t_o;\xi) = \xi$.

We observe now that the matrix $B(t)$ satisfies the conditions of the definition of the number p. For this number, we have $\dim E_{o\mathcal{D}}\ge p$, by Lemma 4, and so $\xi \in E_{o\mathcal{D}}$ may the taken depending on p parameters. If $z(t;\xi) = \tilde{z}(t;\xi)$ for $t \ge T_o$, we have a family $\mathcal{F}$ of solutions $z(t;\xi)$ for (12) depending on p parameters, satisfying $t^{\mu}e^{\rho t}||z(t;\xi)||\to 0$ as $t \to \infty$.

Now let $\sigma > 0$ such that $C_o\sigma + K = (1-\lambda K)R$. From above, we extract $K\eta < (1-\lambda K)R$ and by Theorem C the mapping $H : z(T_o;\xi) \in \mathcal{F}(T_o) \to \xi \in V_\sigma$ is an homeomorphism. To each solution $z(t;\xi)$ of (12), we have the solution $x(t;\xi) = y(t) + t^{\ell}e^{at}z(t;\xi)$, $t \ge T_o$, of (P). Since

$$\frac{t^{\ell}e^{at}z||(t;\xi)||}{t^{-\mu}e^{-\rho t}||y(t)||} = t^{\mu}e^{\rho t}||z(t;\xi)|| \frac{t^{\ell}e^{at}}{||y(t)||},$$

we have by (3) that it satisfies the required condition. Finally, we note that the mapping $z(T_o;\xi) \in \mathcal{F}(T_o) \to x(T_o;\xi) \in E$ is an homeomorphism onto its range. The Theorem is proved.

<u>LEMMA</u> 6. Let the hypotheses of Theorem 1 be satisfied, with $f(t,x)$ linear in x. Let $x(t) \ne 0$ be a solution of (P). Then, there exists a solution $y(t) \ne 0$ of equation (H) such that

$$x(t) = y(t) + o(t^{-\mu}e^{-\rho t}||y(t)||).$$

<u>Proof</u>. Let $U(t) = (y_1(t) \ldots y_n(t))$ be the matrix solution for (H) with $U(t_o) = I$. By using Theorem 1, we get a matrix $V(t) = (x_1(t) \ldots x_n(t))$ of solutions of (P), such that $x_i(t) = y_i(t) + o(t^{-\mu}e^{-\rho t}||y_i(t)||)$, $i = 1, 2, \ldots, n$. By [6, Corollary 1], $V(t)$ is a fundamental matrix for (P). Hence, given the solution $x(t)$ of (P), there are constants $\zeta_1, \ldots, \zeta_n$ such that $x(t)=\zeta_1 x_1(t) + \ldots + \zeta_n x_n(t)$. If $Z = \{i : \zeta_i \neq 0\}$, then, $Z \neq \phi$. Let $y(t) = \sum_{i\in Z} \zeta_i y_i(t)$. This solution of (H) satisfies the required condition.

<u>Theorem</u> 2. Let $A(t)$, s and $f(t,x)$ be as in Theorem 1. If $x(t) \neq 0$ is a solution of (P), then there is a $\tilde{p}$-parameter family of solutions $y(t)$ of (H) defined for $t \geq$ some $\tilde{T}_o \geq t_o$, satisfying $y(t) = x(t) + o(t^{-\mu}e^{-\rho t}||x(t)||)$ as $t \to \infty$. Moreover, the set of initial values $y(\tilde{T}_o)$ is homeomorphic to a subspace of E, of dimension $\geq \tilde{p}$.

<u>Proof</u>. If $x^*(t)$ denotes the complex conjugate transpose of $x(t)$, we consider the linear system

$$(18) \quad \dot{w} = A(t)w + \frac{x^*(t)w}{x^*(t)x(t)}f(t,x(t)).$$

One can easily see that system (18) has $x(t)$ as a solution and satisfies the hypotheses of Lemma 6. So, there is a solution $z(t)\neq 0$ of (H) such that $x(t) = z(t) + o(t^{-\mu}e^{-\rho t}||x(t)||)$. By applying Theorem 1 to solution $z(t)$, we get a $\tilde{p}$-parameter family of solutions $y(t)$ of (H) defined for $t \geq \tilde{T}_o$, such that $y(t)=z(t)+o(t^{-\mu}e^{-\rho t}||z(t)||)$. The last two relations concerning $z(t)$ imply the desired condition on $y(t)$, and the set of initial values $y(\tilde{T}_o)$ is homeomorphic to a subspace of E, of dimension $\geq \tilde{p}$. The proof is complete.

REFERENCES

[1] - CORDUNEANU, C., Sur certains systèmes différentiels non-linéaires. An. Sti. Univ. "Al. I. Cuza". Iasi, Sect.I, <u>6</u> (1960), 257-260.

[2] - FAEDO, S., Proprieta assintotiche delle soluzioni dei sistemi differenziali lineari omogenei, Ann. Mat. Pura Apl. <u>26</u> (1947), 207-215.

[3] - HARTMAN, P. and ONUCHIC, N., On the asymptotic integration of ordinary differential equations, Pacific J. Math. 13(1963), 1193-1207.

[4] - LEVI, E., Sul comportamento asintotico delle soluzioni dei sistemi di equazioni differenziali lineari omogenee, Atti Acad. Naz. Lincei Cl. Sci. Fis. Nat. 8(1950), 465-470; 9(1950), 26-31.

[5] - MASSERA, J.L. and SCHÄFFER, J.J., Linear differential equations and functional analysis, IV. Math. Ann. 139(1960), 287-342.

[6] - ONUCHIC, N., Asymptotic relationships at infinity between the solutions of two systems of ordinary differential equations. J. Diff. Equations, 3(1967), 47-58.

[7] - ONUCHIC, N. and TÁBOAS, P.Z., Qualitative properties of nonlinear ordinary differential equations, Proc. of the Royal Soc. of Edinburgh, 79 A, (1977), 47-58.

[8] - RODRIGUES, H.M., Relative asymptotic equivalence with weight t^{μ}, between two systems of ordinary differential equations, Dynamical Systems, Vol. 2. An International Symposium, Academic Press, Inc., 1976.

[9] - STRAUSS, A. and YORKE, J.A., Perturbations theorems for ordinary differential equations, J. Diff. Equations, 3(1967), 15-30.

[10] - SZMYDT, Z., Sur l'allure asymptotique des intégrales de certains systèmes d'équations différentielles non-linéaires, Ann. Pol. Math., (1965), 253-276.

APPLICATIONS OF THE INTEGRAL AVERAGING BIFURCATION METHOD TO RETARDED FUNCTIONAL DIFFERENTIAL EQUATIONS

by Julio Ruiz-Claeyssen and Bernardo Cockburn

0. Introduction.

The integral averaging bifurcation method as developed by Chow and Mallet-Paret [1] allows to discuss Hopf bifurcation through any center curve of the characteristic equation of a retarded functional differential equation. Averaging procedures has been considered by Hale [5] when discussing bifurcation through a center stable curve.

In this paper we present applications of such method to the equation

$$x'(t) = g(x(t),x(t-r),\alpha)$$

with α a real parameter and g a smooth real function with $g(0,0,\alpha) = 0$.

Hopf bifurcation is assumed to occur at $\alpha = 0$. The direction of bifurcation is then determined by computing a constant K which depends on the nonlinearities and the crossing speed v obtained from the characteristic equation. This information allows to describe the qualitative behavior of the bifurcating solution such as amplitude, period or stability. This is an advantage over fixed point techniques.

The function g is chosen in such a way to include equations of interest in mathematical modelling. The general case could be treated numerically from the formulae derived in [7]. Bifurcation by varying the delay, is discussed for an equation proposed by Lasota and Wazewska and work out by Chow [2] by fixed point techniques. Our computer numerical studies shows an interesting variation of K in terms of the bifurcating delay and involved parameters.

1. Preliminars.

We consider the retarded differential equation

$$(1.1) \qquad x'(t) = g(x(t),x(t-r))$$

where g is a smooth real function and the origen being an isolated equilibrium point. We let the phase space be C the set of continuous real functions defined on the interval $[-r,0]$ with the supremum norm. The state solution $x_t \in C$ is defined by $x_t(\theta) = x(t+\theta)$, $-r \leq \theta \leq 0$, where $r \geq 0$ is fixed. The notation of Hale is followed.

Equation (1.1) is conveniently written

$$(1.2) \qquad x'(t) = ax(t) + bx(t-r) + h(x(t),x(t-r))$$

where $a = g_x(0,0)$, $b = g_y(0,0)$ and $h(x,y)$ contains the nonlinearities, begining with at least quadratic terms. This equation generates an evolution equation for x_t in the space BC of all uniformly continuous functions on $-r \leq \theta < 0$ with at most one jump discontinuity at $\Theta = 0$. This has been accomplished by extending the definition of the infinitesimal generator A of the stongly continuous semigroup associated with the linear part of (1.2). Details are omitted, [1]. We have

$$(1.3) \qquad \frac{d}{dt} x_t = Ax_t + X_o H(x_t)$$

where

$$A\phi = \phi' + X_o[a\phi(0) + b\phi(-r) - \phi'(0)], \qquad \phi \in C^1$$

$$H(\phi) = h(\phi(0),\phi(-r))$$

and $X_o(\theta)$ is the jump function

$$X_o(\theta) = \begin{cases} 1, & \theta = 0 \\ 0, & -r \leq \theta < 0 \end{cases}$$

Hale's Descomposition.

Let us assume that the characteristic equation

(1.4) $\lambda = a + be^{-\lambda r}$

has a simple pair of pure imaginary roots $\lambda_1 = i\omega$, $\lambda_2 = \bar{\lambda}_1$, $\omega > 0$.

We define P to be linear subspace in BC generated by the eigenfunctions $\text{Cos}\omega\theta$ and $\text{Sin}\omega\theta$, that is, $P = \{\phi \in C: \phi = \Phi u,\ u \in R^2\}$ where

$$(1.5)\qquad \Phi(\theta) = \Phi(0)e^{B\theta} = [\text{Cos}\omega\theta \quad \text{Sin}\,\omega\theta]$$

$$B = \begin{bmatrix} 0 & \omega \\ -\omega & 0 \end{bmatrix}.$$

It follows that $A = d\Phi/d\theta = \Phi B$. Let C^* denote the space of continuous real functions defined on the interval $[0,r]$. We define the bilinear functional

$$(1.6)\qquad \langle\psi,\phi\rangle = \psi(0)\phi(0) + b\int_{-r}^{0} \psi(r+\theta)\phi(\theta)d\theta, \quad \phi \in C, \quad \psi \in C^*$$

It has been established by Hale that for the adjoint matrix

$$\Psi^*(s) = e^{-Bs}\Psi(0)$$

the matrix $\langle\Psi^*,\Phi\rangle = [\langle\psi_i,\phi_j\rangle]$ is nonsingular. Then $\psi = \langle\Psi^*,\phi\rangle^{-1}\Psi^*$ will be such that $\langle\Psi,\Phi\rangle = I$, the 2x2 indentity matrix. We have [7]

<u>Lemma</u> 1. The adjoint matrix $\Psi(s)$ is given by

$$(1.7)\qquad \Psi(s) = 2\mu \begin{bmatrix} A\ \text{Cos}\omega s + B\ \text{Sin}\omega s \\ C\ \text{Cos}\omega s + D\ \text{Sin}\omega s \end{bmatrix}$$

where $A = D = (1-ar)$, $C = -B = \omega r$ and $\mu = 1/(\omega^2 r^2 + (1-ar)^2)$.

The space BC can be now descomposed as $BC = P \oplus Q$ where $\phi = \phi^P + \phi^Q$ with $\phi^P = \Phi\langle\Psi,\phi\rangle$ and $\phi^Q = \phi - \phi^P$. We let A_Q denote A restricted to Q. Then $x_t = \Phi u(t) + y_t$ where $u(t) = \langle\Psi,x_t\rangle$, $\Phi u(t) = x_t^P$ and $y_t = x_t^Q$ does not necessarily satisfies $y_t(\theta) = y(t+\theta)$. We let A_Q denote A restricted to Q. With this descomposition, (1.3) becomes

$$(1.8) \quad u'(t) = Bu(t) + \psi(0)H(\phi u(t) + y_t)$$

$$\frac{d}{dt} y_t = A_Q y_t + X_o^Q H(\phi u(t) y_t)$$

where $X_o^Q = X_o - X_o^P$, $X_o^P = \Phi\Psi(0)$.

2. Root Analysis.

We say that a curve C in R^2 is a *center curve* or a *Hopf bifurcation curve* if for any point (a,b) lying in the curve, the characteristic equation

$$(2.1) \quad (\lambda) = \lambda - a - be^{-\lambda r} = 0$$

has a simple pair of pure imaginary roots. A center curve is said to be *stable* if all remaining roots have negative real part. A geometric description of all center curves is shown in Figure 1. The pair (n,m) indicates the number of pure imaginary roots (n) and the rootes with positive real part (m). The analytical description is as follows.

The complex number $\lambda = (y+is)/r+a$ will satisfy (2.1) if and only if the real numbers y and s satisfy the equations

$$(2.2) \quad y = -s \operatorname{Cotg} s$$

$$A = -s \operatorname{Cosec} s \,.\, \exp(-s \operatorname{Cotg} s)$$

where $A = br \exp(-ar)$ and $s \neq k\pi$, k integer. We define

$$G(s) = -s \operatorname{Cotg} s$$

$$G_k^{-1} = \text{inverse of } G \text{ restricted to } (\pi k,(k+1)\pi),\ k \geq 0$$

$$H(s) = -s \operatorname{Cotg}(s)\exp(G(s))$$

$$H_k^{-1} = \text{inverse of } H \text{ restricted to } (\pi k,(k+1)\pi),\ k \geq 0$$

Therefore λ is a pure imaginary root of (2.1) if and only if $A = A_k$ for some integer $k \geq 0$, where A_k is defined from the relations

(2.3) $\bar{s}_k = G_k^{-1}(-ar)$

$A_k = H(\bar{s}_k)$

In this case we have that $\lambda = (\bar{s}_k/r)i$.

Let us define $\bar{A} = -ar\exp(-ar)$ and $\bar{A}_o = A_o$ (resp. $\bar{A}$) if $ar < 1$ (resp. $ar \geq 1$). We thus enunciate the following [3].

<u>Theorem</u>. Let $r \geq 0$ be fixed. Then equation (2.1) has exactly k rootes with nonegative real part if and only if (a,b) lies in the region $R_k(r)$, has exactly two pure imaginary roots if and only if (a,b) lies in the curve $C_k(r)$ for some integer $k \geq 0$. The curves C_k are given by

$$C_o(r) = \{(a,b) : A = A_o, \quad ar < 1\}$$

$$C_{m+1}(r) = \{(a,b) : A = A_{m+1}\}, \quad m = 0,1,2,\ldots$$

The regions R_k have C_k and C_{k+1} as boundaries for $k \geq 2$. R_o has C_o and the line $a + b = 0$ as boundaries, and R_1 has the line $a + b = 0$ and C_1 as boundaries. The curves C_k are asymptotic to the lines $a \pm b = 0$.

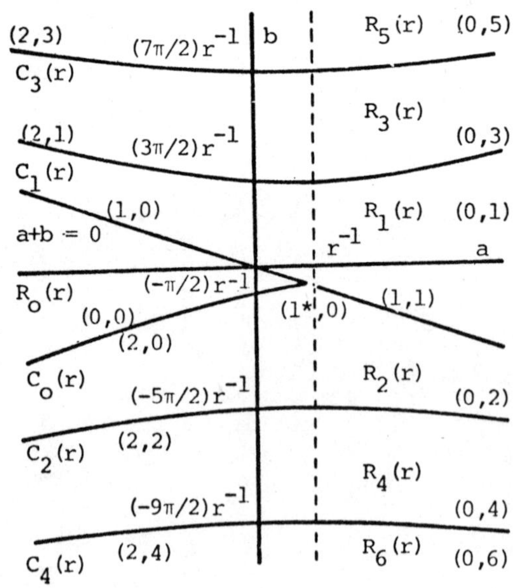

Figure 1

Consider the retarded functional differential equation

$$x'(t) = g(x(t),x(t-r),\alpha)$$

where α is a real parameter, and let $\lambda = a(\alpha) + b(\alpha)e^{-\lambda r}$ be its characteristic equation. Assume that $x = 0$ is an equilibrium point. It follows then that Hopf bifurcation will occur at $\alpha = \alpha_o$ if and only if there is a nonnegative integer k such that $F_k(0) = 0$, $F_k'(0) \neq 0$, where $F_k(\alpha) = G(H_k^{-1}(A(\alpha)) + a(\alpha)r$. This is the condition for which the curve $\Gamma(\alpha) = (a(\alpha)r, b(\alpha)r)$ crosses transversally a center curve $C_k(1) = rC_k(r)$.

The following characteristic equation

$$\lambda = -\sigma - e^{-\beta}e^{-\lambda r} \tag{2.4}$$

corresponds to a model arising in the study of the survival of red blood cells [2]. There β denotes the unique real root of the equation

$$\sigma\beta = e^{-\beta} \tag{2.5}$$

for a given $\sigma > 0$. We have

<u>Lemma</u> 2. Let $0 < \sigma < e^{-1}$ be given. Then there is a sequence r_k such that $(-\sigma r_k, -e^{-\beta}r_k)$ lies on the center curve $C_k(1)$. The sequence is given by

$$r_k = \left| \frac{\mathrm{Cos}_k^{-1}(-1/\beta)}{\omega} \right| \tag{2.6}$$

where Cos_k^{-1} denotes the k-th branch of arcosine and $\omega = \sigma\sqrt{\beta^2 - 1}$.

<u>Proof</u>. From theorem 1 we have that the bifurcation curves $C_k(1)$ are contained in the region $(ar+br)(ar-br) < 0$. Thus for $a = -e^{-\beta}/\beta$ and $b = -e^{-\beta}$ this will be the case if and only if $|b/a| > 1$. This implies $|\beta| > 1$, that is, $0 < \sigma < e^{-1}$. The value (2.6) follows from (2.2) and (2.3).

3. Integral averaging method.

Consider the retarded differential equation

(3.1) $$x'(t) = a(\alpha)x(t) + b(\alpha)x(t-r) + h(x(t),x(t-r),\alpha)$$

where $\Gamma(\alpha) = (a(\alpha),b(\alpha))$ is a smooth curve in R^2 that crosses transversally a center curve C_k at (a,b) for $\alpha = 0$. We assume that $h(x,y,\alpha)$ is continuously differentiable with $h(0,0,\alpha) = 0$. Thus Hopf bifurcation occurs at $x = 0$.

We scale (3.1) by $x \to \varepsilon x$, $\alpha \to \varepsilon\alpha$ and then we set $\alpha = 0$. It turns out that

(3.2) $$x'(t) = ax(t) + bx(t-r) + \varepsilon h(\varepsilon x(t),\varepsilon x(t-r))/\varepsilon^2$$

where $a = a(0)$, $b = b(0)$ and $h(x,y) = h(x,y,0)$. Upon descomposing as in section 1, we have

$$u'(t) = Bu(t) + \varepsilon\Psi(0)H(\Phi u(t) + y_t,\varepsilon)$$

$$\frac{d}{dt} y_t = A_Q y_t + \varepsilon X_o^Q H(\Phi u(t) + y_t,\varepsilon)$$

where $H(\phi,\varepsilon) = h(\varepsilon\phi(0),\varepsilon\phi(-r))/\varepsilon^2$, $\Psi(0) = 2\mu\,\mathrm{col}[A\ C]$, B given by (1.5) and A,C,μ as in lemma 1.

In polar coordinates, the above equations become

$$r' = 2\mu[A\mathrm{Cos}\gamma + C\mathrm{Sin}\gamma]h(\varepsilon(r\mathrm{Cos}\gamma+y_t(0)),\varepsilon(r\mathrm{Cos}(\omega r+\gamma)+y_t(-r)))/\varepsilon$$

$$\gamma' = -\omega + \frac{2\mu}{r}(-A\mathrm{Sin}\gamma + C\,\mathrm{Cos}\gamma)h/\varepsilon$$

$$y_t' = A_Q y_t + \varepsilon X_o^Q(h/\varepsilon^2)$$

where h is as in the equation for the amplitude and $u = \mathrm{col}(r\mathrm{Cos}\gamma, r\mathrm{Sin}\gamma)$. We expand h in Taylor series up to third order terms, and we neglect second order terms in y. Then

$$r' = \varepsilon r^2 C_3(\gamma) + \varepsilon^2 r^3 C_4(\gamma) + \varepsilon r G_2(\gamma)y_t + 0(\varepsilon^3)$$

$$\gamma' = -\omega + \varepsilon r D_3(\gamma) + \varepsilon G_2^*(\gamma)y_t + 0(\varepsilon^2)$$

$$y_t' = A_Q y_t + \varepsilon J(0,0)u^2 + 0(\varepsilon^2)$$

where the term in uy has also been neglected on the last equation. The functions $C_3(\gamma)$, $C_4(\gamma)$, $D_3(\gamma)$ are homegeneous polynomials in $(\mathrm{Cos}\gamma, \mathrm{Sin}\gamma)$, more precisely

$$C_3(\gamma) = \frac{\mu}{b^2} \sum_{k=0}^{3} A_{4-k} \mathrm{Cos}^k\gamma \mathrm{Sin}^{3-k}\gamma$$

$$D_3(\gamma) = \frac{\mu}{b^2} \sum_{k=0}^{3} B_{4-k} \mathrm{Cos}^k\gamma \mathrm{Sin}^{3-k}\gamma$$

$$C_4(\gamma) = \frac{\mu}{b^3} \sum_{k=0}^{4} F_{5-k} \mathrm{Cos}^k\gamma \mathrm{Sin}^{4-k}\gamma$$

The function $G_2(\gamma)$ is a linear functional defined on the space C and $J(0,0)(\mathrm{Cos}\gamma, \mathrm{Sin}\gamma)^2$ is a bilinear form taking values in the subspace Q for each u in R^2.

Let $K = K^* + K^{**}$ where

$$K^* \frac{1}{2\pi} \int_0^{2\pi} [C_4(\gamma) - C_3(\gamma) D_3(\gamma)/\omega] d\gamma$$

$$K^{**} \frac{1}{2\pi} \int_0^{2\pi} w^*(\gamma) J(0,0)(\mathrm{Cos}\gamma, \mathrm{Sin}\gamma)^2 d\gamma$$

and $w^*(\gamma)$ is the unique 2π-periodic solution on the equation

$$G_2(\gamma) - \omega \frac{dw^*}{d\gamma} + w^*(\gamma) A_Q = 0$$

The results of Chow and Mallet-Paret [1] tell us the following.

Let us assume $K \neq 0$. Then there is a unique periodic solution bifurcating from the origin, either $\alpha > 0$ (when $Kv < 0$, $v = \mathrm{Re}\lambda'(0)$) or $\alpha < 0$ (when $Kv > 0$). More precisely, in the original coordinates (u, y, α) with $u = (r\mathrm{Cos}\theta, r\mathrm{Sin}\theta)$ the solution has the form

$$r(t,\varepsilon) = \varepsilon r_o + 0(\varepsilon^2), \qquad r_o = |v/K|^{1/2}$$

$$\theta(t,\varepsilon) = \omega t + 0(\varepsilon)$$

$$y(t,\varepsilon) = 0(\varepsilon^2)$$

$$T(\varepsilon) = \text{period} = 2\pi/\omega + 0(\varepsilon)$$

where $\alpha = -\operatorname{sgn}(Kv)\varepsilon^2$.

We shall now write down in a condensed manner the coefficients that are necessary for computing the constant K. See [7] for details. We expand $h(x,y)$ up to third order terms

$$h(x,y) = \frac{1}{2!}[x^2+2xy+y^2]H_1 + \frac{1}{3!}[x^3+3x^2y+3xy^2+y^3]H_3 + 0(x^4+y^4)$$

where H_2 and H_3 are the column matrices

$$H_2 = \operatorname{col}[h_{xx}(0,0) \quad h_{xy}(0,0) \quad h_{yy}(0,0)]$$

$$H_3 = \operatorname{col}[h_{xxx}(0,0) \quad h_{xxy}(0,0) \quad h_{xyy}(0,0) \quad h_{yyy}(0,0)]$$

and we let $x = \varepsilon(r\operatorname{Cos}\gamma + y_t(0))$, $y = \varepsilon(r\operatorname{Cos}(\omega r+\gamma) + y_t(-r))$. We define the matrices

$$M(X,Y) = \begin{bmatrix} b^2X & -2abX & a^2X \\ b^2Y & -2abY \;\; 2bX & a^2Y-2a\omega Y \\ 0 & 2b\omega Y & \omega^2X-2a\omega Y \\ 0 & 0 & \omega^2Y \end{bmatrix}$$

$$N(X,Y) = \begin{bmatrix} bX & -aX \\ bY & \omega X-aY \\ 0 & \omega Y \end{bmatrix}$$

$$P(A,C) = \begin{bmatrix} Ab^3/3 & -Aab^2 & Aba & Aa^3/3 \\ C/3 & -Cab^2+A\omega b^2 & Ca^2b+2\omega AC & -Ca^3/3+\omega^2aA \\ 0 & C\omega b^2 & A\omega^2b-2ab\omega C & -A\omega^2a-Ca^2 \\ 0 & 0 & Cb\omega^2 & -Ca\omega^2 \;\; A\omega^3/3 \\ 0 & 0 & 0 & -C\;{}^3/3 \end{bmatrix}$$

Let $S = \text{col}[A_1 \quad A_2 \quad A_3 \quad A_4 \quad B_1 \quad B_2 \quad B_3 \quad B_4]$ and $F = [F_1 \quad F_2 \quad F_3 \quad F_4 \quad F_5]$.

Theorem 2. We have the following relations for the coefficients

$$(3.3) \qquad S = \text{col}[M(A,C) \quad M(C,-A)]H_2$$
$$F = P(A,C)H_3$$

and

$$K^* = \frac{\mu}{8b^3}[3(F_1+F_5)+F_3] - \frac{\mu^2}{16b^4\omega}[5A_1B_1+A_1B_3+A_2B_2+A_2B_4+A_3B_1+A_3B_3+A_4B_2+5A_4B_4].$$

Proof. The relations (3.3) are obtained by identifying terms of first and second order in ε once h is expanded in Taylor series. The value of K^* follows by integration. The terms involving F_2 and F_4 have mean value zero.

The linear functional $G_2(\gamma)$ is given by

$$G_2(\gamma) = \frac{2\mu}{b}[M_2(\gamma)\phi(0) + L_2(\gamma)\phi(-r)]$$

where M_2 and L_2 are homogeneous polynomials of second order in $(\text{Cos}\gamma, \text{Sin}\gamma)$. Their coefficients M_{2j} and L_{2j}, $j=1,3$, are given by

$$(3.4) \qquad \mathcal{M} = \text{diag}[N(A,C),\ N(A,C)]H_2$$

where $\mathcal{M} = [M_{21} \quad M_{22} \quad M_{23} \quad L_{21} \quad L_{22} \quad L_{23}]$. Finally, the bilinear functional $J(0,0)(\text{Cos}\gamma,\text{Sin}\gamma)^2 = x_0^Q J_2(\gamma)$ is given by

$$J_2(\gamma) = \frac{1}{2b^2}[\mathcal{B}_1 e^{2i\gamma} + \overline{\mathcal{B}}_1 e^{-2i\gamma} + \mathcal{B}_2]$$

where

$$\mathcal{B}_1 = \frac{J_1 - iJ_2 - J_3}{4}, \qquad \mathcal{B}_2 = \frac{J_1+J_3}{2}$$

Let us write $M_2(\gamma)$ and $L_2(\gamma)$ in complex notation (see $J_2(\gamma)$) and denote their coefficients by $\mathcal{A}_1$, $\mathcal{A}_2$ and $\mathcal{C}_1$, $\mathcal{C}_2$ respectively. We have

Theorem 3. The constant K^{**} is given by

$$K^{**} = \frac{4\mu}{b}[\operatorname{Re} A_1\bar{B}_1(2i\omega - A_Q)^{-1}X_o^Q(0) + C_1\bar{B}_1(2i\omega-A_Q)^{-1}X_o^Q(-r)]$$

$$- \frac{2\mu}{b}[A_2B_2 \quad A_Q^{-1}X_o^Q(0) + C_2B_2 \quad A_Q^{-1}X_o^Q(-r)]$$

Proof. We claim that the unique 2π-periodic solution of the equation

$$G_2(\gamma) + w^*(\gamma) \quad A_Q = \omega\frac{dw^*}{d\gamma}$$

is given by

$$w^*(\gamma) = \frac{1}{\omega}\int_o^{2\pi} G_2(\gamma+s)e^{-A_Q s/\omega}(e^{-2\pi A_Q/\omega} - I)^{-1}ds$$

Therefore

$$K^{**} = \frac{1}{2\pi\omega}\int_o^{2\pi}\int_o^{2\pi} G_2(\gamma+s)E(s,.)ds\ J_2(\gamma)d\gamma$$

where

$$E(s,\theta) = e^{-A_Q s/\omega}(e^{-2\pi\ A_Q/\omega} - I)^{-1}\ X_o^Q(\theta)$$

From (3.4) we obtain

$$K^{**} = \frac{1}{2\pi\omega}\int_o^{2\pi}\int_o^{2\pi}[\frac{2\mu}{b}\ M_2(\gamma+s)E(s,0) + L_2(\gamma+s)E(s,-r)]J_2(\gamma)\ ds\ d\gamma$$

The results follows then by integration.

The values

(3.6) $$(2i\omega-A_q)^{-1}\ X_o^Q(0) = (2i\omega-be^{-2i\omega r}-a)^{-1} + \frac{2\mu}{3}(C + 2iA)$$

$$(2i\omega-A_Q)^{-1}\ X_o^Q(-r) = e^{-2i\omega r}(2i\omega-be^{-2i\omega r}-a)^{-1}$$

$$- \frac{2\mu}{3}[(2iaA + Ca) - (\omega A - 2i\omega C)]$$

$$A_Q^{-1}X_o^Q(0) = -\frac{2\mu C}{\omega} - (a+b)^{-1}$$

$$A_Q^{-1}X_o^Q(-r) = \frac{2\mu(\omega A+aC)}{b\omega} - (a+b)^{-1}$$

are obtained by solving $\phi = (2id - A)^{-1}\psi$, $d = 0,\omega$ subject to the boundary condition $\dot{\phi}(0) = a\phi(0) + b\phi(-r)$ and then by replacing $\psi = X_0^Q = X_0 - X_0^P$, $X_0^P = \Phi\Psi(0)$.

Thus we have obtained the necessary ingredients for computing the value for K. We shall now apply such results to concrete situations.

4. Applications.

In this section we assume that Hopf bifurcation occurs in the equation

$$x'(t) = a(\alpha)x(t) + b(\alpha)x(t-r) + h(x(t),x(t-r),\alpha)$$

at $\alpha = 0$. This means that the curve $\Gamma(\alpha) = (a(\alpha),b(\alpha))$ crosses transversally a center curve C_k at $\alpha = 0$. See Figure 1 in section 2. We let $\lambda(\alpha)$ denote the eigenvalue which crosses the imaginary axis at $\alpha = 0$. The sign of $v = \mathrm{Re}\lambda'(0)$ can be easily determined. For instance, $v > 0$ when $\Gamma(\alpha)$ crosses downwards a center curve C_k with k even. As noted before, the constant K, which determines the direction of bifurcation when $K \neq 0$, only depends on the equation at $\alpha = 0$. This equation will be simply written

$$x'(t) = ax(t) + bx(t-r) + h(x(t),x(t-r)).$$

A. The equation $x'(t) = b(\alpha)x(t-r) + x(t)f(x(t-r),\alpha)$

From the discussion in section 2, it turns out that Hopf bifurcation will occur at $\alpha = 0$ if and only if

$$(4.1) \quad b(0) = (-1)^{N+1}(\tfrac{\pi}{2} + N\pi)/r, \quad b'(0) \neq 0$$

for some nonnegative integer N. Moreover

$$\lambda(0) = i\omega, \quad \omega = |b_N|, \quad b_N = b(0)$$

$$v = \frac{b'(0)b_N r}{1+b_N^2}$$

From now on we let N arbitrary but fixed. We have from Lemma 1

$$\Psi(0) = 2\mu \begin{bmatrix} A \\ C \end{bmatrix} \; ; \quad A = 1, \quad C = \omega r, \quad \mu = \frac{1}{1+\omega^2 r^2}$$

Let $h(x,y) = xf(y)$. Then all their second and third order derivatives vanish at the origin but $h_{xy}(0,0) = f'(0)$ and $h_{xyy}(0,0) = f''(0)$, unless f', f'' vanish too. Thus only the terms in the second column of the matrix $M(A,C)$ as well as the terms in the third column of $P(A,C)$ will contribute. From theorem 2 we obtain

$$(4.2) \qquad K^* = \frac{\mu f''(0)}{8}$$

It follows from (3.4) and (3.5) that

$$A_1 = \frac{M_{21} - iM_{22} - M_{23}}{4} = -\frac{\omega}{4}(\omega r + i)f'(0),$$

$$A_2 = \frac{M_{21} + M_{23}}{2} = \frac{\omega^2 r}{2} f'(0)$$

$$C_1 = \frac{b}{4}(1 - i\omega r)f'(0),$$

$$C_2 = \frac{b}{2}f'(0)$$

$$B_1 = -\frac{ib\omega}{2}f'(0)$$

$$B_2 = 0$$

By using the values (3.6) we obtain from theorem 2

$$(4.3) \qquad K^{**} = -\frac{\mu}{20\omega}(3\omega r + (-1)^{N} 1)(f'(0))^2$$

The value for $K = K^* + K^{**}$ is thus computed from (4.2) and (4.3)

Example. Wright's equation

$$x'(t) = \sigma x(t-r)(1+x(t))$$

is of the type considered. We let $\sigma = b(\alpha) = \alpha + b_N$ and $h(x,y,\alpha) = (b_N+\alpha)xy$. Then

$$K = K^{**} = -\frac{\mu\omega}{20}(3\omega r + (-1)^{N}\ 1)$$

This result agrees with the one obtained by Chow and Mallet-Paret [1].

B. The equation $x'(t) = b(\alpha)x(t-r) + x(t-r)f(x(t),\alpha)$

By symmetry considerations, the value for K^{**} coincides with (4.3). The only difference being with K^* which involves h_{xxy} rather than h_{xyy}. Therefore

$$K = \left(\frac{\mu r b_N}{8}\right) f''(0) + K^{**}$$

where K^{**} is given by (4.3).

Example. The equation

$$x'(t) = \sigma x(t-r)(1-x^2(t))$$

is of the type considered. Here $f(x) = -b_n x^2$ implies $f'(0) = 0$, $f''(0) = -2b_N$ and $b(\alpha) = \alpha + b_N$ implies $b'(0) = 1$. Therefore

$$K = K^* = -\frac{\mu r\omega^2}{4}, \quad v = \mathrm{Re}\lambda'(0) = \mu b_N$$

We obtain then that the bifurcating solution has the form

$$x(t) = \left[\frac{4(\sigma-b_N)}{rb_N}\right]^{\frac{1}{2}} \mathrm{Cos}\, b_N t + 0(\sigma-b_N)$$

Thus bifurcation occurs to the right of b_N for N odd, and the left for N even. Only the solution at $N = 0$ is stable, and it has the form

$$x(t) = -\left[\frac{2\sigma r+\pi}{\pi r}\right]^{\frac{1}{2}} \mathrm{Cos}\,\frac{\pi t}{2} + 0(\sigma+\frac{\pi}{2})$$

C. <u>The equation $x'(t) = b(\alpha)x(t-r) + h(x(t-r),\alpha)$</u>.

For this equation we just simply state the results. We have

$$K^* = \frac{\mu}{8}(b_N r h_{yyy}(0,0) + 2\mu r h_{yy}^2(0,0)$$

Since all derivatives with respect to x vanish for $h(y)$, we obtain

$$K^{**} = \frac{4\mu}{b}(\mathrm{Re}C_1\overline{B}_1(2i\omega-A_Q)^{-1}X_o^Q(\theta r)) - \frac{2\mu}{b}(C_2B_2A_Q^{-1}X_o^Q(-r))$$

$$= \frac{\mu}{20}(3\omega^2 r + 2b_N + 2r\mu\omega^2)h_{yy}^2(0,0)$$

It follows that for $h(y)$ begining with third order terms the value for $K = K^* + K^{**}$ will be

$$\frac{\mu b_N r}{8}\, h_{yyy}(0,0)$$

and

$$\overline{K} = Kv = \frac{\mu 2\omega 2_r}{8}\, h_{yyy}(0,0)$$

will only depend on the sign of the third derivative of h at $(0,0)$. This implies that bifurcation will occur in one side for any N.

D. <u>The equation $x'(t) = b(\alpha)x(t-r) + \sum_{i+j=3} s_{ij}x^i(t)x^j(t-r)$</u>.

This equation was discussed by Kazarinoff et al [6] in order to illustrate their formulae obtained for Hopf bifurcation by using the center manifold technique. Their results are restricted to the case $\Gamma(\alpha)$ crossing the center stable curve C_o. With the integral averaging method such restriction os easily removed. The computations are quite simple because only third order derivatives are involved. This implies that $K^{**} = 0$. From theorem 2 it follows that

$$F_1 = 2b_N^3 D, \quad F_3 = 2b_N\omega^2 E + 2b_N\omega G, \quad F_5 = 2\omega^4 rH$$

where

$$D = s_{30}, \quad E = s_{12}, \quad G = s_{21}, \quad H = s_{03}$$

Therefore

$$(4.6) \qquad K = K^* = \frac{\mu}{4}(3D + E + G + 4b_N rH)$$

<u>Example</u>. The equation

$$x'(t) = -2x(t-r) + \beta x^3(t)$$

comes from a example with two delays discussed in [8]. With the change of variables $t = \tau r$, $x(t) = y(\tau)$, the equation becomes

$$y'(t) = -2ry(t-1) + r\beta y^3(t)$$

which is of the type considered. Here the delay r is taken as the bifurcation parameter. From (4.6)

$$K = \frac{3\mu r_N \beta}{4}$$

where $b_N = -2r_N$. Therefore

$$\overline{K} = Kv = -\frac{3\mu^2 b_N^2}{8}\beta$$

implies that bifurcation occurs to the right of b_N for $\beta > 0$, and the left for $\beta < 0$. The solution is stable for $N = 0$ if $\beta > 0$, otherwise unstable.

E. <u>The equation $x'(t) = a(\alpha)x(t) + b(\alpha)x(t-r) + h(x(t-r),\alpha)$</u>.

For this case all the terms involving a in the matrices $M(A,C)$, $P(A,C)$ and $N(A,C)$ will have to be considered when computing K from theorems 2 and 3. It follows from lemma 1 that

$$A = 1-ar, \qquad C = \omega r, \qquad \mu = (A^2+C^2)^{-1}$$

After computations, it turns out that

$$(4.7) \qquad K^* = \frac{\mu}{8b}(-aA+\omega C)h_{yyy}(0,0) - \frac{\mu^2}{4\omega b^2}[a\omega(A^2-C^2)(a^2-\omega^2)AC]h^2_{yy}(0,0)$$

Since only the derivatives of h with respect to y are to be taken into account, we obtain from (3.4), (3.5) and (3.6) that $A_1 = A_2 = 0$ and

$$C_1 = \frac{be^{i\theta}(A-iC)}{4} h_{yy}(0,0) \ ,$$

$$C_2 = \frac{-aA+\omega C}{2} h_{yy}(0,0)$$

$$B_1 = \frac{b^2 e^{2i\theta}}{4} h_{yy}(0,0)$$

$$B_2 = \frac{b^2}{b} h_{yy}(0,0)$$

where $\theta = \omega r$, $i\theta = ar + bre^{-i\theta}$. Then

(4.8) $$K^{**} = [\frac{\mu r^2 b^2}{4\Delta}[a(A\mathrm{Cos}3\theta - C\mathrm{Sin}3\theta) + b(A(2\mathrm{Cos}2\theta - \mathrm{Cos}\theta) - C(2\mathrm{Sin}2\theta - \mathrm{Sin}\theta)]$$

$$+ \frac{\mu^2}{b}[a(A^2+C^2) + 3\omega AC]]h_{yy}^2(0,0)$$

where

$$\Delta = a^2 r^2 + r^2 ab(2\mathrm{Cos}\theta - \mathrm{Cos}2\theta) + b^2 r^2(5 - 4\mathrm{Cos}\theta)$$

The crossing speed is computed from the characteristic equation $\lambda = a(\alpha) + b(\alpha)e^{-\lambda r}$. It turns out that

(4.9) $$v = \mathrm{Re}\lambda'(0) = \frac{\mu}{b}[ba'(0)(1-ar) + \omega^2 b'(0)r]$$

where $a = a(0)$ and $b = b(0)$.

The constant $\overline{K}$ is thus computed from (4.7), (4.8) and (4.9)

Example. The retarded differential equation

$$u'(t) = -\sigma u(t) + e^{-u(t-r)}$$

is a proposed model of Lasota and Wazewska [2] on the survival of red blood cells. The equilibrium points of this equation are the real roots of the equation

$$\sigma\beta = e^{-\beta}$$

The behavior of solutions near an equilibrium point is more

conveniently described by the equation

$$x'(t) = -\frac{e^{-\beta}}{\beta}x(t) - e^{-\beta}x(t-r) + h(x(t-r),\beta)$$

$$h(y) = e^{-\beta}\sum_{n=2}^{\infty}\frac{(-1)^n y^y}{n!}$$

$$x(t) = u(t)-\beta$$

With the substitution $t = \tau r$, $x(t) = y(\tau)$, this equation becomes

$$y'(t) = \overline{a}(r)y(t) + \overline{b}(r)y(t-1) + rh(y(t-1),\beta)$$

$$\overline{a}(r) = -\frac{e^{-\beta}}{\beta}r$$

$$\overline{b}(r) = -e^{-\beta}r$$

which is of the type considered.

Chow [2] by using fixed point techniques over certain cones, has shown the existence of periodic solutions for $0 < \sigma < e^{-1}$ and $r \geq R$ sufficiently large. We can discuss this equation in terms of Hopf bifurcation by taking the delay as the bifurcation parameter, that is, $r = \alpha + r_k$ where r_k is a bifurcation point.

For a given value of σ (or β), Hopf bifurcation will occur whenever the curve $\Gamma(r) = (-e^{-\beta}r/\beta, -e^{-\beta}r)$ intersects a center curve C_k in the $\overline{a}\,\overline{b}$-plane. This is ilustrated in Figure 2. The curve Λ gives the values of σ and β such that $\sigma\beta = e^{-\beta}$. The intersection of Γ with the center curves will define the sequence of delays r_k characterized in Lemma 2. We should observe that for $\sigma > e^{-1}$ the origin is asymptotically stable and bifurcation can not occur.

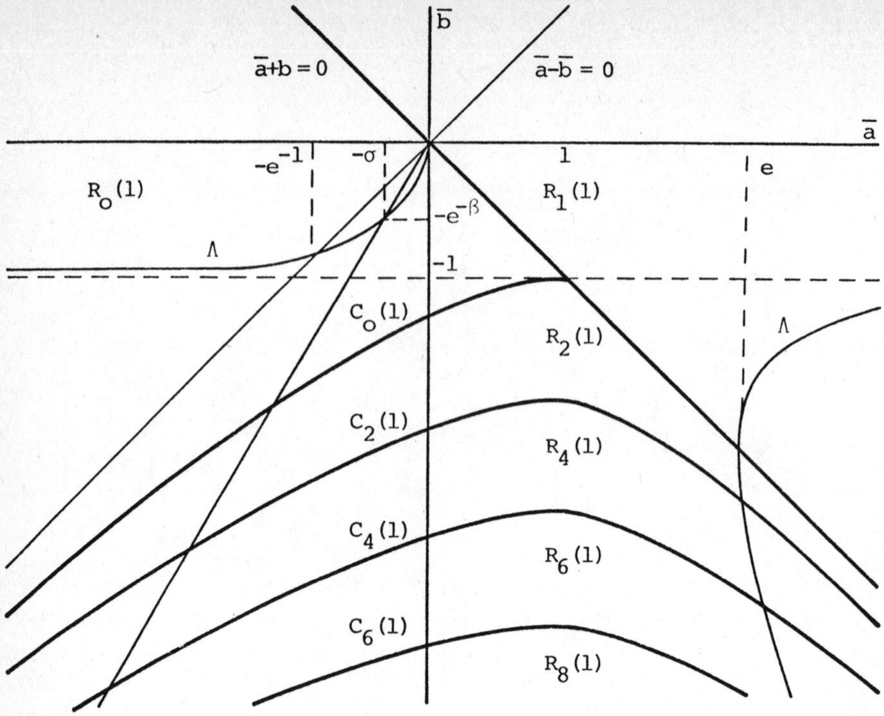

Figure 2

The direction of bifurcation in thus determined from (4.7), (4.8) and (4.9), for the case $a(\alpha) = -e^{-\beta}(\alpha+r_k)/\beta$ $b(\alpha) = -e^{-\beta}(\alpha+r_k)$ with $k \geq 0$ integer. We should observe that the value for K will be sufficient for such purpose because $v = \mathrm{Re}\lambda'(0)$ will have a definite negative sign

$$v = \frac{\mu}{b(0)}[2\sigma e^{-\beta} r_k(1+\sigma r_k) + e^{-\beta}\omega^2 r_k]$$

where $\omega^2 = \sigma^2(\beta^2-1)$, $b(0) = -e^{-\beta}$. Our computer numerical studies reveal the following facts:

i) the values of K for a given σ decrease in absolute value as the delay r_k increases,

ii) there is a value of σ for which K changes its sign. This value is less than $e^{-1/2}$

Thus higher order averaging has to be considered for $K = 0$.

Acknowledgment. We would to acknowledge Professor Mallet-Paret for submitting our paper to the meeting. Thanks John.

REFERENCES

[1] - CHOW, S.-N. and MALLET-PARET, J., Integral averaging and bifurcation, J. Diff. Equations, 26(1977), 112-158.

[2] - CHOW, S.-N., Existence of periodic solutions of autonomous functional differential equations, J. Diff. Equations, 15 (1974), 350-378.

[3] - COCKBURN, B., On the equilibrium points of the equation x'(t) F(x(t),x(t-r)). Proc. IV Lat. Ame. Sch. Math. Tech. Rep., nº 2 (1979), Departamento de Matematicas, Universidad Nacional de Ingenieria, (1979).

[4] - HALE, J.K., Theory of Functional Differential Equations, Springer-Verlag, (1977).

[5] - HALE, J.K., Behavior near constant solutions of functional differential equations, J. Diff. Equations, 15(1974), 278-294.

[6] - KAZARINOFF, N.D., WAN, Y.H. and DRIESCHE, P. Van der, Hopf bifurcation and stability of periodic solutions of differential--difference and integro-differential equations, J. Inst. Math. Appl., (1978), 461-477.

[7] - RUIZ-CLAEYSSEN, J., The integral averaging bifurcation method and the general one delay equation, Tech. Rep. nº 3, (1979), Departamento de Matematicas, Universidad Nacional de Ingenieria.

[8] - RUIZ-CLAEYSSEN, J., Effects of delays on functional differential equations, J. Diff. Equations, (1976), 404-440.

MODULI AND BIFURCATIONS; NON-TRANSVERSAL INTERSECTIONS OF INVARIANT MANIFOLDS OF VECTORFIELDS

by Floris Takens*

1. Introduction.

We consider smooth vectorfields on a manifold M. It is known [1], [2] that generically such vectorfields have only hyperbolic singularities and transversal intersections of stable and unstable manifolds; for the definitions see below. If one considers however one-parameter families, or arcs of such vectorfields, then, for a discrete set of parameter values, one expects to get a non-hyperbolic singularity or a non-transversal intersection of a stable and an unstable manifold. These are special cases of the so called *bifurcation* values (of the parameter); we speak of a bifurcation or a bifurcation value of the parameter if the topological equivalence class, as function of the parameter, is not locally constant. In this paper we are concerned with such vectorfields, occuring in generic one-parameter families, which have a non-transversal intersection of invariant manifolds. In [3] we discussed a kind of such tangencies, or non-transversal intersections, in dimension 4. We saw that some real function of the eigenvalues of the singularities, whose stable and unstable manifolds have a tangency, appears as a topological invariant. Hence in a neighbourhood of this vectorfield there are uncountably many topological equivalence classes. In fact, four is the lowest dimension where such examples occur in generic one-parameter families of vectorfields. In this paper we want to show that in some sense this real invariant is the only one if we restrict our attention to a small neighbourhood of the orbit of tangency. So, although we have uncountably many equivalence classes, we have a nice parametrization of them; such

*) The author acknowledges financial support of the Volkswagen foundation for a stay at I.H.E.S. (Bures-s-Yvette; France) where a part of this research was carried out.

a parametrization with one (or more) real variable(s) is also called *a modulus* (or *moduli*). If we do not restrict our attention to a small neighbourhood of the orbit of tangency, the situation is much harder to describe and our results indicate that in some cases there may even be no finite dimensional parametrization of the toplogical equivalence classes.

An analogous situation has been studied by Palis [4], de Melo [5] and Newhouse, Palis, Takens [6] for diffeomorphisms where there is a tangency between a stable and an unstable manifold. These results are also applicable to vectorfields for the case of tangency of stable and unstable manifolds of closed (non-constant) orbits. It should be mentioned that in our example several of the eigenvalues of the vector field at the relevant singularities are non-real; this is essential: by [3] and [7], no modulus can occur if all the eigenvalues are real.

In order to describe the results more precisely we need some definitions. Let M be a compact manifold without boundary. $\mathfrak{X}(M)$ denotes the set of smooth, is C^∞, vector fields on M. We say that two vector fields $X, Y \in \mathfrak{X}(M)$ are *topologically equivalent* if there is a homeomorphism $h : M \to M$ such that h maps integral curves of X direction preserving to integral curves of Y. If o_X is an integral curve of X and o_Y an integral curve of Y then we say that *X at o_X is topologically equivalent with Y at o_Y* if there are neighbourhoods U_X and U_Y of the closures of o_X and o_Y respectively and a homeomorphism $h : U_X \to U_Y$ which maps o_X to o_Y and which maps integral curves of $X|U_X$ direction preserving to integral curves of $Y|U_Y$. For $X \in \mathfrak{X}(M)$ and $p \in M$ we call p a *singularity* of X if $X(p) = 0$. Such singularity p is called a *hyperbolic singularity* of X if, in local coordinates $x_1, \ldots, x_n$ with $x_i(p) = 0$ in which X has the form $X = \sum_{i=1}^{n} X_i(x_1, \ldots, x_n)\frac{\partial}{\partial x_i}$, all the eigenvalues of $(\frac{\partial X_i}{\partial x_j}(0))$ have non-zero real parts. If p is such a hyperbolic singularity of the vector field X, we define its *stable manifold* $W^s(p)$ or

$W^s(p,X)$, as

$$W^s(p) = \{q \in M \mid \lim_{t\to\infty} X_t(q) = p\},$$

where $t \to X_t(q)$ is the X integral curve starting in q; $W^s(p)$ is a smoothly and injectively immersed submanifold of M whose dimension equals the number of eigenvalues (with multiplicity) of X, or dX, at p with negative real part. The unstable manifold $W^u(p)$, or $W^u(p,X)$, is defined in the same way except we take $\lim_{t\to\infty} X_{-t}(q)$. From [1] and [2] we know that there is a residual set $R \subset \mathcal{X}(M)$ such that for each pair of singularities, $p,q \in M$ of $X \in R$, p and q are hyperbolic and $W^u(p)$ and $W^s(q)$ have only transversal intersections; $p = q$ is not excluded. So for arcs of vector fields, i.e., vector fields X_μ depending on $\mu \in R$ such that $X_\mu(x)$ depends smoothly on (x,μ), we have generically that there is a discrete set $B \subset R$ such that $\bar{\mu} \in B$ if and only if either one of the singularities of $X_{\bar{\mu}}$ is non-hyperbolic or there is one orbit of non-transversal intersection of a stable and an unstable manifold.

We say that an orbit γ of non-transversal intersection of invariant manifolds W^u and W^s of the vector field X is an *orbit of quasi-transversal intersection* if there are local coordinates $x_1,\ldots,x_n$ near some point of γ such that

$$X = \frac{\partial}{\partial x_1}, \qquad \gamma = \{(x_1,\ldots,x_n) \mid x_2 = \ldots = x_n = 0\}$$

$$W^s = \{(x_1,\ldots,x_n) \mid x_2 = \ldots = x_{n-s+1} = 0\}$$

$$W^u = \{(x_1,\ldots,x_n) \mid x_{u+2}=\ldots=x_n=0 \; , \; x_2 = f(x_{n-s+2},\ldots,x_{u+1})\}$$

where:

$n = \dim(M)$;

$s = \dim(W^s)$;

$u = \dim(W^u)$;

f is a homogeneous quadratic function, if $n-s+2 > u+1$, one should read $x_1 = 0$. It is not hard to see that generic arcs of vector fields have the property that for any $\bar{\mu}$, such that X_μ is has an orbit of tangency of a stable and an unstable manifold, there is only one such orbit and at this orbit, the intersection is quasi-transversal. This fact is more or less well known, a proof appears in [6]. In fact we shall use the above description mainly in the case $n = 4$, $u = s = 2$.

Let now $X \in \mathcal{X}(M)$ be a vector field on M with hyperbolic singularities $p,q \in M$, and let $\gamma \in W^u(p,X) \cap W^s(q,X)$ be an orbit of quasi-transversal intersection of $W^u(p,X)$ and $W^s(q,X)$. Then there is a neighbourhood U of X in $\mathcal{X}(M)$ such that:

(a) there are continuous maps $P,Q: U \to M$ such that $P(X) = p$, $Q(X) = q$ and such that $P(X')$, $Q(X')$ are hyperbolic singularities of X';

(b) there is a codimension one submanifold $\Sigma \subset U$ such that $X' \in \Sigma$ if and only if X' has an orbit $\Gamma(X')$ of quasi-transversal intersection of $W^u(P(X'),X')$ and $W^s(Q(X'),X'))$ which is near γ, i.e., $\Gamma(X) = \gamma$.

From the results announced in [7] it follows that if all the eigenvalues of dX at both p and q are real and distinct and if U is small enough, then Σ has an open and dense subset $\tilde{\Sigma} \subset \Sigma$ such that if X' and X'' are in the same connected component of $\tilde{\Sigma}$, then X' at $\Gamma(X')$ is topologically equivalent with X'' at $\Gamma(X'')$. In [3] we proved that this is not true in general if the eigenvalues of dX at p and q are no longer real. To be more precise, consider a vector field X on an n-manifold M, $n = 4$ with two hyperbolic singularities p and q such that

1. dX has in p two non-real eigenvalues with negative real part, $-a \pm ib$ $(a,b > 0)$ and $(n-2)$ eigenvalues with

positive real part;

2. dX has in q two non-real eigenvalues with positive real part, $\alpha \pm i\beta$ $(\alpha,\beta > 0)$ and $(n-2)$ eigenvalues with negative real part;

3. $W^u(p)$ and $W^s(q)$ have one orbit γ of quasi-transversal intersection.

In this case we can find a neighbourhood U of X in $\chi(M)$ as above with $P,Q:U \to M$, $\Sigma \subset U$ and Γ. By [8] there is an open and dense subset $\tilde{\Sigma} \subset \Sigma$ containing vector fields $X' \in \Sigma$ for which X' can be locally linearized near $P(X')$ and $Q(X')$ by a change of coordinates which is at least C^2. We denote by $a,b,\alpha,\beta:U \to R$ the functions such that $a(X') \pm i\cdot b(X')$ are the contracting eigenvalues of dX' at $P(X')$ and $\alpha(X') \pm i\beta(X')$ are the expanding eigenvalues at $Q(X')$. Finally $m:U \to R$ is defined by

$$m(X') = \frac{a(X')}{b(X')} \cdot \frac{\beta(X')}{\alpha(X')} .$$

In the above terminology the main result in [3] was that for $X',X'' \in \Sigma$ and $m(X') \neq m(X'')$, X' and X'' are not topologically equivalent. In this paper we prove:

<u>Theorem</u>. If $X',X'' \in \tilde{\Sigma}$ and $m(X') = m(X'')$ then X' at $\Gamma(X')$ is topologically equivalent with X'' at $\Gamma(X'')$.

In the final section, we discuss some generalization and state some conjectures and open problems.

It should be pointed out that there is a close relation between the results in this paper and those of de Melo [5] showing that in dimension two the topological invariant for tangencies, of stable and unstable manifolds of diffeomorphisms (as defined in [4]), is "complete" if we restrict to a neighbourhood of the orbit with tangency and may not be complete as a topological invariant for the global bifurcation. In de Melo's paper even the exact number of parameters, or moduli, needed to parametrize the topological equivalence classes of all nearby bifurcations, is determined in some cases.

2. Invariant foliations near singularities.

Since all the singularities of vector fields we shall use in this paper have the property that they can be C^2-linearized (see the introduction) we shall restrict in this section to linear vector fields on R^n. We consider such a linear hyperbolic vector field X on R^n which has a 2-dimensional stable manifold W^s, with non--real stable eigenvalues $-a \pm i\cdot b$, $a,b > 0$. We also assume that for $x \in W^s$, $\langle x,X(x)\rangle < 0$ and for $x \in W^u$, $\langle x,X(x)\rangle > 0$; W^u denotes the unstable manifold of X (of dimension $n-2$) and $\langle\ ,\ \rangle$ the standard inner product on R^n.

We denote by π_u,π_s the canonical linear projections of R^n on W^u, respectively W^s, i.e., $\pi_u(W^s) = \pi_s(W^u) = 0$. Let S be a smooth co-dimension one submanifold transversal to X and W^u such that $S \cap W^u$ is a fundamental domain of $X|W^u$. We shall consider C^o co-dimension one foliations $\mathcal{F}$ of $R^n\backslash W^u$ or of $U\backslash W^u$, U a neighbourhood of W^u, such that

1^e the leaves of $\mathcal{F}|R^n - (W^u \cup W^s)$ are smooth;

2^e X is everywhere tangent to the leaves of $\mathcal{F}$; $\mathcal{F}|W^s - \{0\}$ consists of integral curves of X.

We call such foliations *orbit-unstable* foliations.

We want to show that certain foliations of S can be extended to orbit-unstable foliations. For an open map $f:S \to W^s$, mapping $S \cap W^u$ to 0, we define the induced partition $\mathcal{F}_f$ of a neigehbourhood of W^u by the following requirements:

(a) if p belongs to a member F of $\mathcal{F}_f$, then also the X-integral curve through p;

(b) $p,q \in S$ belong to the same member of F_f if and only if $f(p)$ and $f(q)$ belong to the same integral curve of $X|W^s$;

(c) $p \in S$ and $f(p)$ belong to the same member of F_f.

Proprosition (2.1). For a differentiable (at least C^1) map $f:S \to W^s$ as above, which has in each point of $S \cap W^u$ the same 1-jet as $\pi_s|S$, the partition F_f induces an orbit-stable foliation on the complement of W^u.

Proof: Let $\tilde{S}$ be another smooth co-dimension one submanifold, transversal to X and transversally intersecting W^u in a fundamental domain. For $f:S \to W^u$ as in the assumptions of the proposition, we want to show there is also $\tilde{f}:\tilde{S} \to W^u$ as in these same assumptions such that the partitions F_f and $F_{\tilde{f}}$ are equal. Let $P:\tilde{S} \to S$ and $T:\tilde{S} \to R$ be defined as follows: for $\tilde{x} \in \tilde{S}$, $P(\tilde{x})$ is the intersection of the X integral curve through $\tilde{x}$ with S and $(X_{T(\tilde{x})})(\tilde{x}) = P(\tilde{x})$; these maps are of course only defined in a neighbourhood of $\tilde{S} \cap W^u$, also the equallity of F_f and $F_{\tilde{f}}$ should be understood as equallity in some neighbourhood of W^u. We define now $\tilde{f}:\tilde{S} \to W^s$ by $\tilde{f}(\tilde{x}) = (X_{-T(\tilde{x})})(f \circ P(\tilde{x}))$; this has clearly the required properties.

From the above argument it follows that without restricting generality, we may assume that S has some special form: we assume that $\pi_u(S) = S$.

Let Θ be an orbit of $X|W^s$, $\Theta \neq 0$. We want to show that the leaf of the foliation, determined by f, containing Θ is a C^0 manifold which is differentiable outside W^s. We parametrize $\overline{\Theta}$ by arclength starting in 0 (so we now think of $\Theta(\sigma)$ as a point of this orbit, $\Theta(0) = 0$). $f^{-1}(\overline{\Theta})$ can be written as $\bigcup_{r \in S \cap W^u} \ell(r)$ where $\ell(r)$ is the curve $\pi^{-1}(r) \cap f^{-1}(\overline{\Theta})$ which can be parametrized so that

$$(\ell(r))(\sigma) = (r, \Theta(\sigma) + g(r,\sigma)) \in W^u \oplus W^s, \quad \sigma \geq 0,$$

with $g(r,\sigma) = 0(\sigma)$ uniformly in r. The leaf through $f^{-1}(\Theta)$ consists of the X-integral curves through $f^{-1}(\Theta)$; outside W^s this is clearly a differentiable manifold. Except for Θ it consists of lines $X_{-t}(\overset{o}{\ell}(r))$ which can be parametrized as

$$(X_{-t}(r), \Theta(\sigma) + G(t,r,\sigma)), \ \sigma > 0.$$

For $t \to \infty$, $X_{-t}(r)$ and $G(t,r,\sigma)$ go to zero, uniformly in r,σ. For $G(t,r,\sigma)$ this follows from the special form of X and $g(r,\sigma) = o(\sigma)$. From this parametrization of the curves $X_{-t}(\overset{o}{\ell}(r))$, and hence of the leaf through $f^{-1}(\Theta)$ it follows that this leaf is a C^o manifold containing Θ.

Proposition (2.2). Let X and X' be linear vector fields on R^n satisfying all the assumption which we made in this section on X. Let S, S' be smooth co-dimension one submanifolds transversal to X, respectively to X', and to $W^u(X)$, $W^u(X')$ so that $W^u(X) \cap S$, $W^u(X') \cap S'$, is a fundamental domain for $X|W^u(X)$, $X'|W^u(X')$. $D \subset W^u(X)$, $D' \subset W^u(X')$, denotes the closed disc of those points whose forward X, X', integral curve meets S, S'. Let F, F', be orbit-unstable foliations of X, X'. Then any homeomorphism $h: S \to S'$, mapping leaves of $F|S$ to leaves of $F'|S'$, can be extended to a topological equivalence $H: U \to U'$ between $X|U$ and $X'|U'$; U, U'; neighbourhoods of D, D', in R^n.

Proof. We first define the notion of a Liapunov function for X. This is a smooth function $L: R^n \to R$ such that $L(0) = 0$ and $X(L) > 0$ except in 0. For example $L(x) = ||\pi_u(x)||^2 - ||\pi_s(x)||^2$ is a Liapunov function.

We choose for X and X' Liapunov functions L, L' such that $L(S) = L'(S') = 1$. Now we define H on a neighbourhood of D, except on $W^s - \{0\}$, by the following two requirements.

H maps integral curves of X to integral curves of X';

$L = L' \circ H$.

Since h maps leafs of $F \cap S$ to leaves of $F' \cap S'$, H maps leaves of F to leaves of F'. Hence H has a unique extension to W^s.

3. Joined linearizations of invariant foliations.

Let us assume, as in the introduction, that X is a vector field on a 4-dimensional manifold M with hyperbolic singularities p, q, satisfying the conditions 1, 2, and 3 in the introduction. Assume furthermore that X admits, both near p and near q, a C^2--linearization. Then there are projections on $W^s(p)$ and $W^u(q)$, defined by these linearizations, which we denote by $\tilde{\pi}_s$ and $\tilde{\pi}_u$; they are defined on some neighbourhood of $W^u(p)$ and $W^s(q)$ respectively, are at least C^2 and commute with the action of the vector field. We take a smooth co-dimension one (3-dimensional) submanifold S, transversal to the orbit γ of non-transversal intersection of $W^u(p)$ and $W^s(q)$. We denote $\bar{\pi}_s = \tilde{\pi}_u|S$.

From the previous section it is clear that $\check{\pi}_s : S \to W^s(p)$ and $\check{\pi}_u : S \to W^u(q)$ are C^1 maps with, along $S \cap W^u(p)$ respectively $S \cap W^s(q)$ the same 1-jet as $\bar{\pi}_s$, $\bar{\pi}_u$, then there is an orbit--unstable foliation F_p for X at p and an orbit-stable foliation F_q for X at q such that the leaves of $F_q \cap S$, $F_q \cap S$, are inverse images under the map $\check{\pi}_s$, $\check{\pi}_u$, of X-integral curves in $W^s(p)$, $W^u(q)$. An *orbit-stable* foliation of X is an orbit-unstable foliation of $-X$; see section 2.

We would like to be able to choose $\check{\pi}_s$ and $\check{\pi}_u$ so that, for appropriate coordinates in S near $\gamma \cap S$, both $\check{\pi}_s$ and $\check{\pi}_u$ are linear with respect to the linear structure on $W^s(p)$ and $W^u(q)$, induced by the above linearizations. In fact we need somewhat less: we only need that the foliations, defined by $\check{\pi}_s$ and $\check{\pi}_u$ in S are "linear", i.e., like foliations defined by linear projection. (Note that it is easy to linearize one foliation since X is linearizable near p and q, but that it is harder to linearize two of them simultaneously!)

Proposition (3.1). There are C^1-maps $\check{\pi}_s$, $\check{\pi}_u$ from S to $W^s(p)$, $W^u(q)$ as above and there are C^1-maps $p_s: S \to W^s(p)$, $p_u: S \to W^u(q)$, such that p_s and $\check{\pi}_s$ define the same foliation in $S \setminus (S \cap W^u(p))$ and also p_u and $\check{\pi}_u$ define same foliation in $S \setminus (S \cap W^s(q))$ and such that there are C^1-coordinates x_1, x_2, x_3 on S near $S \cap \gamma$, with respect to which p_u and p_s are both linear.

Proof. We make the local coordinates x_1, x_2, x_3 so that

(a) $S \cap \gamma = 0$;

(b) $\check{\pi}_s^{-1}(0) = \bar{\pi}_s^{-1}(0)$ is the x_2-axis and $\check{\pi}_u^{-1}(0) = \bar{\pi}_u^{-1}(0)$ is the x_1-axis.

Let $L \subset S$ be a differentiable 2-manifold containing $\bar{\pi}_s^{-1}(0)$ and $\bar{\pi}_u^{-1}(0)$, at least near $\gamma \cap S$. We want to arrange that $L = \{x_3 = 0\}$. For this we need to "modify" $\bar{\pi}_s$ and $\bar{\pi}_u$.

First we note that if λ is a (smooth) function on S then the map $p_s^\lambda: S \to W^s(p)$, defined by $p_s^\lambda(r) = X_{\lambda(r)}(\bar{\pi}_s(r))$ for $r \in S$, induces the same foliation in S as $\bar{\pi}_s$.

Let $e^1(p)$, $e^2(p)$ be a basis of $T_p(W^s(p))$. Then there is a C^1-function $\lambda: S \to R$ such that for each $r \in \bar{\pi}_s^{-1}(0)$, $d(p_s^\lambda)(T_r(L))$ is the linear subspace generated be $e^1(p)$. Also we choose a basis $e^1(q)$, $e^2(q)$ of $T_q(W^u(q))$ and a C^1-function $\sigma: S \to R$ such that for each $r \in \bar{\pi}_s^{-1}(0)$, $d(p_u^\sigma)(T_r(L))$ is the linear subspace of $T_q(W^u(q))$ spanned by $e^1(q)$. It follows that there are C^1-coordinates x_1, x_2 on L such that

$$d(p_s^\lambda)\left(\frac{\partial}{\partial x_1}\right) = e^1(p) \quad \text{along the } x_2\text{-axis and}$$

$$d(p_u^\sigma)\left(\frac{\partial}{\partial x_2}\right) = e^1(q) \quad \text{along the } x_1\text{.axis.}$$

Once we have this, it is not hard to extended x_1 and x_2 to a neighbourhood of L and to define the third coordinate x_3 such that $x_3|L \equiv 0$ and such that

$$d(p_s^\lambda)(\frac{\partial}{\partial x_3}) = c_s.e^2(p) \text{ along the } x_2\text{-axis and}$$

$$d(p_u^\sigma)(\frac{\partial}{\partial x_3}) = c_u.e^2(p) \text{ along the } x_1\text{-axis; } c_s \text{ and } c_u$$

are constants.

Using these coordinates, we define p_s and p_u to be the linear maps, having along the x_2-axis respectively the x_1-axis the same 1-jet as p_s^λ, p_u^σ. Finally $\check{\pi}_s$ and $\check{\pi}_u$ are defined by

$$\check{\pi}_s(r) = X_{-\lambda(r)}(p_s(r)) \text{ and } \check{\pi}_u(r) = X_{-\sigma(r)}(p_u(r)).$$

4. Topological classification of pairs of spiral foliations.

We say that F is a *linear spiral foliation* of R^3 if, for some basis v_1, v_2, v_3, each leaf of F has the form

$$\{A.e^{-\frac{\gamma}{2\pi}t}.\text{sint}.v_1 + A.e^{-\frac{\gamma}{2\pi}t}.\text{cost}.v_2 + s.v_3 \mid s,t \in \mathbb{R}\}$$

where $A > 0$ is a constant depending on the leaf and $\gamma > 0$ is independent of the leaf. In this case the axis $\ell(F)$ is the line through 0 in the v_3 direction and the contraction coefficient $c(F)$ is $e^{-\gamma}$ (see also section 3 and section 4 of [3]).

Proposition (4.1). Let F_1, F_2, F_1' and F_2' be four linear spiral foliations of R^3 with $\ell(F_1) \neq \ell(F_2)$ and $\ell(F_1') \neq \ell(F_2')$. There is a homeomorphism $H:R^3 \to R^3$ sending leaves of F_1 to leaves of F_1' and leaves of F_2 if and only if

$$(*) \quad \frac{\ell n(c(F_1))}{\ell n(c(F_2))} = \frac{\ell n(c(F_1'))}{\ell n(c(F_2'))}$$

Proof. For a pair of linear spiral foliations like F_1 and F_2 (i.e., with $\ell(F_1) \neq \ell(F_2)$), there is a line m consisting of points where the leaves of F_1 and F_2 are tangent. There are two maps $T_1, T_2: m \to m$ with the property that for $p \in m$, $T_i(p)$ is the intersection of the F_i leave through p with m, between p and zero and nearest to p. From the above notation we see that T_i is a linear contraction with a factor $c(F_i)$.

Starting with the foliations F_1', F_2' one analogously defines m', T_1' and T_2'. If there is a homeomorphism H as in our proposition, then it has to map m to m' and it has to conjugate T_i with T_i'.

Lemma (4.2). Let c_1, c_2, c_1' and c_2' be positive real numbers, smaller than one. $T_i, T_i': R \to R$ are linear contractions with a factor c_1, c_i'. There is a homeomorphism $h: R \to R$ such that $hT_i = T_i'h$ for $i = 1,2$ if and only if

$$(**) \quad \frac{\ln c_1}{\ln c_2} = \frac{\ln c_1'}{\ln c_2'} .$$

Proof. Let $n_{1,i}$, $n_{2,i}$ be sequences of integers such that $[n_{1,i} \cdot \ln c_1 - n_{2,i} \cdot \ln c_2] \to 0$ for $i \to \infty$. Then for $o \neq p \in R$, $T_1^{n_{1,i}} \circ T_2^{n_{2,i}}(p)$ tends to p. So if there is a homeomorphism h as above, then also $T_1'^{n_{1,i}} \circ T_2'^{n_{2,i}}(h(p))$ tends to $h(p)$. Then we have $n_{1,i} \ln c_1' - n_{2,i} \ln c_2' \to 0$ and hence (**).

On the other hand, if (**) holds we define $\beta = \dfrac{\ln c_1'}{\ln c_1} = \dfrac{\ln c_2'}{\ln c_2}$ and take for $h: h(t) = \dfrac{t}{|t|} \cdot |t|^{\beta}$. This proves the lemma.

From the above lemma we see that (*) is a necessary condition for the existence of the homeomorphism H. If (*) is satisfied, we obtain the required homeomorphism as follows: from a half line of m (i.e., one side of 0) to a half line of m' we take the homeomorphism h provided by lemma (4.2). Then we construct one-dimensional foliations X_1, X_2, X_1' and X_2' the leaves of which are straight lines parallel to $\ell(F_1)$,

$\ell(F_2)$, $\ell(F_1')$ and $\ell(F_2')$ respectively. We want H to map also leaves of X_1, X_2 to leaves of X_1', X_2'. We first extend h to a plane P_1 formed by the X_2 leaves through m. In this plane the extension is unique:

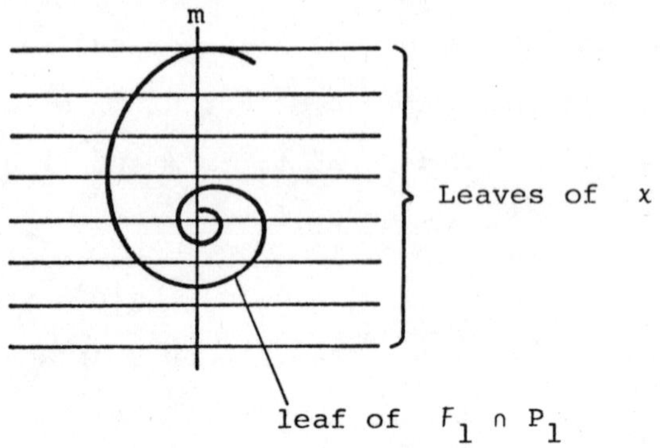

Figure 1

since we have to preserve F_1, h extends uniquely to all of m. Since the leaves of X_2 and $F_1 \cap P_1$ are only tangent along m, there is now a unique extension of h to P_1, mapping the leaves of these foliations to leaves of X_2', $F_1' \cap P_1'$.

The extension of h to P_2, formed by the X_1 leaves through m goes in the same way. Finally, the extension from $P_1 \cup P_2$ over all of R^3 is determined by the rule that H maps X_i leaves to X_i' leaves. This homeomorphism H indeed maps leaves of F_i to leaves of F_i'.

5. Proof of the main theorem.

Let $X', X'' \in \tilde{\Sigma}$ with $m(X') = m(X'')$; see the introduction. We denote $P(X') = p'$, $P(X'') = p''$, $Q(X') = q'$, $Q(X'') = q''$, $\Gamma(X') = \gamma'$ and $\Gamma(X'') = \gamma''$. Let S', S'' be co-dimension one submanifolds transverse to γ', γ''. We choose orbit-unstable foliations F_1', F_1'' of X', X'' near $W^u(p', X')$, $W^u(p'',X'')$ and orbit-stable foliations F_2', F_2'' of X', X'' near $W^s(q',X')$,

$W^s(q'',X'')$ so that there is a common linearization of $F_1' \cap S'$ and $F_2' \cap S'$ and a common linearization of $F_1'' \cap S''$ and $F_2'' \cap S''$ as in section 3. Since $m(X') = m(X'')$, there is a homeomorphism from S' to S'' mapping leaves of $S' \cap F_1'$ to leaves of $S'' \cap F_1''$ and leaves of $S' \cap F_2'$ to leaves of $S'' \cap F_2''$.

We extend S' in two ways: $S_u' \supset S'$ is a co-dimension one submanifold, transverse to X' and intersecting $W^u(p',X')$ in a fundamental domain; $S_s' \supset S'$ intersects $W^s(q',X')$ in a fundamental domain. S_u'' and S_s'' are similarly defined. We extend the homeomorphism from S_u' to S_u'' such that leaves of $F_1' \cap S_u'$ are mapped to leaves of $F_1'' \cap S_u''$. We similarly extend to S_s'. By the result of section 2, the homeomorphism from $S' \cup S_u' \cup S_s'$ to $S'' \cup S_u'' \cup S_s''$ can now be extended to a topological equivalence from X' at γ' to X'' at γ''.

6. Some further remarks.

1. In the case that $\dim(M) > 4$, we may expect our result still holds: Consider a vector field X on a manifold M of dimension $n > 4$ with hyperbolic singularities p,q satisfying the conditions 1, 2 and 3 in the introduction. We define P, Q, Γ, m, Σ and $\tilde{\Sigma}$ similarly. I expect that one can prove, extending somewhat the considerations of the present paper, that also in this case we have for $X', X'' \in \tilde{\Sigma}$ that if $m(X') = m(X'')$, then X' at $\Gamma(X')$ is topologically equivalent with X'' at $\Gamma(X'')$. However in this case where $\dim(M) > 4$ I do not see any way of proving the *conjecture* that $m(X')$ is a *topological invariant* of X' at $\Gamma(X')$, or, in other words, that if X', X'' are in $\tilde{\Sigma}$ and X' at $\Gamma(X')$ is topologically equivalent with X'' at $\Gamma(X'')$, then $m(X') = m(X'')$.

2. One may consider generic 1-parameter families of vector fields X_μ (μ R is the parameter on which X depends) on a 4-dimensional manifold M such that for $\mu = 0$, $X_o \in \Sigma$; $\tilde{\Sigma}$ as before. Let X_μ' be

a nearby 1-parameter family of vector fields on M such that also $X_o' \in \tilde{\Sigma}$ and such that $m(X_o) = m(X_o')$. In this case one might want to extend the topological equivalence h of X_o at $\Gamma(X_o)$ with X_o' at $\Gamma(X_o')$ to a topological equivalence of the 1-parameter family X_μ at $(\Gamma(X_o),0)$ to the 1-parameter family X_μ' at $(\Gamma(X_o'),0)$; for the definitions see [3]. I conjecture that there is in general no such extension.

3. The topolopogical equivalence considered in this paper are in general quite rigid.

Claim. Let $X',X'' \in \tilde{\Sigma}$ be as in the introduction $(\dim(X) = 4)$ with $m(X') = m(X'')$ irrational. A topological equivalence h from X' at $\Gamma(X')$ to X'' at $\Gamma(X'')$ induces a bijection H_s between orbits in $W^s(P(X'),X')$ and orbits in $W^s(P(X''),X'')$. This bijection is completely fixed as soon as the immage of one orbit is fixed. The same holds for the induced bijection between orbits in $W^u(Q(X'),X')$ and the orbits in $W^u(Q(X''),X'')$.

Remark. As in the case of tangencies between stable and unstable manifolds of diffeomorphisms in dimension two, this leads to moduli of stability if we consider the vector fields globally; see de Melo [5].

Sketch of the proof. The proof of this claim is heavily based on the spiral constructions in [3]. Let F_1, F_2 be two linear spiral foliations in R^3 with non-coinciding axis and with contraction factors c_1, c_2. Let L_i be a leaf of F_i. $L_i^{\alpha i}$, $c_i < \alpha_i \leq 1$ denotes $\{x \in R^3 \mid \exists\, \alpha \in [\alpha_i,1],\ \exists\, r \in L_i$ such that $\alpha.r = x\}$. $L_i^{\alpha i}$ is a union of leaves of F_i. With the methods of [3] one can show that

1^e if $\alpha_1.\alpha_2 > \sqrt{c_1.c_2}$ then any continuous path $\lambda:[0,1]\to \overline{L_1^{\alpha_1} \cap L_2^{\alpha_2}}$ with $\lambda(0) = 0$, is constant;

2^e if $\alpha_1 . \alpha_2 < \sqrt{c_1 . c_2}$ then there is a continuous non-constant path $\lambda : [0,1] \to \overline{L_1^{\alpha_1} \cap L_2^{\alpha_2}}$ with $\lambda(0) = 0$.

To apply this to a vector field $X' \in \tilde{\Sigma}$, we take $c_1 = e^{a(X') . \frac{2\pi}{b(X')}}$ (note that $a(X') < 0$; we may assume that $b(X'') > 0$) and $c_2 = e^{-\alpha(X') . \frac{2\pi}{\beta(X')}}$ (here $\alpha(X') > 0$ and we assume that $\beta(X') > 0$).

Consider in $W^s(p',X')$, $p' = P(X')$, linearizing coordinates (they are unique up to a linear transformation [9]). For $r \in W^s(p',X')$, $r \neq p'$, and $c_1 < \alpha_1 \leq 1$ we consider all orbits in $W^s(p',X')$ which pass through a point $\alpha . r$, $\alpha_1 \leq \alpha \leq 1$ (scalar multiplication in the linearizing coordinates). The union of these orbits is called an *orbit interval* with *ratio* α_1. Analogously we define an orbit interval with ratio $c_2 < \alpha_2 \leq 1$ in $W^u(q',X')$. Let S' be again a codimension one submanifold of M transverse to $\gamma' = \Gamma(X')$.

If $\alpha_1 . \alpha_2 > \sqrt{c_1 . c_2}$ then, for any orbit interval O_1 in $W^s(p',X')$ with ratio α_1 and orbit interval O_2 in $W^u(q',X')$ with ratio α_2 there are X'-invariant neighbourhoods U_1, U_2 of O_1, O_2 in M, so small that any continuous path $\lambda : [0,1] \to S \cap U_1 \cap U_2$ with $\lambda(0) = S \cap \gamma'$, is constant.

If $\alpha_1 . \alpha_2 < \sqrt{c_1 . c_2}$ then, for any orbit interval O_1 in $W^s(p',X')$ with ratio α_1 and any orbit interval O_2 in $W^u(q',X')$ with ratio α_2 and X'-invariant neighbourhoods U_1, U_2 of O_1, O_2, there is a continuous non-constant path $\lambda : [0,1] \to \overline{S \cap U_1 \cap U_2}$ with $\lambda(0) = S \cap \gamma'$.

This last statement gives a topological characterization of certain "pairs of ratios", namely the pairs (α_1, α_2) with $\alpha_1 . \alpha_2 = \sqrt{c_1 . c_2}$. Using this and the irrationality of $m(X') = \frac{\ell n(c_1)}{\ell n(c_2)}$, we find a topological characterization of the ratio of any orbit intervals. From this the claim follows.

REFERENCES

[1] - KUPKA, I., Contribution à la theorie des champs génériques, in Contributions to differential equations, Vol. 2, (1963), 457-484 and Vol. 3, (1964), 411-420.

[2] - SMALE, S., Stable manifolds for differential equations and diffeomorphisms, Ann. Scuola Norm. Sup., Pisa, 18, (1963), 97-116.

[3] - TAKENS, F., Global phenomena in bifurcations of dynamical systems with simple recurrence, to appear in the proceedings of the annual congress (1978) of the German Mathematical Society D.M.V.

[4] - PALIS, J., A differentiable invariant of topological conjugacies and moduli of stability, Astérisque 51, (1978). 335-346.

[5] - MELO, W. de, Moduli of stability of two dimensional diffeomorphisms, to appear in Topology.

[6] - NEWHOUSE, S., PALIS, J., and TAKENS, F., Stable families of diffeomorphisms, to appear.

[7] - PALIS, J., Moduli of stability and bifurcation theory, to appear in the proceedings of the international congress in Helsinki, (1978).

[8] - STERNBERG, S., On the structure of local homeomorphisms of Euclidean n-space II, Amer. J. Math. 80, (1958), 623-631.

[9] - STERNBERG, S., Local contractions and a theorem of Poincaré, Amer. J. Math. 79, (1957), 809-824.

STABILITY PROPERTIES IN ALMOST PERIODIC SYSTEMS OF FUNCTIONAL DIFFERENTIAL EQUATIONS

by Taro Yoshizawa

1. Introduction.

Nonlinear ordinary differential equations of the second order which are periodic and whose solutions satisfy some boundedness condition (dissipative or ultimate boundedness) were studied by Levinson [28] in 1944. In 1950, Massera [29] showed that for scalar ω-periodic equations and for linear ω-periodic systems, the existence of a bounded solution implies the existence of an ω-periodic solution, and he showed also that for 2-dimensional ω-periodic systems, there exists an ω-periodic solution if all solutions exist in the future and one of them is bounded. For higher dimensional systems, this is not true generally. However, we can show that if solutions are ultimately bounded, then there exists an ω-periodic solution, by applying Browder's fixed point theorem [3], (cf. [40]). For scalar functional differential equations, the existence of a bounded solution does not necessarily imply the existence of a periodic solution, see [4], [14] for finite delay and [4] for infinite delay. For linear ω-periodic systems with finite delay, there exists an ω-periodic solution if the system has a bounded solution [5], [13]. This fact happens also for linear systems with infinite delay whose phase spaces satisfy some conditions [5]. For general ω-periodic systems with finite delay in which the solution map T is completely continuous, Jones [21] and Yoshizawa [40] showed that T has a fixed point by using Browder's theorem under condition. In this case, ω is assumed to be greater than the delay because of complete continuity of T, and we can show the existence of an ω-periodic solution if solutions are uniformly bounded and ultimately bounded. By discussing fixed point theorems in a Banach space, Hale and Lopes [18] showed that there exists an ω-periodic solution if solutions are ultimately bounded, where it should be noticed that they do not need the condition that ω is greater than the delay.

Even for scalar almost periodic equations, the boundedness of all solutions does not necessarily imply the existence of almost periodic solutions, see [32], and also uniformly ultimate boundedness does not assure the existence of almost periodic solutions, see [11]. Thus we need additional conditions in discussing the existence of almost periodic solutions connecting with boundedness. Favard [9] obtained results on the existence under some kind of separation condition for linear systems, and Amerio [1] generalized some of Favard's results to nonlinear systems. Further extensions have been made by Seifert and Fink, see [10], [43] for the references. Another direction is to assume that bounded solutions have some kind of stability properties, uniform stability, uniformly asymptotic stability, total stability and so on. In this direction, Miller [30], Seifert [37], and Sell [38] have discussed the existence of an almost periodic solution by using the theory of dynamical systems and hence the uniqueness of solutions is assumed. On the other hand, Coppel [7] and Yoshizawa [41] have studied the same question through asymptotically almost periodic functions introduced by Fréchet [12] without assuming the uniqueness. Probably, the first result through asymptotically almost periodic functions appeared in a paper of Reuter [33] about a differential equation of the second order. Halanay [13] also utilized properties of asymptotically almost periodic functions for a quasi-linear system. Recently, Sacker and Sell [35] have discussed the lifting properties of skew-product flow so that both theories by separation and by stability are consequences of the same general principle arising in the study of skew-product flows. We can unify both theories also by discussing relationships between separatedness and some kind of stability property.

Recently, some of those works have been extended to functional differential equations with infinite delays. In the theory of functional differential equations with finite delay, the development of a general qualitative theory is not too sensitive to the choice for the space of initial data for a solution. But for the infinite delay, this choice is never very clear. Hale and Kato [17] have developed a qualitative theory for fundamental theorems and stability by requiring that the phase space satisfies only some general qualitative properties. If we assume that the right-hand side of an equation with finite delay is completely continuous, after one delay interval, the states of solutions belong to some compact set, but for an equation with infinite delay, this is not verified generally. The treatments for infinite delay are more complicated than for finite delay. Hino [19, 20] has studied the existence of an almost periodic solution by considering stability properties in the hull, and Sawano [36] has discussed Liapunov type theorem for a linear system and extended results by Hale [15] and Yoshizawa [39] for finite delays.

2. Almost periodic systems.

For a functional differential equation, the space of initial data is not locally compact, and hence we note here some remarks on almost periodic functions. Let $f(t,\phi)$ be a continuous function defined on $R \times D$ with values in R^n, where $R = (-\infty,\infty)$ and D is an open set in a separable Banach space X. $f(t,\phi)$ is said to be almost periodic in t uniformly for $\phi \in D$, if for any $\varepsilon > 0$ and any compact set S in D, there exists an $\ell(\varepsilon,S) > 0$ such that every interval of length $\ell(\varepsilon,S)$ contains a τ for which $|f(t+\tau,\phi) - f(t,\phi)| \leq \varepsilon$ for all $t \in R$ and all $\phi \in S$. Such a number τ is called an ε-translation number of f on S. A continuous function $f(t,\phi)$ is almost periodic in t uniformly for $\phi \in D$ if and only if for any sequence $\{\tau_k\}$ there exists a subsequence $\{\tau_{k_j}\}$ such that $f(t+\tau_{k_j},\phi)$ converges uniformly on $R \times S$, S any compact set in D, as $j \to \infty$. To make the statement for this convergence short, we say $f(t+\tau_{k_j},\phi)$ converges c-uniformly on $R \times D$. If $f(t,\phi)$ is almost periodic in t uniformly for $\phi \in D$ and $f(t+\tau_k,\phi) \to g(t,\phi)$ c-uniformly on $R \times D$, so is $g(t,\phi)$, and we can find a sequence $\{\sigma_m\}$ such that $\sigma_m \to \infty$ (or $\sigma_m \to -\infty$) as $m \to \infty$ and $f(t+\sigma_m,\phi) \to g(t,\phi)$ c-uniformly on $R \times D$ (cf. [10],[43]).

Let $f(t,\phi)$ be almost periodic in t uniformly for $\phi \in D$. Then the hull $H(f)$ of f is a set of all functions $g(t,\phi)$ such that $\lim_{k\to\infty} f(t+\tau_k,\phi) = g(t,\phi)$ for some sequence $\{\tau_k\}$. Since the space X is separable, we can see that the set of real number λ such that

$$a(\lambda,\phi) = \lim_{T\to\infty} \frac{1}{T}\int_0^T f(t,\phi)e^{-i\lambda t}dt, \quad i = \sqrt{-1},$$

is not identically zero for $\phi \in D$ is a countable set, see [43]. Thus we can define the module of f denoted by $m(f)$ in the usual way. Then, for f and g which are almost periodic in t uniformly for $\phi \in D$, $m(g) \subset m(f)$ if and only if for any sequence $\{\tau_k\}$ of real numbers for which $\{f(t+\tau_k,\phi)\}$ converges c-uniformly on $R \times D$, the sequence $\{g(t+\tau_k,\phi)\}$ also converges c-uniformly on $R \times D$.

Now we shall give the definition of asymptotically almost periodic functions. Let $f(t)$ be a continuous function defined on $I = [0,\infty)$ with values in R^n. $f(t)$ is said to be asymptotically almost periodic if it is a sum of an almost periodic function $p(t)$ and a continuous function $q(t)$ defined on I which tends to zero as $t \to \infty$, that is, $f(t) = p(t) + q(t)$. Then $f(t)$ is asymptotically almost periodic if and only if for any sequence $\{\tau_k\}$ such that $\tau_k \to \infty$ as $k \to \infty$, there exists a subsequence $\{\tau_{k_j}\}$ such that $\{f(t+\tau_{k_j})\}$ converges uniformly on I as $j \to \infty$ (cf. [12], [43]).

Suppose $0 \le r \le \infty$ is given. If $x : [\sigma-r, \sigma+A) \to R^n$, $A > 0$, is a given function, let $x_t : [-r,0] \to R^n$ for each $t \in [\sigma, \sigma+A)$ be defined by $x_t(\theta) = x(t+\theta)$, $-r \le \theta \le 0$. Now assume $r < \infty$ and let C be the space of continuous functions from $[-r,0] \to R^n$ with the uniform norm, that is, for $\phi \in C$, $|\phi| = \sup\{|\phi(\theta)|;\ -r\le\theta\le 0\}$. We use the same symbol $|\cdot|$ for the norm in C and for the vector norm in R^n. Denote by C_α the set of $\phi \in C$ such that $|\phi| < \alpha$, and consider a system of functional differential equations with finite delay

(1) $\quad \dot{x}(t) = f(t,x_t)$,

where $\dot{x}(t)$ is the right-hand derivative by t and $x_t \in C$. We assume that $f(t,\phi)$ in (1) is continuous on $R \times C_\beta$, $0 < \beta \le \infty$, and $f(t,\phi)$ is almost periodic in t uniformly for $\phi \in C_\beta$. For the fundamental theorems on functional differential equations, see [16]. Then the basic theorem is the following.

<u>Theorem</u> 1. Suppose that system (1) has a solution $\xi(t)$ defined on I such that $|\xi_t| \le \alpha < \beta$ for all $t \ge 0$. If $\xi(t)$ is asymptotically almost periodic, then system (1) has an almost periodic solution.

<u>Proof</u>. Since $\xi(t)$ is asymptotically almost periodic, it has the decomposition $\xi(t) = p(t) + q(t)$, where $p(t)$ is almost periodic and $q(t) \to 0$ as $t \to \infty$. Since $\xi(t)$ and $p(t)$ are uniformly continuous, there exists a compact set $S \subset C_\beta$ such that $\xi_t \in S$, $t \ge 0$, and $p_t \in S$ for all $t \in R$. Let $\{\tau_k\}$ be a sequence such that $\tau_k \to \infty$ as $k \to \infty$ and that $f(t+\tau_k,\phi) \to f(t,\phi)$ uniformly on $R \times S$ and $p(t+\tau_k) \to p^*(t)$ uniformly on R, where $p^*(t)$ is almost periodic. Set $\xi^k(t) = \xi(t+\tau_k)$. Then $\xi^k(t)$ is defined on $-\tau_k - r \le t < \infty$, $\xi^k_t \in S$ for all $t \ge -\tau_k$ and for all k, and $\xi^k(t)$ is a solution of $\dot{x}(t) = f(t+\tau_k, x_t)$. Since $\xi(t+\tau_k) = p(t+\tau_k) + q(t+\tau_k)$ and $q(t) \to 0$ as $t \to \infty$, we can see that $\xi^k(t) \to p^*(t)$ uniformly on any compact interval $[-N,N]$,

$N > 0$, as $k \to \infty$. Therefore we can see that $p^*(t)$ is an almost periodic solution of (1).

3. Stability properties in almost periodic systems.

In this section, we shall discuss relationships between stability properties of a bounded solution which imply the existence of an almost periodic solution of system (1).

Definition 1. Let K be a given compact set in C_β and $\xi(t)$ be a solution of (1) such that $\xi_t \in K$ for all $t \geq 0$. For $g \in H(f)$ and $h \in H(f)$, define $\rho(g,h;K)$ by

$$\rho(g,h;K) = \sup\{|g(t,\phi) - h(t,\phi)|;\ t \in R,\ \phi \in K\}.$$

The solution $\xi(t)$ of (1) is said to be stable under disturbances from $H(f)$ with respect to K, if for any $\varepsilon > 0$ there is a $\delta(\varepsilon) > 0$ such that $|\xi_{t+\tau} - x_t(0,\psi,g)| \leq \varepsilon$ for $t \geq 0$, whenever $g \in H(f)$, $|\xi_\tau - \psi| \leq \delta(\varepsilon)$ and $\rho(f^\tau,g;K) \leq \delta(\varepsilon)$ for some $\tau \geq 0$, and $\psi \in K$, where $f^\tau(t,\phi) = f(t+\tau,\phi)$ and $x(0,\psi,g)$ is a solution of

(2) $\quad \dot{x}(t) = g(t,x_t)$

such that $x_o = \psi$.

Remark 1. This definition is equivalent to the following: $\xi(t)$ is stable under disturbances from $H(f)$ with respect to K, if for any $\varepsilon > 0$ there is a $\delta(\varepsilon) > 0$ such that for any $\tau \geq 0$, the solution $x(\tau,\psi,g)$, $x_\tau = \psi$, of (2) satisfies $|\xi_t - x_t(\tau,\psi,g)| \leq \varepsilon$ for all $t \geq \tau$, whenever $g \in H(f)$, $\psi \in K$, $|\xi_\tau - \psi| \leq \delta(\varepsilon)$, and $\rho(f,g;K) \leq \delta(\varepsilon)$.

This kind of stability was introduced by Sell [38] which is equivalent to Σ-stability introduced by Seifert [37].

Theorem 2. Let K be a compact set in C_β. If system (1) has a solution $\xi(t)$ such that $\xi_t \in K$ for all $t \geq 0$ and if $\xi(t)$ is stable under disturbances from $H(f)$ with respect to K, then

$\xi(t)$ is asymptotically almost periodic, and consequently system (1) has an almost periodic solution.

<u>Proof</u>. Let $\{\tau_k\}$ be a sequence such that $\tau_k \to \infty$ as $k \to \infty$, and set $\xi^k(t) = \xi(t+\tau_k)$. Then $\xi^k(t)$ is a solution of

(3) $\dot{x}(t) = f(t+\tau_k, x_t)$

and $\xi_o^k = \xi_{\tau_k}$ and $\xi_t^k \in K$ for all $t \geq 0$ and for all k. Moreover, it is clear that $\xi^k(t)$ is stable under disturbances from $H(f^{\tau_k})$ with respect to K with the same δ as for $\xi(t)$. There is a subsequence of $\{\tau_k\}$, which we shall denote by $\{\tau_k\}$ again, such that $f(t+\tau_k,\phi)$ converges uniformly on $R \times K$ as $k \to \infty$, and hence there exists an integer $k_o(\varepsilon) > 0$ such that if $m \geq k \geq k_o(\varepsilon)$,

$$|f(t+\tau_k,\phi) - f(t+\tau_m,\phi)| \leq \delta(\varepsilon) \quad \text{on} \quad R \times K,$$

where δ is the number in Definition 1. Thus $\rho(f^{\tau_k}, f^{\tau_m}; K) \leq \delta(\varepsilon)$ if $m \geq k \geq k_o(\varepsilon)$. Since $\xi_o^k \in K$, we can assume that $|\xi_o^k - \xi_o^m| \leq \delta(\varepsilon)$ if $m \geq k \geq k_o(\varepsilon)$. $\xi^m(t)$ is a solution of $\dot{x}(t) = f(t+\tau_m, x_t)$ such that $\xi_o^m \in K$ and since $\xi^k(t)$ is stable under disturbances from $H(f^{\tau_k})$ with respect to K and $f^{\tau_m} \in H(f^{\tau_k})$, we have $|\xi_t^k - \xi_t^m| \leq \varepsilon$ for all $t \geq 0$ if $m \geq k \geq k_o(\varepsilon)$. This implies $|\xi(t+\tau_k) - \xi(t+\tau_m)| \leq \varepsilon$ for all $t \geq 0$ if $m \geq k \geq k_o(\varepsilon)$, which shows that $\xi(t)$ is asymptotically almost periodic. The existence of an almost periodic solution follows from Theorem 1.

Now, denoting by $C(I,R^n)$ the set of all continuous functions on I with values in R^n, let B be a Banach space $\subset C(I,R^n)$ with the norm $|\cdot|_B$, and consider a general system

(4) $\dot{x}(t) = F(t,x_t)$,

where $F(t,\phi)$ is continuous on $I \times C_\beta$ and $F(t,\phi)$ is assumed to take closed bounded sets of $I \times C$ into closed bounded sets in R^n. Let $\xi(t)$ be a solution of (4) such that $|\xi_t| \leq \alpha < \beta$ for all $t \geq 0$. We denote by $x(\sigma,\psi,h)$ a solution through (σ,ψ) of

(5) $\dot{x}(t) = F(t,x_t) + h(t), \quad h \in B.$

Definition 2. The solution $\xi(t)$ of (4) is said to be stable under B perturbations (called BS), if for any $\varepsilon > 0$ there exists a $\delta(\varepsilon) > 0$ such that $|\xi_t - x_t(\sigma,\psi,h)| < \varepsilon$ for all $t \geq \sigma$, whenever $\sigma \geq 0$, $|\xi_\sigma - \psi| < \delta(\varepsilon)$ and $|h|_B < \delta(\varepsilon)$.

Let T, L, and M be the Banach spaces $\subset C(I,R^n)$ with the norms $|\cdot|_T$, $|\cdot|_L$ and $|\cdot|_M$, respectively, where

$$|h|_T = \sup_{t\geq 0} |h(t)|, \quad |h|_L = \int_0^\infty |h(t)|dt \text{ and } |h|_M = \sup_{t\geq 0}\int_t^{t+1} |h(s)|ds.$$

Then stability under B perturbations with $B = \{0\}$, $B = T$, $B = L$ or $B = M$ corresponds to uniform stability (US), total stability (TS), integral stability (IS) or M-stability (MS).

Definition 3. The solution $\xi(t)$ of (4) is said to be attracting under B perturbations, if there exists a $\delta_o > 0$ such that for any $\varepsilon > 0$ there exist $\tau(\varepsilon) > 0$ and $\gamma(\varepsilon) > 0$ such that $|\xi_\sigma - \psi| < \delta_o$ implies $|\xi_t - x_t(\sigma,\psi,h)| < \varepsilon$ for $t \geq \sigma+\tau(\varepsilon)$, whenever $\sigma \geq 0$ and $|h|_B < \gamma(\varepsilon)$.

Definition 4. The solution $\xi(t)$ of (4) is said to be asymptotically stable under B perturbations (called BAS), if it is stable under B perturbations and is attracting under B perturbations. If $B = \{0\}$, $B = T$, $B = L$, or $B = M$, this gives the definition of uniformly asymptotic stability (UAS), totally asymptotic stability (TAS), integrally asymptotic stability (IAS), or M-asymptotic stability (MAS).

Moreover, we shall give the following two definitions, which characterize asymptotic stability under B perturbations.

Definition 5. We say that $\xi(t)$ has uniform continuous dependence under B perturbations, if for any $\varepsilon > 0$ and any $T > 0$ there exists an $\eta_1(\varepsilon) > 0$, which is independent of T, and an $\eta_2(\varepsilon,T) > 0$ such that if $\sigma \geq 0$, $|\xi_\sigma - \psi| < \eta_1(\varepsilon)$ and

$|h|_B < \eta_2(\varepsilon,T)$, then $|\xi_t - x_t(\sigma,\psi,h)| < \varepsilon$ on $\sigma \le t \le \sigma+T$.

Definition 6. We say that $\xi(t)$ has uniform finite time attracting under B perturbations, if there exists an $\eta_o > 0$ such that for any $\varepsilon > 0$, there exist $\tau_1(\varepsilon) > 0$ and $\eta_3(\varepsilon) > 0$ such that $x(\sigma,\psi,h)$ is continuable on $[\sigma, \sigma+\tau_1(\varepsilon)]$ and

$$|\xi_{\sigma+\tau_1(\varepsilon)} - x_{\sigma+\tau_1(\varepsilon)}(\sigma,\psi,h)| < \varepsilon,$$

whenever $\sigma \ge 0$, $|\xi_\sigma - \psi| < \eta_o$, and $|h|_B < \eta_3(\varepsilon)$.

Theorem 3. If solution $\xi(t)$ of (4) has uniform continuous dependence under B perturbations and uniform finite time attracting under B perturbations, then $\xi(t)$ is asymptotically stable under B perturbations. The converse is evident from the definitions.

Proof. Let η_o, η_1, η_2, η_3 and τ_1 be the numbers in Definition 5 and 6. Let $\rho(\varepsilon) = \min\{\eta_o, \eta_1(\varepsilon)\}$ and $\delta(\varepsilon) = \min\{\rho(\varepsilon), \eta_2(\varepsilon,\tau_1(\rho(\varepsilon))), \eta_3(\rho(\varepsilon))\}$. Then, if $\sigma \ge 0$, $|\xi_\sigma - \psi| < \delta(\varepsilon)$ and $|h|_B < \delta(\varepsilon)$, we have $|\xi_t - x_t| < \varepsilon$ on $\sigma \le t \le \sigma+\tau_1(\rho(\varepsilon))$, where $x = x(\sigma,\psi,h)$, by uniform continuous dependence under B perturbations. Since $|\xi_\sigma - \psi| < \rho(\varepsilon) \le \eta_o$ and $|h|_B < \delta(\varepsilon) \le \eta_3(\varepsilon)$, x is continuable on $[\sigma, \sigma+\tau_1(\rho(\varepsilon))]$ and $|\xi_{\sigma+\tau_1(\rho(\varepsilon))} - x_{\sigma+\tau_1(\rho(\varepsilon))}| < \rho(\varepsilon)$. On the interval $\sigma+\tau_1(\rho(\varepsilon)) \le t \le \sigma+2\tau_1(\rho(\varepsilon))$, $|h|_B < \delta(\varepsilon) \le \eta_2(\varepsilon,\tau_1(\rho(\varepsilon)))$ and $|\xi_{\sigma+\tau_1(\rho(\varepsilon))} - x_{\sigma+\tau_1(\rho(\varepsilon))}| < \rho(\varepsilon)$, and hence we have $|\xi_t - x_t| < \varepsilon$ on $[\sigma+\tau_1(\rho(\varepsilon)), \sigma+2\tau_1(\rho(\varepsilon))]$ and $|\xi_{\sigma+2\tau_1(\rho(\varepsilon))} - x_{\sigma+2\tau_1(\rho(\varepsilon))}| < \rho(\varepsilon)$. Repeating the process, if $|\xi_\sigma - \psi| < \delta(\varepsilon)$ and $|h|_B < \delta(\varepsilon)$, $|\xi_t - x_t| < \varepsilon$ for $t \ge \sigma$, which shows that $\xi(t)$ is BS.

Now let $\delta_o = \eta_o$. Then for $\delta(\varepsilon)$ above, there exists a $\tau_1(\delta(\varepsilon))$ and an $\eta_3(\delta(\varepsilon))$ such that if $\sigma \ge 0$, $|\xi_\sigma - \psi| < \delta_o$, and $|h|_B < \eta_3(\delta(\varepsilon))$, $x = x(\sigma,\psi,h)$ is continuable on $[\sigma, \sigma+\tau_1(\delta(\varepsilon))]$ and $|\xi_{\sigma+\tau_1(\delta(\varepsilon))} - x_{\sigma+\tau_1(\delta(\varepsilon))}| < \delta\ \varepsilon$. Therefore $|\xi_t - x_t| < \varepsilon$ for $t \ge \sigma+\tau_1(\delta(\varepsilon))$ by BS. This shows that $\xi(t)$ is attracting under B perturbations. This completes the proof.

By constructing Liapunov function, Chow and Yorke [6] showed that IAS $\leftrightarrow$ MAS for ordinary differential systems, but their method is not applicable to functional differential equations. Since we can see that solution $\xi(t)$ of (4) has uniform continuous dependence under M perturbations if it is integrally stable and that $\xi(t)$ has uniform finite time attracting under M perturbations if it is integrally attracting, we have the following.

<u>Theorem</u> 4. If the solution $\xi(t)$ of (4) is IAS, then it is MAS, and consequently IAS is equivalent to MAS. Moreover, evidently MAS $\rightarrow$ TAS.

Now we shall consider the almost periodic system (1), where we assume that for each γ, $0 \le \gamma < \beta$, there is an $L(\gamma) > 0$ such that $|f(t,\phi)| \le L(\gamma)$ for all $t \in R$ and all ϕ such that $|\phi| \le \gamma$. For the solution $\xi(t)$ of (1) defined on I, $(\eta,g) \in H(\xi,f)$ means that there exists a sequence $\{\tau_k\}$, $\tau_k \ge 0$, such that $f(t+\tau_k,\phi) \rightarrow g(t,\phi)$ c-uniformly on $R \times C_\beta$ as $k \rightarrow \infty$ and $\xi(t+\tau_k) \rightarrow \eta(t)$ uniformly on any compact set in I as $k \rightarrow \infty$ (we say simply $\xi(t+\tau_k) \rightarrow \eta(t)$ locally uniformly on I). Then clearly, $\eta(t)$ is a solution of (2).

<u>Definition</u> 7. The solution $\xi(t)$ is said to be stable in the hull under B perturbations (called BSH), if for every $(\eta,g) \in H(\xi,f)$, $\eta(t)$ is BS and the number δ in the definition of BS can be chosen independently of (η,g). $\xi(t)$ is said to be asymptotically stable in the hull under B perturbations (BASH), if for any $(\eta,g) \in H(\xi,f)$, $\eta(t)$ is BAS and we can choose common numbers δ, δ_o, τ and γ in the definition of BAS.

<u>Theorem</u> 5. If the solution $\xi(t)$ of (1) such that $|\xi_t| \le \alpha < \beta$ for all $t \ge 0$ is TAS, then it is TASH, and consequently it is UASH. If $\xi(t)$ is MAS, then it is MASH, and consequently it is IASH and also TASH.

If solution $\xi(t)$ is uniformly stable in the hull, it has uniform continuous dependence under T perturbations, and if $\xi(t)$ is UASH, it has uniform finite time attracting under T perturbations. Therefore UASH $\to$ TAS (see [22]). For ordinary differential systems, we can see that UASH $\to$ MAS [27]. For functional differential equations, in order to obtain a corresponding result, we change the definitions which depend on continuity properties of perturbation functions. Let C_1 be the set of all continuous functions μ defined on $[0,r]$, where r is the delay, such that $\mu(0) = 0$, $\mu(s) > 0$ for $s > 0$ and μ is increasing. Denote by $B(\mu)$ the set of all $h \in B$ such that

$$\left|\int_t^{t'} h(s)ds\right| \le \mu(|t-t'|) \quad \text{if} \quad |t-t'| \le r.$$

For the case $r = 0$, that is, ordinary differential systems, $B(\mu) = B$ for any $\mu \in C_1$. In all definitions above, we assume that those numbers in the definition may depend on μ. For example, $\xi(t)$ is stable under B perturbations, if for any $\mu \in C_1$ and any $\varepsilon > 0$, there exists a $\delta(\varepsilon,\mu) > 0$ such that $|\xi_t - x_t(\sigma,\psi,h)| < \varepsilon$ for all $t \ge \sigma$, whenever $\sigma \ge 0$, $|\xi_\sigma - \psi| < \delta(\varepsilon,\mu)$, $h \in B(\mu)$, and $|h|_B < \delta(\varepsilon,\mu)$. For total stability, both definitions are equivalent since it is sufficient to consider only a special $\mu(s) = s$.

<u>Lemma</u> 1. Suppose that for any $(\eta,g) \in H(\xi,f)$, $\eta(t)$ is unique for initial conditions. Let ε and τ^* be positive constants. Then there exists a $\delta_1(\varepsilon,\tau^*) > 0$ such that if $\sigma \ge 0$, $|\xi_\sigma - \psi| < \delta_1(\varepsilon,\tau^*)$, and $|h|_M < \delta_1(\varepsilon,\tau^*)$, then $|\xi_t - x_t(\sigma,\psi,h)| < \varepsilon$ on $[\sigma, \sigma+\tau^*]$, where $x(\sigma,\psi,h)$ is a solution of

(6) $\dot{x}(t) = f(t,x_t) + h(t), \quad h \in M,$

such that $x_\sigma = \psi$.

This lemma can be proved by the same idea as in the proof of Lemma 6 in [41].

<u>Theorem</u> 6. If the solution $\xi(t)$ of (1) such that $|\xi_t| \le \alpha < \beta$, $t \ge 0$, is uniformly stable in the hull, it has uniform continuous dependence under M perturbations (in the sense of the new definition). Moreover, if $\xi(t)$ is UASH, it has uniform finite time attracting under M perturbations (in the new sense). Thus UASH $\to$ MAS (in the new sense).

<u>Proof</u>. Let $K = K(\mu)$ be a compact set in C_β such that $\xi_t \in K$ for all $t \ge 0$ and that

$$\{\phi \in C;\ |\phi| \le \frac{\alpha+\beta}{2},\ |\phi(\theta)-\phi(\theta')| \le \mu(|\theta-\theta'|) + L|\theta-\theta'|,\ \theta,\theta' \in [-r,0]\} \subset K,$$

where $L = L(\frac{\alpha+\beta}{2})$ and in the case $\beta = \infty$, we assume $\beta = \alpha+1$. We can assume $\varepsilon < \frac{\beta-\alpha}{2}$.

First of all, we shall show that there exists an $\eta_2'(\varepsilon,T,\mu) > 0$ such that if $\sigma \ge 0$, $|x_\sigma-\xi_\sigma| < \frac{1}{2}\delta(\frac{\varepsilon}{2})$, $x_\sigma \in K$, $h \in B(\mu)$, and $|h|_M < \eta_2'(\varepsilon,T,\mu)$, then $|\xi_t-x_t| < \varepsilon$ on $[\sigma,\sigma+T]$, where x is a solution of (6) and δ is the number for uniform stability in the hull. Suppose that there is no $\eta_2'(\varepsilon,T,\mu)$. Then there are sequences $\{t_k\}$, $t_k \ge 0$, $\{h_k(t)\}$, $\{x^k(t)\}$ and $\{\tau_k\}$, $t_k \le \tau_k \le t_k+T$, such that $h_k \in B(\mu)$, $|h_k|_M < \frac{1}{k}$, $|\xi_{t_k}-x^k_{t_k}| < \frac{1}{2}\delta(\frac{\varepsilon}{2})$, $x^k_{t_k} \in K$ and that

$$|\xi_{\tau_k}-x^k_{\tau_k}| = \varepsilon,\quad |\xi_t-x^k_t| < \varepsilon \quad \text{on} \quad [t_k,\tau_k),$$

where x^k is a solution of

$$\dot{x}(t) = f(t,x_t) + h_k(t).$$

We can assume that $\tau_k-t_k \to \tau$, $0 < \tau \le T$, $\xi(t+t_k) \to \zeta(t)$ locally uniformly on I and $f(t+t_k,\phi) \to g(t,\phi)$ c-uniformly on $R \times C_\beta$.

If we set $y^k(t) = x^k(t+t_k)$, then $y^k_o = x^k_{t_k} \in K$ and $y^k(t)$ is a solution of

$$\dot{y}(t) = f(t+t_k,y_t) + h_k(t+t_k),$$

which is defined on $[0,\tau_k-t_k]$. Taking a subsequence, if necessary, we can assume that $y^k_o \to \psi \in K$, because K is compact. Moreover, $H_k(t) = \int_o^t h_k(s+t_k)ds$ is uniformly bounded and equicontinuous on

$[0,\tau]$ since $|h_k|_M \to 0$ as $k \to \infty$. Since $|y^k(t)| < \frac{\alpha+\beta}{2}$, $\{y^k(t)\}$ is uniformly bounded and equicontinuous on $[-r,\tau]$ if k is sufficiently large. Furthermore, $H_k(t) \to 0$ uniformly on $[0,\tau]$ and $y_t^k \in K$ for all $t \in [0, \tau_k - t_k]$, and hence a subsequence of $\{y^k(t)\}$ converges to $y(t)$ uniformly on $[-r,\tau]$ and $y(t)$ is a solution of (2) such that $y_o = \psi \in K$. However, $\xi_{t_k} \to \zeta_o$, $y_o^k \to \psi = y_o$ and $|\xi_{t_k} - x_{t_k}^k| < \frac{1}{2}\delta(\frac{\varepsilon}{2})$. Therefore $|\zeta_o - y_o| \le \frac{1}{2}\delta(\frac{\varepsilon}{2}) < \delta(\frac{\varepsilon}{2})$. But $|\zeta_\tau - y_\tau| = \varepsilon$ since $|\xi_{\tau_k} - x_{\tau_k}^k| = \varepsilon$. This contradicts the uniform stability of $\zeta(t)$. Thus there exists an $\eta_2'(\varepsilon,T,\mu)$.

On the other hand, whatever x_σ is, it follows from Lemma 1 that if

$$|\xi_\sigma - x_\sigma| < \delta_1(\tfrac{1}{2}\delta(\tfrac{\varepsilon}{2}),r) \quad \text{and} \quad |h|_M < \delta_1(\tfrac{1}{2}\delta(\tfrac{\varepsilon}{2}),r), \quad h \in B(\mu),$$

then $|\xi_{\sigma+r} - x_{\sigma+r}| < \frac{1}{2}\delta(\frac{\varepsilon}{2})$. Since x is a solution of (6) and $h \in B(\mu)$, $x_{\sigma+r} \in K$. Now let

$\eta_1(\varepsilon,\mu) = \delta_1(\frac{1}{2}\delta(\frac{\varepsilon}{2}),r)$ and $\eta_2(\varepsilon,T,\mu) = \min\{\delta_1(\frac{1}{2}\delta(\frac{\varepsilon}{2}),r),\ \eta_2'(\varepsilon,T,\mu)\}$.

Then, if $|\xi_\sigma - \psi| < \eta_1(\varepsilon,\mu)$, $h \in B(\mu)$ and $|h|_M < \eta_2(\varepsilon,T,\mu)$, $|\xi_t - x_t(\sigma,\psi,h)| < \frac{1}{2}\delta(\frac{\varepsilon}{2})$ on $[\sigma,\sigma+r]$ and $x_{\sigma+r}(\sigma,\psi,h) \in K$. Moreover, $|\xi_{\sigma+r} - x_{\sigma+r}(\sigma,\psi,h)| < \frac{1}{2}\delta(\frac{\varepsilon}{2})$ and $|h|_M < \eta_2'(\varepsilon,T,\mu)$, which implies $|\xi_t - x_t(\sigma,\psi,h)| < \varepsilon$ on $[\sigma+r,\sigma+r+T]$. This shows that if $\sigma \ge 0$, $h \in B(\mu)$, $|h|_M < \eta_2(\varepsilon,T,\mu)$ and $|\xi_\sigma - \psi| < \eta_1(\varepsilon,\mu)$, we have $|\xi_t - x_t(\sigma,\psi,h)| < \varepsilon$ on $[\sigma,\sigma+T]$.

Next we shall show that if $\xi(t)$ is UASH, it has uniform finite time attracting under M perturbations, that is, for any $\mu \in C_1$ there exists an $\eta_o(\mu) > 0$ such that for any $\varepsilon > 0$ there are a $\tau_1(\varepsilon,\mu)$ and an $\eta_3(\varepsilon,\mu) > 0$ such that $x = x(\sigma,\psi,h)$ is continuable on $[\sigma, \sigma+\tau_1(\varepsilon,\mu)]$ and

$$|\xi_{\sigma+\tau_1(\varepsilon,\mu)} - x_{\sigma+\tau_1(\varepsilon,\mu)}| < \varepsilon,$$

whenever $\sigma \ge 0$, $|\xi_\sigma - \psi| < \eta_o(\mu)$, $h \in B(\mu)$, and $|h|_M < \eta_3(\varepsilon,\mu)$. Let δ, δ_o and τ be the numbers for UASH. As was seen, $\xi(t)$ has uniform continuous dependence under M perturbations. Let η_1

and η_2 be the numbers for that. Let $\tau_1(\varepsilon) = \tau(\frac{\varepsilon}{2})+r$. Suppose that for each $\mu \in C_1$,

$$|\xi_\sigma-\psi| < \eta_1(\frac{\beta-\alpha}{2},\mu),\ |h|_M < \eta_2(\frac{\beta-\alpha}{2},\tau_1(\varepsilon),\mu),\ h \in B(\mu).$$

Then x is continuable to $\sigma+\tau_1(\varepsilon)$ by uniform continuous dependence. Setting $\eta_o' = \min\{\delta_o,\ \eta_1(\frac{\beta-\alpha}{2},\ \mu)\}$, we can find a positive number $\eta_3'(\varepsilon,\mu) \le \eta_2(\frac{\beta-\alpha}{2},\tau_1(\varepsilon),\mu)$ such that if $\sigma \ge 0$, $|\xi_\sigma-\psi| < \eta_o'$, $\psi \in K(\mu)$, $h \in B(\mu)$, and $|h|_M < \eta_3'(\varepsilon,\mu)$, we have $|\xi_{\sigma+\tau(\frac{\varepsilon}{2})}-x_{\sigma+\tau(\frac{\varepsilon}{2})}| < \varepsilon$. This can be proved by the same idea as in the proof of the existence of $\eta_2'(\varepsilon,T,\mu)$.

Now let $\eta_o(\mu) = \delta_1(\eta_o',r)$ and

$$\eta_3(\varepsilon,\mu) = \min\{\delta_1(\eta_o',r),\ \eta_2(\frac{\beta-\alpha}{2},\tau_1(\varepsilon),\mu),\ \eta_3'(\varepsilon,\mu)\}.$$

Then, if $\sigma \ge 0$, $|\xi_\sigma-\psi| < \eta_o(\mu)$, $h \in B(\mu)$ and $|h|_M < \eta_3(\varepsilon,\mu)$, we have

$$|\xi_t-x_t| < \eta_o' \quad \text{on} \quad [\sigma,\sigma+r],$$

$$x_{\sigma+r} \in K(\mu)$$

and

$$|\xi_{\sigma+r+\tau(\frac{\varepsilon}{2})}-x_{\sigma+r+\tau(\frac{\varepsilon}{2})}| < \varepsilon.$$

Letting $\tau_1(\varepsilon) = \tau(\frac{\varepsilon}{2})+r$, this shows that $\xi(t)$ has uniform finite time attracting under M perturbations.

Thus, using the result corresponding to Theorem 3, $\xi(t)$ is MAS (in the new sense) if $\xi(t)$ is UASH.

Remark 2. If system (1) is periodic or if for each $g \in H(f)$ solutions of (2) are unique for initial conditions, we have UAS $\leftrightarrow$ UASH, see [22], [41]. For almost periodic systems without uniqueness, UAS does not necessarily imply UASH [22].

In the sequel, we always assume that $f(t,\phi)$ in almost periodic system (1) satisfies the condition that for each γ, $0 \le \gamma < \beta$, there is an $L(\gamma)$ such that $|f(t,\phi)| \le L(\gamma)$ for all $t \in R$ and ϕ such that $|\phi| \le \gamma$. Let $\xi(t)$ be a solution of (1) such that

$|\xi_t| \le \alpha < \beta$ for all $t \in I$, and let K be a compact set in C_β such that $\xi_t \in K$ for all $t \ge 0$ and that

$$\{\phi \in C;\ |\phi| \le \tfrac{\alpha+\beta}{2},\ |\phi(\theta)-\phi(\theta')| \le L|\theta-\theta'| \text{ on } [-r,0]\} \subset K,$$

where $L = L(\frac{\alpha+\beta}{2})$. If $\beta = \infty$, we let $\beta = \alpha+1$.

Theorem 7. If $\xi(t)$ is stable under disturbances from $H(f)$ with respect to K, then $\xi(t)$ is uniformly stable for $t \ge 0$. For a periodic system, we have the converse.

Proof. Let $\sigma \ge 0$, $\varepsilon < \frac{\beta-\alpha}{2}$, and let $x(\sigma,\psi)$ be a solution of (1) such that $|\xi_\sigma-\psi| < \delta(\varepsilon)$, $\psi \in K$, where δ is the number in Definition 1. Then $y(t) = x(t+\sigma,\sigma,\psi)$ is a solution of $\dot{x}(t)=f(t+\sigma,x_t)$ and $y_0 = \psi$. Since $\xi(t)$ is stable under disturbances from $H(f)$ with respect to K and $\rho(f^\sigma,f^\sigma;K) = 0$, we have $|\xi_{\sigma+t}-y_t| < \varepsilon$ for $t \ge 0$, that is, if $|\xi_\sigma-\psi| < \delta(\varepsilon)$, $\psi \in K$, we have $|\xi_t-x_t| < \varepsilon$ for all $t \ge \sigma$. On the other hand, it is easily seen that for any $(\eta,g) \in H(\xi,f)$, $\eta(t)$ is stable under disturbances from $H(g)$ with respect to K with the same δ as for $\xi(t)$. Thus it follows from the above that for any $(\eta,g) \in H(\xi,f)$, $\eta(t)$ is unique for initial conditions. Therefore, by Lemma 1, there exists a $\delta_1(\delta(\varepsilon)) > 0$ such that if $\sigma \ge 0$ and $|\xi_\sigma-\psi| < \delta_1(\delta(\varepsilon))$, then $|\xi_t-x_t(\sigma,\psi)| < \delta(\varepsilon)$ for all $t \in [\sigma,\sigma+r]$, and moreover, $x_{\sigma+r}(\sigma,\psi) \in K$. Therefore $|\xi_t-x_t(\sigma,\psi)| < \varepsilon$ for $t \ge \sigma+r$. This shows that $|\xi_t-x_t(\sigma,\psi)| < \varepsilon$ for all $t \ge \sigma$ if $|\xi_\sigma-\psi| < \delta_1(\delta(\varepsilon))$, which proves uniform stability.

Now we consider the case where $f(t,\phi)$ is periodic in t, that is, $f(t+\omega,\phi) = f(t,\phi)$, $\omega > 0$, on $R \times C_\beta$ and we shall show that if $\xi(t)$ is uniformly stable, then $\xi(t)$ is stable under disturbances from $H(f)$ with respect to K. If f is autonomous on K, this is evident. In the case where f is not autonomous on $R \times K$, there is a smallest positive period ω^* of $f(t,\phi)$ on $R \times K$ and we can see that for any $g \in H(f)$ and any $\tau \ge 0$, there is a

$\sigma = \sigma(\tau,g,K)$ such that $\tau-\frac{\omega^*}{2} \le \sigma \le \tau+\frac{\omega^*}{2}$ and $g(t,\phi) = f(t+\sigma,\phi)$ on $R \times K$. For such a σ, we can see that for any $\varepsilon > 0$ there exists a $\gamma_1(\varepsilon) > 0$ such that if $\tau \ge 0$, $g \in H(f)$ and $\rho(f^\tau,g;K) \le \gamma_1(\varepsilon)$, then $|\tau-\sigma| < \varepsilon$.

Let δ be the number for uniform stability of $\xi(t)$. We can assume $\varepsilon < \frac{\beta-\alpha}{2}$. Since $\xi_t \in K$ for all $t \ge 0$, there is a $\lambda(\varepsilon) > 0$ such that $|\xi(t)-\xi(t')| < \frac{\delta(\varepsilon)}{2}$ for t, $t' \in [-r,\infty)$ if $|t-t'| < \lambda(\varepsilon)$. Then there is a $\gamma(\varepsilon) > 0$ such that if $\tau \ge 0$, $g \in H(f)$ and $\rho(f^\tau,g;K) \le \gamma(\varepsilon)$, then

(7) $\quad |\tau-\sigma| < \lambda(\varepsilon),$

where we can assume $\gamma(\varepsilon) < \frac{\delta(\varepsilon)}{2}$. Moreover $g(t,\phi) = f(t+\sigma,\phi)$ on $R \times K$. Let $\eta(t) = \xi(t+\tau)$. Then $\eta(t)$ is a solution of

(8) $\quad \dot{x}(t) = f(t+\tau,x_t).$

Letting $|\xi_\tau-\psi| \le \gamma(\varepsilon)$, $\psi \in K$ and $g \in H(f)$ such that $\rho(f^\tau,g;K) \le \gamma(\varepsilon)$, consider a solution $x = x(0,\psi)$ of (2). As long as x exists, $x_t \in K$, and hence x is a solution of

(9) $\quad \dot{x}(t) = f(t+\sigma,x_t), \quad \sigma = \sigma(\tau,g,K),$

and we have (7). If we set $y(t) = \eta(t+\sigma-\tau)$, then $y(t) = \xi(t+\sigma)$. First of all, we assume $\sigma \ge 0$. Then $y(t)$ is a solution of (9) such that $y_o = \xi_\sigma$ and $y(t)$ is uniformly stable with the same pair $(\varepsilon,\delta(\varepsilon))$ as for $\xi(t)$. Since we have (7), $|\xi_\sigma-\xi_\tau| < \frac{\delta(\varepsilon)}{2}$, and hence $|y_o-\psi| < \delta(\varepsilon)$. Thus the uniform stability of $y(t)$ implies that $|y_t-x_t| < \varepsilon$ for $t \ge 0$. Moreover, (7) implies $|y_t-\eta_t| < \frac{\delta(\varepsilon)}{2} < \varepsilon$ for $t \ge 0$. Thus we have $|\eta_t-x_t| < 2\varepsilon$, or $|\xi_{t+\tau}-x_t| < 2\varepsilon$ for $t \ge 0$. Now consider the case where $\sigma < 0$, and consequently $\tau-\sigma > 0$. If we set $z(t) = x(t+\tau-\sigma)$, $z(t)$ is a solution of (8) such that $z_o = x_{\tau-\sigma}(0,\psi)$. Since we have (7) and $\psi \in K$, $|x_{\tau-\sigma}-\psi| < \frac{\delta(\varepsilon)}{2}$. Thus we have $|\eta_o-z_o| < \delta(\varepsilon)$, which implies $|\eta_t-z_t| < \varepsilon$ for $t \ge 0$. Moreover, $|z_t-x_t| < \frac{\delta(\varepsilon)}{2} < \varepsilon$ for

$t \geq 0$, and thus $|\eta_t - x_t| < 2\varepsilon$, or $|\xi_{t+\tau} - x_t| < 2\varepsilon$ for $t \geq 0$. This completes the proof.

From the Definitions and Remark 1, the following theorem is almost evident.

<u>Theorem</u> 8. If the solution $\xi(t)$ of (1) is totally stable, then it is stable under disturbances from $H(f)$ with respect to K, and consequently system (1) has an almost periodic solution.

<u>Remark</u> 3. By Theorem 6 and Remark 2, for a periodic system, UAS $\rightarrow$ TS, and for an almost periodic system, UAS $\rightarrow$ TS if for any $g \in H(f)$ solutions of (2) are unique for initial conditions. Therefore, in these cases, systems have almost periodic solutions. Kato and Sibuya [25] constructed an almost periodic equation which has a uniformly asymptotically stable solution bounded for $t \geq 0$ but has no almost periodic solution.

4. <u>Separation conditions</u>.

First of all, we consider Favard's conditions. Consider linear systems

$$(10) \quad \dot{x}(t) = A(t, x_t)$$

and

$$(11) \quad \dot{x}(t) = A(t, x_t) + h(t),$$

where $A(t,\phi)$ is continuous in $(t,\phi) \in R \times C$, linear in ϕ and almost periodic in t uniformly for $\phi \in C$, and $h(t)$ is almost periodic. Notice that there exists an $L > 0$ such that $|A(t,\phi)| \leq L|\phi|$ for all $t \in R$ and $\phi \in C$. Corresponding to Favard's conditions, we consider the following conditions:

(a) For any $B \in H(A)$, every nontrivial solution x bounded on R of

$$(12) \quad \dot{x}(t) = B(t, x_t)$$

satisfies $\inf_{t \in R} |x_t| > 0$.

(b) For any $B \in H(A)$, the bounded solution on R of (12) is only $x \equiv 0$.

Recently, for ordinary differential equations Sacker and Sell [34] have shown that if Favard's condition (b) is satisfied, then $\dot{x} = A(t)x$ admits an exponential dichotomy on R, see also [24], [42]. Then for every $B \in H(A)$, $\dot{x} = B(t)x$ admits also an exponential dichotomy, and conversely if the system admits an exponential dichotomy, then the bounded solution on R is only $x \equiv 0$. Therefore Favard's condition (b) is equivalent to saying that the system admits an exponential dichotomy, and hence, system $\dot{x} = A(t)x + h(t)$ has a bounded solution automatically. See also [8].

Theorem 9. For system (11), suppose that condition (a) is satisfied. If system (11) has a solution bounded on I, then system (11) has an almost periodic solution and its module is contained in $m(A,h)$.

This theorem has been proved by Kato [23] by considering a minimal solution with respect to the norm $|\cdot|_*$ in C defined by

$$|\phi|_* = \left(\int_{-r}^{0} |\phi(s)|^2 ds\right)^{1/2}.$$

Theorem 10. For system (11), suppose that condition (b) is satisfied. If system (11) has a solution bounded on I, then the solution of (11) bounded on R is almost periodic and its module is contained in $m(A,h)$.

This theorem follows from general separation condition by Amerio [1]. We say that system (1) satisfies separation condition in S, $S = \{\phi;\ |\phi| \le \alpha < \beta\}$, if for each $g \in H(f)$ there exists a $\lambda(g) > 0$ such that if x, y are distinct solutions of (2) such that $x_t \in S$, $y_t \in S$ for all $t \in R$, then $|x_t - y_t| \ge \lambda(g)$ for all $t \in R$.

Theorem 11. Suppose that system (1) satisfies separation condition in S. Then a solution $\xi(t)$ of (1) such that $\xi_t \in S$ for all

$t \geq 0$ is asymptotically almost periodic, and consequently system (1) has an almost periodic solution.

This theorem can be proved by the same argument as in ordinary differential equations by applying the following lemma which is a version of Bochner's theorem [2].

Lemma 2. Let $f(t,\phi)$ be almost periodic in t uniformly for $\phi \in D$, where D is an open set in a separable Banach space. Then for any two sequences $\{\alpha_k'\}$ and $\{\beta_k'\}$, there exist subsequences $\{\alpha_s\} = \{\alpha_{k_s}'\}$ and $\{\beta_s\} = \{\beta_{k_s}'\}$ for a common sequence of indices $\{k_s\}$ such that

$$\lim_{m\to\infty}\{\lim_{n\to\infty} f(t+\alpha_n+\beta_m,\phi)\} = \lim_{s\to\infty} f(t+\alpha_s+\beta_s,\phi),$$

where each of the three limits exist c-uniformly on $R \times D$.

5. Stability and separation condition.

For the closed bounded set $S = \{\phi \in C;\ |\phi| \leq \alpha < \beta\}$, $A(f,S)$ denotes the family of solutions x of (1) such that $x_t \in S$, $t \geq \sigma$, for some $\sigma \in R$. For $x \in A(f,S)$, let σ_x be the infimum of σ, and σ_x may be $-\infty$. $B(f,S)$ denotes the family of solutions x of (1) such that $x_t \in S$ for all $t \in R$. The following stabilities were considered by Nakajima [31] for ordinary differential equations.

Definition 8. $\xi \in B(f,S)$ is conditionally totally stable (called CTS) in S, if for any $\varepsilon > 0$ there exists a $\delta(\varepsilon) > 0$ such that $|\xi_t - y_t(h)| < \varepsilon$ for all $t \geq \sigma$, whenever $y(h) \in A(f+h,S)$, $|y_\sigma(h)-\xi_\sigma| < \delta(\varepsilon)$ at some $\sigma > \sigma_{y(h)}$, and $|h(t)| < \delta(\varepsilon)$ on $[\sigma,\infty)$, where $h(t)$ is continuous. System (1) is said to be conditionally totally stable in S, if every $x \in B(f,S)$ is CTS in S.

Theorem 12. If system (1) satisfies separation condition in S, then for each $g \in H(f)$, system (2) is CTS in S. Moreover, we can choose the number δ so that $\delta(\varepsilon)$ depends only on ε and is independent of g and solutions.

Proof. We shall prove that for any $\varepsilon > 0$ there is a $\delta(\varepsilon) > 0$

such that for any $g \in H(f)$ and $x \in B(g,S)$, $|x_t-y_t| < \varepsilon$ for all $t \geq \sigma$, whenever $y \in A(g+h,S)$, $|x_\sigma-y_\sigma| < \delta(\varepsilon)$ for some $\sigma > \sigma_y$, and $|h(t)| < \delta(\varepsilon)$ on $[\sigma,\infty)$. Suppose not. Then there exists an $\varepsilon > 0$ and sequences $g_k \in H(f)$, $h_k(t)$, $x^k \in B(g_k,S)$, $y^k \in A(g_k+h_k,S)$, t_k, τ_k, $\tau_k > t_k$, such that $|h_k(t)| < \frac{1}{k}$ on $t_k \leq t < \infty$, $|x^k_{t_k}-y^k_{t_k}| < \frac{1}{k}$, $t_k > \sigma_{y^k}$, and $|x^k_{\tau_k}-y^k_{\tau_k}| = \varepsilon$. If system (1) satisfies separation condition, we can choose a positive number $\lambda_o = \lambda(g)$ independent of g. Here we can assume $\varepsilon \leq \frac{\lambda_o}{2}$. Set $u^k(t) = x^k(t+\tau_k)$ and $v^k(t) = y^k(t+\tau_k)$. Then $u^k(t)$ and $v^k(t)$ are solutions of

$$\dot{x}(t) = g_k(t+\tau_k,x_t)$$

and

$$\dot{x}(t) = g_k(t+\tau_k,x_t) + h_k(t+\tau_k),$$

respectively, and $u^k_t \in S$ for all $t \in R$ and $v^k_t \in S$ for $t \geq t_k - \tau_k$ $(t_k - \tau_k \leq 0)$.

There exists a compact set K in C such that $x^k_t \in K$ and $u^k_t \in K$ for all $t \in R$ and all k. Taking a subsequence, we can assume that $x^k_{t_k} \to \phi \in K$ as $k \to \infty$, and hence $y^k_{t_k} \to \phi$ as $k \to \infty$. Then the set $\{y^k_{t_k};\ k = 1, 2, \ldots;\phi\}$ is compact. Thus $\{y^k(t_k+\theta)\}$, $-r \leq \theta \leq 0$, is uniformly bounded and equicontinuous. Also $|\dot{y}^k(t+\tau_k)| \leq L(\alpha) + 1$ for $t \geq t_k - \tau_k$. Therefore there exists a compact set $K_1 \subset S$ such that $v^k_t \in K_1$ for all $t \geq t_k - \tau_k$ and $u^k_t \in K_1$ for all $t \in R$. Since $g_k(t+\tau_k,\phi) \in H(f)$ and $H(f)$ is compact in the sense of c-uniform convergence, $\{g_k(t+\tau_k,\phi)\}$ has a subsequence, which we denote by $\{g_k(t+\tau_k,\phi)\}$ again, such that

$$g_k(t+\tau_k,\phi) \to p(t,\phi) \quad \text{c-uniformly on } R \times C_\beta \quad \text{as } k \to \infty,$$

and $p \in H(f)$. Moreover, we can assume $t_k - \tau_k \to \tau$ as $k \to \infty$, where τ can be $-\infty$. In the case where $\tau = -\infty$, taking a subsequence if necessary, $v^k(t) \to \eta(t)$ locally uniformly on R as $k \to \infty$, and $\eta \in B(p,S)$. On the other hand, there exists a $\xi(t)$

such that $u^k(t) \to \xi(t)$ locally uniformly on R and $\xi \in B(p,S)$. But $|\xi_o - \eta_o| = \lim_{k\to\infty} |x^k_{\tau_k} - y^k_{\tau_k}| = \varepsilon > 0$ and hence $|\xi_t - \eta_t| \geq \lambda_o$ for all $t \in R$. This contradicts $\varepsilon \leq \frac{\lambda_o}{2}$. In the case where $\tau > -\infty$, we can find $\eta \in A(p,S)$ and $\xi \in B(p,S)$ such that $\eta_\tau = \xi_\tau$. Then if we define

$$\eta^*(t) = \begin{cases} \eta(t) & \text{for } t \geq \tau \\ \xi(t) & \text{for } t < \tau, \end{cases}$$

$\eta^* \in B(p,S)$ and $\xi \in B(p,S)$ are distinct solutions, because $|\eta^*_o - \xi_o| = |\eta_o - \xi_o| = \varepsilon > 0$. Notice that $\tau \leq 0$. Thus we have also a contradition. This completes the proof.

<u>Theorem</u> 13. Suppose that system (1) has a solution $\xi(t)$ such that $\xi_t \in S$ for all $t \in R$. If $\xi(t)$ is CTS in S, then $\xi(t)$ is asymptotically almost periodic, and consequently system (1) has an almost periodic solution.

<u>Proof</u>. First of all, we shall show that for any $c > 0$, $\eta(t) = \xi(t+c)$ is CTS in S with the same $(\varepsilon, \delta(\varepsilon))$ as for $\xi(t)$. $\eta(t)$ is a solution of $\dot{x}(t) = f(t+c, x_t)$ and $\eta_t \in S$ for all $t \in R$. For $\sigma \in R$ and ψ such that $|\eta_\sigma - \psi| < \delta(\varepsilon)$, consider a solution $y(h)$ of

$$\dot{y}(t) = f(t+c, y_t) + h(t)$$

such that $y_t(h) \in S$ for all $t \geq \sigma$, where $|h(t)| < \delta(\varepsilon)$ for $t \geq \sigma$. If we set $z(t) = y(t-c)$, $z(t)$ is defined for $t \geq \sigma+c$ and $z_t \in S$ for $t \geq \sigma+c$, and $z(t)$ is a solution of

$$\dot{x}(t) = f(t, x_t) + h(t-c).$$

Moreover, $z_{\sigma+c} = y_\sigma = \psi$, $|z_{\sigma+c} - \xi_{\sigma+c}| = |\psi - \eta_\sigma| < \delta(\varepsilon)$, and $|h(t-c)| < \delta(\varepsilon)$ for $t \geq \sigma+c$. Since $\xi(t)$ is CTS in S, we have $|z_t - \xi_t| < \varepsilon$ for $t \geq \sigma+c$, and hence $|y_t - \eta_t| < \varepsilon$ for $t \geq \sigma$, which shows that $\eta(t)$ is CTS in S.

Now let $\{\tau_k\}$ be a sequence such that $\tau_k \to \infty$ as $k \to \infty$. Set $\xi^k(t) = \xi(t+\tau_k)$. Then $\xi^k(t)$ is a solution of $\dot{x}(t) = f(t+\tau_k, x_t)$

and $\xi_o^k = \xi_{\tau_k}$. Moreover, $\xi^k(t)$ is CTS in S. Clearly, there exists a compact set K in C_β such that $\xi_t^k \in K$ for all $t \in R$ and all k, and hence we can assume that ξ_o^k converges and also $f(t+\tau_k,\phi) \to g(t,\phi)$ uniformly on $R \times K$ as $k \to \infty$. Thus for any $\varepsilon > 0$ there exists a $k_o(\varepsilon)$ such that if $k,\ m \geq k_o(\varepsilon)$,

(13) $|\xi_o^k - \xi_o^m| < \delta(\varepsilon)$

and

(14) $|f(++\tau_k,\phi) - f(t+\tau_m,\phi)| < \delta(\varepsilon)$ for all $t \in R$ and $\phi \in K$.

$\xi^m(t)$ is a solution of

$$\dot{y}(t) = f(t+\tau_k, y_t) + f(t+\tau_m, \xi_t^m) - f(t+\tau_k, \xi_t^m)$$

and $\xi_t^m \in S$ for $t \geq 0$. Since $\xi^k(t)$ is CTS in S, (13) and (14) imply that if $k,\ m \geq k_o(\varepsilon)$, $|\xi_t^k - \xi_t^m| < \varepsilon$ for all $t \geq 0$. In other words, if $k,\ m \geq k_o(\varepsilon)$,

$$|\xi(t+\tau_k) - \xi(t+\tau_m)| < \varepsilon \quad \text{for all } t \geq 0,$$

which shows that $\xi(t+\tau_k)$ is uniformly convergent on $[0,\infty)$. This proves that $\xi(t)$ is asymptotically almost periodic.

REFERENCES

[1] - AMERIO, L., Soluzioni quasi-periodiche, o limitate, di sistemi differenziali non lineari quasi-periodici, o limitati, Ann. Mat. Pura Appl., 39(1955), 97-119.

[2] - BOCHNER, S., A new approach to almost periodicity, Proc. Nat. Acad. Sci. U.S.A., 48(1962), 2039-2043.

[3] - BROWDER, F.E., On a generalization of the Schauder fixed point theorem, Duke Math. J., 26(1959), 291-303.

[4] - CHOW, S.N., Remarks on one-dimensional delay-differential equations, J. Math. Anal. Appl., 41(1973), 426-429.

[5] - CHOW, S.N. and HALE, J.K., Strong limit-compact maps, Funkcial. Ekvac., 17(1974), 31-38.

[6] - CHOW, S.N. and YORKE, J.A., Lyapunov theory and perturbation of stable and asymptotically stable systems, J. Differential Eqs., 15(1974), 308-321.

[7] - COPPEL, W.A., Almost periodic properties of ordinary differential equations, Ann. Mat. Pura Appl., 76(1967), 27-49.

[8] - COPPEL, W.A., Dichotomies in Stability Theory, Lecture Notes in Mathematics 629, Springer-Verlag, 1978.

[9] - FAVARD, J., Leçons sur les Fonctions Presque-périodiques, Gauthier-Villars, Paris, 1933.

[10] - FINK, A.M., Almost Periodic Differential Equations, Lecture Notes in Mathematics 377, Springer-Verlag, 1974.

[11] - FINK, A.M. and FREDERICKSON, P.O., Ultimate boundedness does not imply almost periodicity, J. Differential Eqs., 9(1971), 280-284.

[12] - FRÉCHET, M., Les fonctions asymptotiquement presque-périodiques, Rev. Scientifique, 79(1941), 341-354.

[13] - HALANAY, A., Differential Equations; Stability, Oscillation, Time Lags, Academic Press, New York, 1966.

[14] - HALANAY, A. and YORKE, J.A., Some new results and problems in the theory of differential-delay equation, SIAM Review, 13(1971), 55-80.

[15] - HALE, J.K., Periodic and almost periodic solutions of functional-differential equations, Arch. Rational Mech. Anal., 15(1964), 289-304.

[16] - HALE, J.K., Theory of Functional Differential Equations, Applied Math. Sciences, Vol. 3, 2nd ed., Springer-Verlag, 1977.

[17] - HALE, J.K. and KATO, J., Phase space for retarded equations with infinite delay, Funkcial. Ekvac., 21(1978), 11-41.

[18] - HALE, J.K and LOPES, O.F., Fixed point theorems and dissipative processes, J. Differential Eqs., 13(1973), 391-402.

[19] - HINO, Y., Stability and existence of almost periodic solutions of some functional differential equations, Tohoku Math. J., 28(1976), 389-409.

[20] - HINO, Y., Favard's separation theorem in functional differential equations with infinite retardations, Tohoku Math. J., 30(1978), 1-12.

[21] - JONES, G.S., Stability and asymptotic fixed point theory, Proc. Nat. Acad. Sci., U.S.A., 59(1965), 1262-1264.

[22] - KATO, J., Uniformly asymptotic stability and total stability, Tohoku Math. J., 22(1970), 254-269.

[23] - KATO, J., Favard's separation theorem in functional differential equations, Funkcial. Ekvac., 18(1975), 85-92.

[24] - KATO, J. and NAKAJIMA, F., On Sacker-Sell's theorem for a linear skew product flow, Tohoku Math. J., 28(1976), 79-88.

[25] - KATO, J. and SIBUYA, Y., Catastrophic deformation of a flow and non-existence of almost periodic solutions, J. Faculty of Sci., Univ. of Tokyo, Ser. IA, 24(1977), 267-280.

[26] - KATO, J. and YOSHIZAWA, T., A relationship between uniformly asymptotic stability and total stability, Funkcial. Ekvac., 12(1969), 233-238.

[27] - KATO, J. and YOSHIZAWA, T., Stability under the perturbation by a class of functions, Dynamical Systems, An International Symposium, Vol. 2, Academic Press, 1976, 217-222.

[28] - LEVINSON, N., Transformation theory of nonlinear differential equations of second order, Ann. of Math., 45(1944), 723-737. Correction, Ann. of Math., 49(1948), 738.

[29] - MASSERA, J.L., The existence of periodic solutions of systems of differential equations, Duke Math. J., 17(1950), 457-475.

[30] - MILLER, R.K., Almost periodic differential equations as dynamical systems with applications to the existence of a.p. solutions, J. Differential Eqs., 1(1965), 337-345.

[31] - NAKAJIMA, F., Separation conditions and stability properties in almost periodic systems, Tohoku Math. J., 26(1974), 305-314.

[32] - OPIAL, Z., Sur une équation différentielle presque-périodique sans solution presque-périodique, Bull. Acad. Polon. Sci. Ser. Sci. Math. Astron. Phys., 9(1961), 673-676.

[33] - REUTER, G.E.H., On certain non-linear differential equations with almost periodic solutions, J. London Math. Soc., 26(1951), 215-221.

[34] - SACKER, R.J. and SELL, G.R., Existence of dichotomies and invariant splittings for linear differential systems 1, J. Differential Eqs., 15(1974), 429-458.

[35] - SACKER, R.J. and SELL, G.R., Lifting properties in skew-product flow with applications to differential equations, Memoirs Amer. Math. Soc., 190(1977).

[36] - SAWANO, K., Exponentially asymptotic stability for functional differential equations with infinite retardations, Tohoku Math. J., 31(1979).

[37] - SEIFERT, G., Almost periodic solutions for almost periodic systems of ordinary differential equations, J. Differential Eqs., 2(1966), 305-319.

[38] - SELL, G.R., Nonautonomous differential equations and topological dynamics I, II, Trans. Amer. Math. Soc., 127(1967), 241-262, 263-283.

[39] - YOSHIZAWA, T., Extreme stability and almost periodic solutions of functional-differential equations, Arch. Rational Mech. Anal., 17(1964), 148-170.

[40] - YOSHIZAWA, T., Stability Theory by Liapunov's Second Method, The Mathematical Society of Japan, Tokyo, 1966.

[41] - YOSHIZAWA, T., Asymptotically almost periodic solutions of an almost periodic systems, Funkcial. Ekvac., 12(1969), 23-40.

[42] - YOSHIZAWA, T., Favard's condition in linear almost periodic systems, International Conference on Differential Equations, Academic Press, 1975, 787-799.

[43] - YOSHIZAWA, T., Stability Theory and the Existence of Periodic Solutions and Almost Periodic Solutions, Applied Math. Sciences, Vol. 14, Springer-Verlag, 1975.

Vol. 640: J. L. Dupont, Curvature and Characteristic Classes. X, 175 pages. 1978.

Vol. 641: Séminaire d'Algèbre Paul Dubreil, Proceedings Paris 1976–1977. Edité par M. P. Malliavin. IV, 367 pages. 1978.

Vol. 642: Theory and Applications of Graphs, Proceedings, Michigan 1976. Edited by Y. Alavi and D. R. Lick. XIV, 635 pages. 1978.

Vol. 643: M. Davis, Multiaxial Actions on Manifolds. VI, 141 pages. 1978.

Vol. 644: Vector Space Measures and Applications I, Proceedings 1977. Edited by R. M. Aron and S. Dineen. VIII, 451 pages. 1978.

Vol. 645: Vector Space Measures and Applications II, Proceedings 1977. Edited by R. M. Aron and S. Dineen. VIII, 218 pages. 1978.

Vol. 646: O. Tammi, Extremum Problems for Bounded Univalent Functions. VIII, 313 pages. 1978.

Vol. 647: L. J. Ratliff, Jr., Chain Conjectures in Ring Theory. VIII, 133 pages. 1978.

Vol. 648: Nonlinear Partial Differential Equations and Applications, Proceedings, Indiana 1976–1977. Edited by J. M. Chadam. VI, 206 pages. 1978.

Vol. 649: Séminaire de Probabilités XII, Proceedings, Strasbourg, 1976–1977. Edité par C. Dellacherie, P. A. Meyer et M. Weil. VIII, 805 pages. 1978.

Vol. 650: C*-Algebras and Applications to Physics. Proceedings 1977. Edited by H. Araki and R. V. Kadison. V, 192 pages. 1978.

Vol. 651: P. W. Michor, Functors and Categories of Banach Spaces. VI, 99 pages. 1978.

Vol. 652: Differential Topology, Foliations and Gelfand-Fuks-Cohomology, Proceedings 1976. Edited by P. A. Schweitzer. XIV, 252 pages. 1978.

Vol. 653: Locally Interacting Systems and Their Application in Biology. Proceedings, 1976. Edited by R. L. Dobrushin, V. I. Kryukov and A. L. Toom. XI, 202 pages. 1978.

Vol. 654: J. P. Buhler, Icosahedral Golois Representations. III, 143 pages. 1978.

Vol. 655: R. Baeza, Quadratic Forms Over Semilocal Rings. VI, 199 pages. 1978.

Vol. 656: Probability Theory on Vector Spaces. Proceedings, 1977. Edited by A. Weron. VIII, 274 pages. 1978.

Vol. 657: Geometric Applications of Homotopy Theory I, Proceedings 1977. Edited by M. G. Barratt and M. E. Mahowald. VIII, 459 pages. 1978.

Vol. 658: Geometric Applications of Homotopy Theory II, Proceedings 1977. Edited by M. G. Barratt and M. E. Mahowald. VIII, 487 pages. 1978.

Vol. 659: Bruckner, Differentiation of Real Functions. X, 247 pages. 1978.

Vol. 660: Equations aux Dérivée Partielles. Proceedings, 1977. Edité par Pham The Lai. VI, 216 pages. 1978.

Vol. 661: P. T. Johnstone, R. Paré, R. D. Rosebrugh, D. Schumacher, R. J. Wood, and G. C. Wraith, Indexed Categories and Their Applications. VII, 260 pages. 1978.

Vol. 662: Akin, The Metric Theory of Banach Manifolds. XIX, 306 pages. 1978.

Vol. 663: J. F. Berglund, H. D. Junghenn, P. Milnes, Compact Right Topological Semigroups and Generalizations of Almost Periodicity. X, 243 pages. 1978.

Vol. 664: Algebraic and Geometric Topology, Proceedings, 1977. Edited by K. C. Millett. XI, 240 pages. 1978.

Vol. 665: Journées d'Analyse Non Linéaire. Proceedings, 1977. Edité par P. Bénilan et J. Robert. VIII, 256 pages. 1978.

Vol. 666: B. Beauzamy, Espaces d'Interpolation Réels: Topologie et Géometrie. X, 104 pages. 1978.

Vol. 667: J. Gilewicz, Approximants de Padé. XIV, 511 pages. 1978.

Vol. 668: The Structure of Attractors in Dynamical Systems. Proceedings, 1977. Edited by J. C. Martin, N. G. Markley and W. Perrizo. VI, 264 pages. 1978.

Vol. 669: Higher Set Theory. Proceedings, 1977. Edited by G. H. Müller and D. S. Scott. XII, 476 pages. 1978.

Vol. 670: Fonctions de Plusieurs Variables Complexes III, Proceedings, 1977. Edité par F. Norguet. XII, 394 pages. 1978.

Vol. 671: R. T. Smythe and J. C. Wierman, First-Passage Perculation on the Square Lattice. VIII, 196 pages. 1978.

Vol. 672: R. L. Taylor, Stochastic Convergence of Weighted Sums of Random Elements in Linear Spaces. VII, 216 pages. 1978.

Vol. 673: Algebraic Topology, Proceedings 1977. Edited by P. Hoffman, R. Piccinini and D. Sjerve. VI, 278 pages. 1978.

Vol. 674: Z. Fiedorowicz and S. Priddy, Homology of Classical Groups Over Finite Fields and Their Associated Infinite Loop Spaces. VI, 434 pages. 1978.

Vol. 675: J. Galambos and S. Kotz, Characterizations of Probability Distributions. VIII, 169 pages. 1978.

Vol. 676: Differential Geometrical Methods in Mathematical Physics II, Proceedings, 1977. Edited by K. Bleuler, H. R. Petry and A. Reetz. VI, 626 pages. 1978.

Vol. 677: Séminaire Bourbaki, vol. 1976/77, Exposés 489–506. IV, 264 pages. 1978.

Vol. 678: D. Dacunha-Castelle, H. Heyer et B. Roynette. Ecole d'Eté de Probabilités de Saint-Flour. VII-1977. Edité par P. L. Hennequin. IX, 379 pages. 1978.

Vol. 679: Numerical Treatment of Differential Equations in Applications, Proceedings, 1977. Edited by R. Ansorge and W. Törnig. IX, 163 pages. 1978.

Vol. 680: Mathematical Control Theory, Proceedings, 1977. Edited by W. A. Coppel. IX, 257 pages. 1978.

Vol. 681: Séminaire de Théorie du Potentiel Paris, No. 3, Directeurs: M. Brelot, G. Choquet et J. Deny. Rédacteurs: F. Hirsch et G. Mokobodzki. VII, 294 pages. 1978.

Vol. 682: G. D. James, The Representation Theory of the Symmetric Groups. V, 156 pages. 1978.

Vol. 683: Variétés Analytiques Compactes, Proceedings, 1977. Edité par Y. Hervier et A. Hirschowitz. V, 248 pages. 1978.

Vol. 684: E. E. Rosinger, Distributions and Nonlinear Partial Differential Equations. XI, 146 pages. 1978.

Vol. 685: Knot Theory, Proceedings, 1977. Edited by J. C. Hausmann. VII, 311 pages. 1978.

Vol. 686: Combinatorial Mathematics, Proceedings, 1977. Edited by D. A. Holton and J. Seberry. IX, 353 pages. 1978.

Vol. 687: Algebraic Geometry, Proceedings, 1977. Edited by L. D. Olson. V, 244 pages. 1978.

Vol. 688: J. Dydak and J. Segal, Shape Theory. VI, 150 pages. 1978.

Vol. 689: Cabal Seminar 76–77, Proceedings, 1976–77. Edited by A.S. Kechris and Y. N. Moschovakis. V, 282 pages. 1978.

Vol. 690: W. J. J. Rey, Robust Statistical Methods. VI, 128 pages. 1978.

Vol. 691: G. Viennot, Algèbres de Lie Libres et Monoïdes Libres. III, 124 pages. 1978.

Vol. 692: T. Husain and S. M. Khaleelulla, Barrelledness in Topological and Ordered Vector Spaces. IX, 258 pages. 1978.

Vol. 693: Hilbert Space Operators, Proceedings, 1977. Edited by J. M. Bachar Jr. and D. W. Hadwin. VIII, 184 pages. 1978.

Vol. 694: Séminaire Pierre Lelong – Henri Skoda (Analyse) Année 1976/77. VII, 334 pages. 1978.

Vol. 695: Measure Theory Applications to Stochastic Analysis, Proceedings, 1977. Edited by G. Kallianpur and D. Kölzow. XII, 261 pages. 1978.

Vol. 696: P. J. Feinsilver, Special Functions, Probability Semigroups, and Hamiltonian Flows. VI, 112 pages. 1978.

Vol. 697: Topics in Algebra, Proceedings, 1978. Edited by M. F. Newman. XI, 229 pages. 1978.

Vol. 698: E. Grosswald, Bessel Polynomials. XIV, 182 pages. 1978.

Vol. 699: R. E. Greene and H.-H. Wu, Function Theory on Manifolds Which Possess a Pole. III, 215 pages. 1979.

Vol. 700: Module Theory, Proceedings, 1977. Edited by C. Faith and S. Wiegand. X, 239 pages. 1979.

Vol. 701: Functional Analysis Methods in Numerical Analysis, Proceedings, 1977. Edited by M. Zuhair Nashed. VII, 333 pages. 1979.

Vol. 702: Yuri N. Bibikov, Local Theory of Nonlinear Analytic Ordinary Differential Equations. IX, 147 pages. 1979.

Vol. 703: Equadiff IV, Proceedings, 1977. Edited by J. Fábera. XIX, 441 pages. 1979.

Vol. 704: Computing Methods in Applied Sciences and Engineering, 1977, I. Proceedings, 1977. Edited by R. Glowinski and J. L. Lions. VI, 391 pages. 1979.

Vol. 705: O. Forster und K. Knorr, Konstruktion verseller Familien kompakter komplexer Räume. VII, 141 Seiten. 1979.

Vol. 706: Probability Measures on Groups, Proceedings, 1978. Edited by H. Heyer. XIII, 348 pages. 1979.

Vol. 707: R. Zielke, Discontinuous Čebyšev Systems. VI, 111 pages. 1979.

Vol. 708: J. P. Jouanolou, Equations de Pfaff algébriques. V, 255 pages. 1979.

Vol. 709: Probability in Banach Spaces II. Proceedings, 1978. Edited by A. Beck. V, 205 pages. 1979.

Vol. 710: Séminaire Bourbaki vol. 1977/78, Exposés 507–524. IV, 328 pages. 1979.

Vol. 711: Asymptotic Analysis. Edited by F. Verhulst. V, 240 pages. 1979.

Vol. 712: Equations Différentielles et Systèmes de Pfaff dans le Champ Complexe. Edité par R. Gérard et J.-P. Ramis. V, 364 pages. 1979.

Vol. 713: Séminaire de Théorie du Potentiel, Paris No. 4. Edité par F. Hirsch et G. Mokobodzki. VII, 281 pages. 1979.

Vol. 714: J. Jacod, Calcul Stochastique et Problèmes de Martingales. X, 539 pages. 1979.

Vol. 715: Inder Bir S. Passi, Group Rings and Their Augmentation Ideals. VI, 137 pages. 1979.

Vol. 716: M. A. Scheunert, The Theory of Lie Superalgebras. X, 271 pages. 1979.

Vol. 717: Grosser, Bidualräume und Vervollständigungen von Banachmoduln. III, 209 pages. 1979.

Vol. 718: J. Ferrante and C. W. Rackoff, The Computational Complexity of Logical Theories. X, 243 pages. 1979.

Vol. 719: Categorial Topology, Proceedings, 1978. Edited by H. Herrlich and G. Preuß. XII, 420 pages. 1979.

Vol. 720: E. Dubinsky, The Structure of Nuclear Fréchet Spaces. V, 187 pages. 1979.

Vol. 721: Séminaire de Probabilités XIII. Proceedings, Strasbourg, 1977/78. Edité par C. Dellacherie, P. A. Meyer et M. Weil. VII, 647 pages. 1979.

Vol. 722: Topology of Low-Dimensional Manifolds. Proceedings, 1977. Edited by R. Fenn. VI, 154 pages. 1979.

Vol. 723: W. Brandal, Commutative Rings whose Finitely Generated Modules Decompose. II, 116 pages. 1979.

Vol. 724: D. Griffeath, Additive and Cancellative Interacting Particle Systems. V, 108 pages. 1979.

Vol. 725: Algèbres d'Opérateurs. Proceedings, 1978. Edité par P. de la Harpe. VII, 309 pages. 1979.

Vol. 726: Y.-C. Wong, Schwartz Spaces, Nuclear Spaces and Tensor Products. VI, 418 pages. 1979.

Vol. 727: Y. Saito, Spectral Representations for Schrödinger Operators With Long-Range Potentials. V, 149 pages. 1979.

Vol. 728: Non-Commutative Harmonic Analysis. Proceedings, 1978. Edited by J. Carmona and M. Vergne. V, 244 pages. 1979.

Vol. 729: Ergodic Theory. Proceedings, 1978. Edited by M. Denker and K. Jacobs. XII, 209 pages. 1979.

Vol. 730: Functional Differential Equations and Approximation of Fixed Points. Proceedings, 1978. Edited by H.-O. Peitgen and H.-O. Walther. XV, 503 pages. 1979.

Vol. 731: Y. Nakagami and M. Takesaki, Duality for Crossed Products of von Neumann Algebras. IX, 139 pages. 1979.

Vol. 732: Algebraic Geometry. Proceedings, 1978. Edited by K. Lønsted. IV, 658 pages. 1979.

Vol. 733: F. Bloom, Modern Differential Geometric Techniques in the Theory of Continuous Distributions of Dislocations. XII, 206 pages. 1979.

Vol. 734: Ring Theory, Waterloo, 1978. Proceedings, 1978. Edited by D. Handelman and J. Lawrence. XI, 352 pages. 1979.

Vol. 735: B. Aupetit, Propriétés Spectrales des Algèbres de Banach. XII, 192 pages. 1979.

Vol. 736: E. Behrends, M-Structure and the Banach-Stone Theorem. X, 217 pages. 1979.

Vol. 737: Volterra Equations. Proceedings 1978. Edited by S.-O. Londen and O. J. Staffans. VIII, 314 pages. 1979.

Vol. 738: P. E. Conner, Differentiable Periodic Maps. 2nd edition, IV, 181 pages. 1979.

Vol. 739: Analyse Harmonique sur les Groupes de Lie II. Proceedings, 1976–78. Edited by P. Eymard et al. VI, 646 pages. 1979.

Vol. 740: Séminaire d'Algèbre Paul Dubreil. Proceedings, 1977–78. Edited by M.-P. Malliavin. V, 456 pages. 1979.

Vol. 741: Algebraic Topology, Waterloo 1978. Proceedings. Edited by P. Hoffman and V. Snaith. XI, 655 pages. 1979.

Vol. 742: K. Clancey, Seminormal Operators. VII, 125 pages. 1979.

Vol. 743: Romanian-Finnish Seminar on Complex Analysis. Proceedings, 1976. Edited by C. Andreian Cazacu et al. XVI, 713 pages. 1979.

Vol. 744: I. Reiner and K. W. Roggenkamp, Integral Representations. VIII, 275 pages. 1979.

Vol. 745: D. K. Haley, Equational Compactness in Rings. III, 167 pages. 1979.

Vol. 746: P. Hoffman, τ-Rings and Wreath Product Representations. V, 148 pages. 1979.

Vol. 747: Complex Analysis, Joensuu 1978. Proceedings, 1978. Edited by I. Laine, O. Lehto and T. Sorvali. XV, 450 pages. 1979.

Vol. 748: Combinatorial Mathematics VI. Proceedings, 1978. Edited by A. F. Horadam and W. D. Wallis. IX, 206 pages. 1979.

Vol. 749: V. Girault and P.-A. Raviart, Finite Element Approximation of the Navier-Stokes Equations. VII, 200 pages. 1979.

Vol. 750: J. C. Jantzen, Moduln mit einem höchsten Gewicht. III, 195 Seiten. 1979.

Vol. 751: Number Theory, Carbondale 1979. Proceedings. Edited by M. B. Nathanson. V, 342 pages. 1979.

Vol. 752: M. Barr, *-Autonomous Categories. VI, 140 pages. 1979.

Vol. 753: Applications of Sheaves. Proceedings, 1977. Edited by M. Fourman, C. Mulvey and D. Scott. XIV, 779 pages. 1979.

Vol. 754: O. A. Laudal, Formal Moduli of Algebraic Structures. III, 161 pages. 1979.

Vol. 755: Global Analysis. Proceedings, 1978. Edited by M. Grmela and J. E. Marsden. VII, 377 pages. 1979.

Vol. 756: H. O. Cordes, Elliptic Pseudo-Differential Operators – An Abstract Theory. IX, 331 pages. 1979.

Vol. 757: Smoothing Techniques for Curve Estimation. Proceedings, 1979. Edited by Th. Gasser and M. Rosenblatt. V, 245 pages. 1979.

Vol. 758: C. Năstăsescu and F. Van Oystaeyen; Graded and Filtered Rings and Modules. X, 148 pages. 1979.